# Tri-Color Phosphors for White Light-Emitting Diodes

Wang Zhengliang, Liang Hongbin, Zhou Qiang, Luo Lijun,
Tang Huaijun, Gong Menglian

SCIENCE PRESS
Beijing

Responsible Editor: Zheng Shufang

**图书在版编目（CIP）数据**

白光 LED 用三基色荧光粉研究进展 = Tri-Color Phosphors for White Light-Emitting Diodes：英文 / 汪正良等著. —北京：科学出版社，2023.7

ISBN 978-7-03-075541-4

Ⅰ. ①白… Ⅱ. ①汪… Ⅲ. ①三基色荧光粉-研究-英文 Ⅳ. ①TN104.3

中国国家版本馆 CIP 数据核字（2023）第 084975 号


Published by Science Press
16 Donghuangchenggen North Street
Beijing 100717, P. R. China

Printed in Chengdu

# Preface

Since Isamu Akasaki, Hiroshi Amano and Shuji Nakamura won the 2014 Nobel Prize in Physics for developing bright and energy-saving white light sources fabricated the first commercial white light-emitting diodes (w-LEDs) using the blue-emitting LED in 1997, more and more interest focused on GaN-based w-LEDs, due to their energy-saving, luminous efficiency, compactness, reliability, long lifetime and environmental friendliness. In general, the commercial w-LEDs were mainly fabricated by the "phosphor-converted" method. For example, the w-LED could be obtained by combining a 465 nm blue-emitting InGaN LED with a broad-band yellow-emitting phosphor, e.g. YAG. However, this type of w-LED suffers from a low color rendering index ($R_a$<80) and high color temperature (CCT>4500 K), because of the scarcity of red emission in its electroluminescent spectrum. To produce wide color-gamut w-LEDs with high $R_a$ and low CCT, some fascinating red phosphors were introduced into such w-LEDs-based on YAG. Meanwhile, highly efficient w-LEDs devices with high $R_a$ and low CCT can also be obtained by coating red, green and blue phosphors on near UV or UV LED chips. Because this wavelength can offer a higher efficiency solid-state lighting. To build highly efficient w-LEDs devices, the luminescent efficiency of phosphors is a critical factor. Hence, it is important to find new tri-color phosphors for w-LEDs.

According to the composition of phosphors, the tri-color phosphors for GaN/InGaN based w-LEDs can be divided into two categories: inorganic phosphors and organic phosphors. The inorganic phosphors contain oxides/composite oxides, halides, sulfides, silicates, phosphates, aluminates, and so on. The organic phosphors are mainly composed of rare-earth ions ($RE^{3+}$) activated $\beta$-diketone complexes. Besides, cationic iridium (III) complexes have been reported for solid-state lighting. Compared with organic phosphors, inorganic phosphors show higher chemical stability and thermal stability. From the viewpoint of applications, the phosphors for w-LEDs are still inorganic ones.

The photo-luminescent (PL) properties of the phosphors are not only influenced by the composition of the host but also by the luminescent centers. For example, $Eu^{3+}$-activated phosphors exhibit red emission and $Tb^{3+}$-activated phosphors exhibit green emission. Meanwhile, the luminescent efficiency can be optimized by a sensitizer, which can absorb energy and transfer the energy to the luminescent center. For example, the green phosphor $LaPO_4:Ce^{3+},Tb^{3+}$ can be obtained by the energy transfer from $Ce^{3+}$ to $Tb^{3+}$.

To obtain high-efficient phosphors, many methods were adopted to prepare these phosphors. The conventional solid-state reaction at high temperature was often used to prepare phosphors. For example, the commercial yellow phosphor YAG was prepared via the solid-state reaction. However, this method suffers from a big particle-size and irregular morphology. To prepare phosphors with uniform size and morphology, many wet-chemical methods such as the sol-gel method, the microemulsion method, the combustion method, the hydrothermal method and the precipitation method were developed. In consideration of practice cost, many commercial phosphors are still prepared by the conventional solid-state method.

In this book, we discuss the photoluminescent (PL) properties of some tri-color phosphors with different compositions, different luminescent centers and different preparation methods, according to our previous work. Meanwhile, we investigate the application on LED devices of these phosphors. The book is divided into four chapters according to the colors of the tri-color phosphors. In chapter I, we mainly investigate the preparation, PL properties and applications of $Eu^{3+}$-activated red phosphors based on molybdates and tungstates. Meanwhile, we also study the PL properties of some organic phosphors, such as $Eu^{3+}$ $\beta$-diketone complexes and cationic iridium (III) complexes. In chapter II, the preparation, PL properties and applications of a series of $Mn^{4+}$-activated fluorides are discussed in detail. In Chapter III, we describe the preparation, PL properties and applications of some green phosphors, such as $Tb^{3+}$ doped molybdates/tungstates and $Eu^{2+}$ doped silicates. In the last chapter, some blue phosphors for w-LEDs have been introduced.

A lot of books and literatures are referred to in this book. We pay the deepest respect and gratitude to those who concern about this book and give some significant comments. We appreciate the grants from the National Natural Science Foundation of China, Yunnan Province Government, Guangdong Province Government, State Ethnic Affairs Commission and Ministry of Education of Yunnan province. With their financial supports, the authors can research in the phosphors field. We also thank the graduate and undergraduate students a lot.

Due to the limited expertise, there may be some deficiencies in this book. Please do not hesitate to point them out.

Wang Zhengliang

December 2022

# Contents

# Chapter I  $Eu^{3+}$-activated red phosphors for white light-emitting diodes

## 1.1  $Eu^{3+}$-activated double molybdates red phosphors for near-UV LED chips

### 1.1.1  Introduction

Since Nakamura et al. fabricated a blue-emitting GaN light-emitting diode (LED) in 1993 and the first commercial w-LED solid-state lighting was developed using this blue-emitting LED in 1997, more and more interest paid on GaN-based white light-emitting diodes (w-LEDs). Such w-LEDs have many advantages, such as high efficiency, long lifetime, environmental friendliness, and so on. Therefore, they attracted increasing attention to replacing traditional solid-state lighting. This technology is expected to reduce energy consumption in the lighting sector by 15% in 2020 and by 40% in 2030—saving 261 TWh in 2030 alone, and its worth is over US$26 billion in savings at today's energy prices. The worth is equivalent to the current total energy consumption of about 24 million homes in the United States. Furthermore, with the given mix of power plants, these energy savings would reduce greenhouse gas emissions by roughly 180 million tons of $CO_2$. In October 2014, Isamu, Hiroshi and Shuji jointly won the Nobel Prize in Physics because of their successful and creative work in "the invention of efficient blue light-emitting diodes which has enabled bright and energy-saving white light sources" since the early 1990s.

At present, these w-LEDs are obtained mainly by combining the 465 nm blue-emitting InGaN LEDs with broad-band yellow-emitting phosphor, such as $Y_3Al_5O_{12}:Ce^{3+}$ (YAG). Additionally, white light can be produced by varieties of other approaches, such as by combining discrete blue, green and red LEDs or by mixing blue, green and red phosphors in one LED. Presently, the emission bands of LEDs are shifted to the near-UV range of around 400 nm, because this wavelength can offer a higher efficiency solid-state lighting. The current phosphors for near-UV GaN-based LEDs are $BaMgAl_{10}O_{17}:Eu^{2+}$ for blue, ZnS:($Cu^{+}$, $Al^{3+}$) for green, and $Y_2O_2S:Eu^{3+}$ for red. However, the efficiency of the $Y_2O_2S:Eu^{3+}$ red phosphor is about 8 times less than that of the blue and green phosphors, and the lifetime of $Y_2O_2S:Eu^{3+}$ is inadequate under UV irradiation. Therefore, it is urgent to find new red phosphors which can be excited in the near-UV range around 400 nm.

A suitable red-emitting UV-LED phosphor should meet the following necessary conditions in general: the host is stable, the phosphor exhibits strong and broad absorption around 400 nm (LED emission wavelength), and the phosphor shows strong emission under 400 nm excitation as well as with the chromaticity coordinates near the NTSC (National Television Standard Committee) standard values. To find novel efficient red-emitting LED phosphors, in this chapter, $Eu^{3+}$-activated double molybdates and $NaLa(MoO_4)_2$:$xEu^{3+}$ were prepared in terms of the following considerations. Firstly, the double molybdates $AB(MoO_4)_2$ (A = $Li^+$, $Na^+$, $K^+$, $Rb^+$, $Cs^+$; B = trivalent rare-earth ions) which share scheelite-like ($CaWO_4$) iso-structure, show excellent thermal and hydrolytic stability and were considered to be efficient luminescent hosts. Secondly, it is well known that $Eu^{3+}$ ions exhibit strong absorption at about 395 nm in many host lattices, which is close to the emission wavelength (400 nm) of GaN-based LED. Thirdly, $Eu^{3+}$ ions are reported to occupy the lattice sites without inversion symmetry in these host lattices, and it is expected that the emission of $Eu^{3+}$ will be mainly originated from the $^5D_0$-$^7F_2$ transition in the host lattices, hence the phosphor is expected to be with appropriate CIE chromaticity coordinates.

In this section, a series of double molybdates doped $Eu^{3+}$ were prepared by different methods. And their structures, luminescent properties and applications are investigated in detail.

### 1.1.2 Preparation method

#### 1.1.2.1 Solid-state reaction at a high temperature

All reagents are analytical grade and without further purification in this work. These reagents include $(NH_4)_6Mo_7O_{24}\cdot 4H_2O$, $NaHCO_3$, $Li_2CO_3$, $K_2CO_3$, $Bi_2O_3$, $Al_2O_3$, $(NH_4)_2SO_4$, $(NH_4)_2HPO_4$, $La_2O_3$, $Gd_2O_3$, $Eu_2O_3$, $Tb_4O_7$, $Sm_2O_3$, $Dy_2O_3$, $Pr_6O_{11}$, $CeO_2$ and $Y_2O_3$. The polycrystalline samples have been prepared by the solid-state technique at a high temperature. The stoichiometric mixtures were ground and pre-fired at 500 ℃ for 4 h and then heated at different temperatures (600-800 ℃) for 4 h.

#### 1.1.2.2 Pechini method

The starting materials include $Eu_2O_3$, $Sm_2O_3$, $HNO_3$, $(NH_4)_6Mo_7O_{24}\cdot 4H_2O$, $NaHCO_3$, citric acid and ethylene glycol. Firstly, 2 mmol $(NH_4)_6 Mo_7O_{24}\cdot 4H_2O$ were dissolved in an aqueous solution. Secondly, $Eu_2O_3$ (1.1824 g) and $Sm_2O_3$ (0.0488 g) were dissolved with the nitric acid aqueous solution. Then citric acid (1.4700 g) and ethylene glycol (0.3500 g) were added into the aqueous solution; the pH was adjusted to about 5.0 with $NH_3\cdot H_2O$. Then the $(NH_4)_6Mo_7O_{24}\cdot 4H_2O$ aqueous solution was added into the solution containing $Eu^{3+}$ and $Sm^{3+}$ ions. The aqueous solution kept at 60 ℃ for 8 h under continuous stirring.

At last, the transparent gel was heated at 150 ℃, and the brown resin was obtained. By firing the precursor at different temperatures of 500 ℃/600 ℃/700 ℃/800 ℃, the white phosphors were obtained.

#### 1.1.2.3 Combustion method

The starting materials include $Eu_2O_3$, $HNO_3$, $(NH_4)_6Mo_7O_{24}\cdot 4H_2O$, $NaHCO_3$, EDTA and urea. Firstly, 2 mmol $(NH_4)_6Mo_7O_{24}\cdot 4H_2O$ were dissolved in an aqueous solution. Secondly, $Eu_2O_3$ (1.2317 g) was dissolved with the nitric acid aqueous solution and EDTA (2.5000 g); $NaHCO_3$ (0.5880 g) were added into the aqueous solution; the pH was adjusted to 5.0 with $NH_3\cdot H_2O$/citric acid. Then the $(NH_4)_6Mo_7O_{24}\cdot 4H_2O$ aqueous solution was added to the $Eu^{3+}$-containing solution. The transparent gel was formed after an hour's continuous stirring at 60 ℃, and the urea (1.2012 g) was added to the transparent gel. The transparent gel was heated, and an auto-combustion process took place accompanied by the evolution of a brown fume. Finally, the fluffy precursor was obtained, and it was treated by calcination at various temperatures for half an hour.

#### 1.1.2.4 Fabrication of LED

The single-color LEDs were fabricated by combining InGaN chips with as-prepared phosphors. Firstly, the mixtures of phosphors and epoxy resin (the mass ratio is 1∶1) were coated on InGaN chips and solidified. Then the devices were packaged with epoxy resin and solidified at 150 ℃ for 1 h. At last, the red LEDs were obtained.

### 1.1.3 Structures, morphologies, PL properties and applications

#### 1.1.3.1 $NaLa(MoO_4)_2{:}xEu^{3+}$

The XRD patterns of $NaLa(MoO_4)_2{:}xEu^{3+}$ ($x = 0$, 0.10, 0.20, 0.30, 0.40, 0.50, 0.60, 0.70, 0.80, 0.90, 1.00) were measured. As examples, the XRD patterns of $NaLa(MoO_4)_2{:}xEu^{3+}$ ($x = 0.02$, 0.10, 1) are shown in Fig.1-1(a). These patterns are in agreement with the JCPDS card [$NaLa(MoO_4)_2$, No.24-1103]. This result indicates that the as-prepared red phosphors are of one single phase with the iso-scheelite structure. With the increase of the $Eu^{3+}$ content, the $d$ values exhibit a slight red-shift, and it is due to the distinct ionic radii between $La^{3+}$ and $Eu^{3+}$.

Fig.1-1(b) is the crystal structure of $NaLa(MoO_4)_2$ with the tetragonal structure ($I41/a$, $a = b = 5.344$ Å, $c = 11.730$ Å, $V = 335.28$ Å$^3$). In this tetragonal scheelite structure, $Mo^{6+}$ is coordinated by four $O^{2-}$ in a tetrahedral site, $La^{+}$ and $Na^{+}$ are disordered in the same site and $La^{3+}/Na^{+}$ is coordinated by eight $O^{2-}$. Due to the similar ionic radii as well as same valence between $La^{3+}$ and $Eu^{3+}$, $Eu^{3+}$ will occupy the crystal site of $La^{3+}$ in $NaLa(MoO_4)_2{:}xEu^{3+}$.

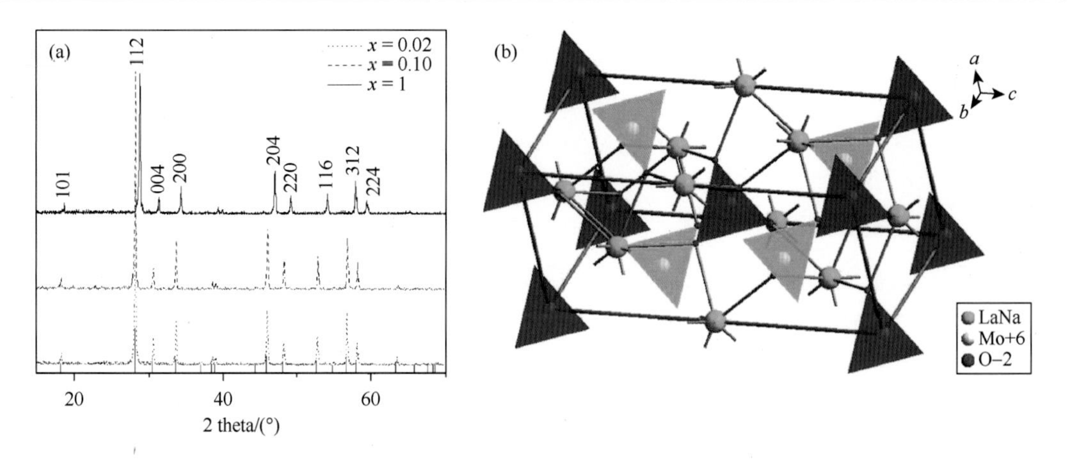


Fig.1-1 (a) The XRD patterns of $NaLa(MoO_4)_2{:}xEu^{3+}$($x$ = 0.02, 0.10, 1);
(b) The crystal structure of $NaLa(MoO_4)_2$

The excitation and emission spectra of $NaLa(MoO_4)_2{:}xEu^{3+}$ samples ($x$ = 0, 0.10, 0.20, 0.30, 0.40, 0.50, 0.60, 0.70, 0.80, 0.90, 1.00) were measured. As examples, the spectra of $NaLa(MoO_4)_2{:}xEu^{3+}$ for $x$ = 0.02, 0.10, 0.50, 1.00 are shown in Fig.1-2. The dominant broad excitation bands peaking at 300 nm in Fig.1-2(a) are attributable to the O→Mo charge transfer (CT) transition, and it seems that the charge transfer band (CTB) position of $NaLa(MoO_4)_2{:}0.10Eu^{3+}$ exhibits a somewhat (about 5 nm) blue-shift compared with that of $NaEu(MoO_4)_2$. The sharp lines are intra-configurational 4f-4f transitions of $Eu^{3+}$ in the host lattices, and the $^7F_0{\to}^5L_6$ and $^7F_0{\to}^5D_2$ transitions at 395 nm and 465 nm are two of the strongest absorptions. Comparing the relative intensity of the $^7F_0{\to}^5L_6$ transition with that of the $^7F_0{\to}^5D_2$ transition in $NaLa(MoO_4)_2{:}0.10Eu^{3+}$ and $NaEu(MoO_4)_2$, it can be found that the $^5D_2$ level tends to be populated in the diluted system $NaLa(MoO_4)_2{:}0.10Eu^{3+}$, while the case is true for $^5L_6$ in pure compound $NaEu(MoO_4)_2$.

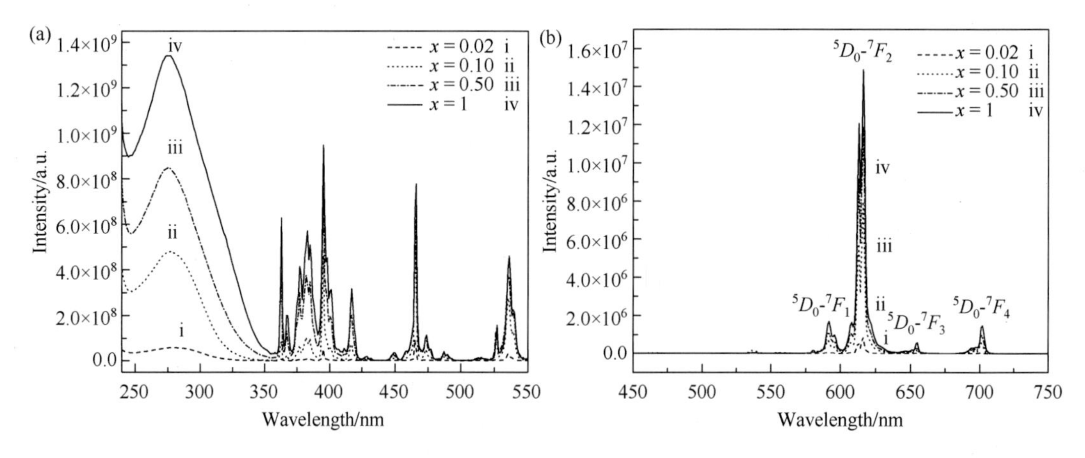


Fig.1-2 The excitation (a, $\lambda_{em}$ = 616 nm) and emission (b, $\lambda_{ex}$ = 395 nm) spectra of $NaLa(MoO_4)_2{:}xEu^{3+}$($x$ = 0.02, 0.10, 0.50, 1)

Fig.1-2(b) represents the emission spectra under the excitation at 395 nm. The main emission line is the $^5D_0 \rightarrow ^7F_2$ transition of $Eu^{3+}$ at 615 nm, other transitions from the excited levels of $^5D_J$ to $^7F_J$ ground states, such as $^5D_0 \rightarrow ^7F_J$ and $^5D_1 \rightarrow ^7F_J$ transitions are very weak, and they are beneficial for a phosphor with good CIE chromaticity coordinates. Moreover, the results imply that $Eu^{3+}$ ions occupy the lattice sites without inversion symmetry, and they are in good agreement with the results of the crystal structure.

Fig.1-3 is the concentration dependence of the relative emission intensity of $Eu^{3+}$ $^5D_0 \rightarrow ^7F_2$ transition in $NaLa(MoO_4)_2$:$xEu^{3+}$. The $^5D_0 \rightarrow ^7F_2$ emission intensity of $NaEu(MoO_4)_2$ is about 4 times higher than that of $NaLa(MoO_4)_2$:0.10$Eu^{3+}$ and no concentration quenching is observed in $NaLa(MoO_4)_2$:$xEu^{3+}$. The CIE chromaticity coordinates of the NTSC for red are $x = 0.67$, $y = 0.33$, and the CIE values for the phosphor $NaEu(MoO_4)_2$ are calculated in terms of the emission spectrum, showing the values are very close to those of NTSC.

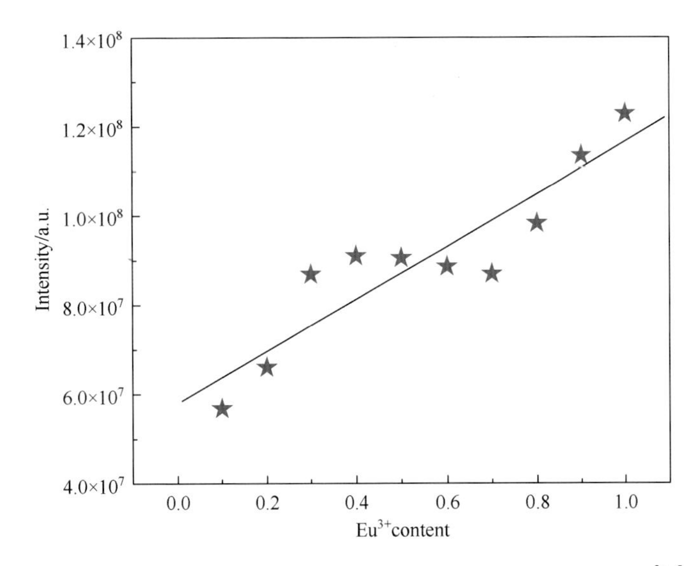


Fig.1-3 Concentration dependence of the relative emission intensity of $Eu^{3+}$ $^5D_0 \rightarrow ^7F_2$ transition in $NaLa(MoO_4)_2$:$xEu^{3+}$ ($\lambda_{ex} = 395$ nm)

Fig.1-4 shows the decay curves for the $^5D_0 \rightarrow ^7F_2$ transition of the $Eu^{3+}$ in the red phosphors $NaLa(MoO_4)_2$:$xEu^{3+}$($x = 0.02$, 0.10, 0.50, 1), which are well fitted into a single-exponential function. The lifetime values are about 433 μs, 424 μs, 374 μs and 419 μs, respectively.

Our purpose is to obtain a highly efficient red component for LED, so a series of red light-emitting LEDs were fabricated with the obtained red phosphors. Fig.1-5 is the electroluminescent (EL) spectrum of the $NaEu(MoO_4)_2$-InGaN-based LED under 20 mA current excitation. The near-UV peak at 395 nm is due to the emission of the LED chip, and the set of red emissions are due to the f-f transitions of $Eu^{3+}$ in $NaEu(MoO_4)_2$. The bright red light from the LED can be observed by naked eyes (Fig.1-5).

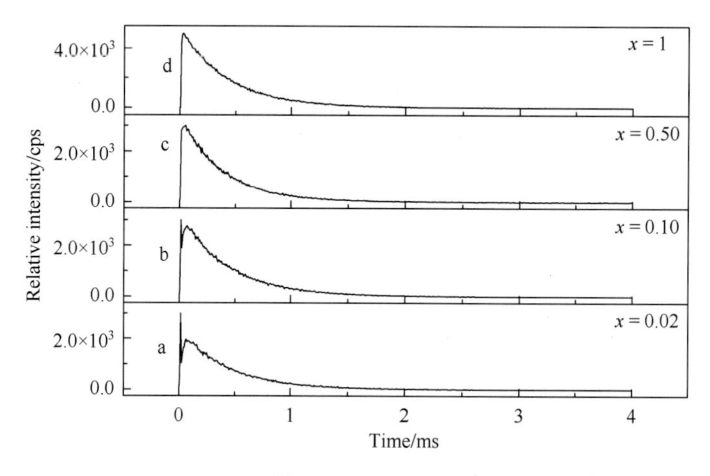


Fig.1-4　The decay curves (a-d) of the $Eu^{3+}$ emission of $NaLa(MoO_4)_2$:$xEu^{3+}$ ($x$ = 0.02, 0.10, 0.50, 1) ($\lambda_{ex}$ = 395 nm; $\lambda_{em}$ = 616 nm)

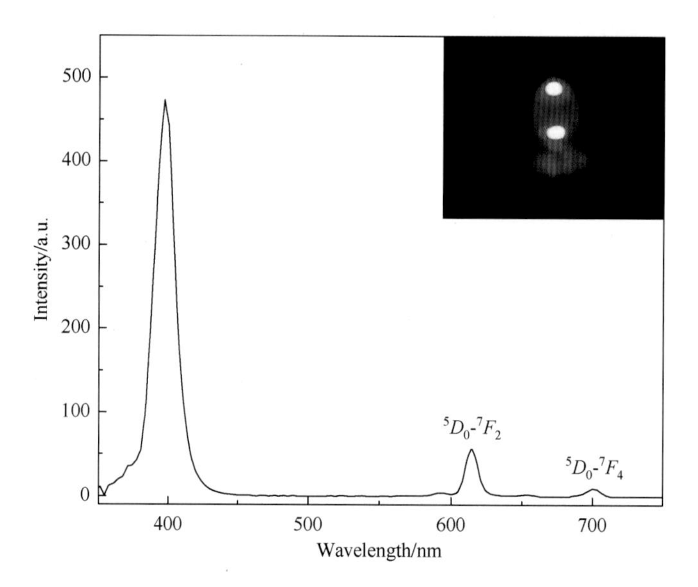


Fig.1-5　The EL spectrum of the $NaEu(MoO_4)_2$-InGaN-based LED under 20 mA current excitation. The inserted figure is the photograph of the red $NaEu(MoO_4)_2$-InGaN LED under 20 mA current excitation

### 1.1.3.2　$NaEu(MoO_4)_2$:$xSm^{3+}$

As mentioned above, the phosphors $NaLa(MoO_4)_2$:$xEu^{3+}$ exhibit a strong absorption line at about 395 nm. $Sm^{3+}$ ions show the main absorption around 405 nm in most host lattices. Therefore, a phosphor with broadened and strengthened absorptions around 400 nm will probably be obtained by an appropriate $Sm^{3+}$-$Eu^{3+}$ co-doping. In general, because the trivalent rare-earth ions share similar ionic radii, it was assumed that it entered the same lattice sites when different rare-earth ions were doped in a specific host lattice. In view of this consideration, we investigated the spectroscopic properties of $NaLa(MoO_4)$:$x$$Sm^{3+}$ ($x$ = 0, 0.02, 0.04, 0.06, 0.08, 0.10, 0.12) and the corresponding results are shown in Fig.1-6.

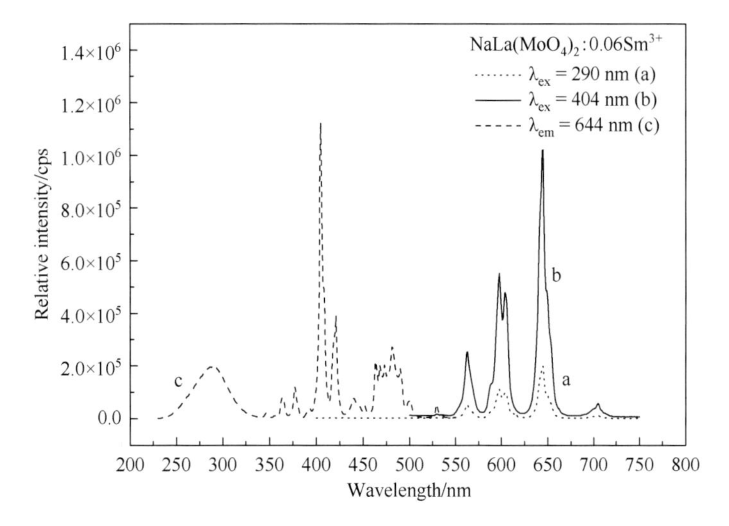


Fig.1-6 The excitation and emission spectra of $NaLa(MoO_4)_2$:0.06$Sm^{3+}$

The broad excitation band peaking at about 290 nm in curve c is assignable to the O→Mo CT absorption as described above. The sharp lines in the 340-550 nm range are f-f transitions of $Sm^{3+}$ in the host lattice, and the main absorption peaking at 404 nm is due to the $^6H_{5/2}\rightarrow{}^4K_{11/2}$ transition of $Sm^{3+}$.

Curves a and b represent the emissions under 290 nm and 404 nm excitation, respectively. The spectra show typical $Sm^{3+}$ f-f transitions. There are three main peaks at 563 nm, 597 nm and 644 nm are due to the $^4G_{5/2}\rightarrow{}^6H_{5/2}$, $^4G_{5/2}\rightarrow{}^6H_{7/2}$, and $^4G_{5/2}\rightarrow{}^6H_{9/2}$ transitions, respectively. The weak emission upon 290 nm excitation indicates that the energy-transfer from the O→Mo CT state to $Sm^{3+}$ 4f levels is not highly efficient.

Our purpose is to find efficient red-emitting phosphors that show stronger absorption at 400 nm. Hence, the luminescent properties of phosphors $NaEu(MoO_4)_2$:$x$$Sm^{3+}$ ($x$ = 0, 0.02, 0.04, 0.06, 0.08, 0.10, 0.12) are investigated. Curve c in Fig.1-7 is the excitation spectrum of $NaEu(MoO_4)_2$:0.08$Sm^{3+}$. Both $Eu^{3+}$ and $Sm^{3+}$ f-f transition absorptions are observed and each excitation line in Fig.1-7(c) is broader than that in Fig.1-6(c) because of the formation of solid solution in $Sm^{3+}$-$Eu^{3+}$ co-doped system. This results in the absorption around 400 nm for the sample $NaEu(MoO_4)_2$:0.08$Sm^{3+}$ is broader than that of $NaEu(MoO_4)_2$.

Curves a and b in Fig.1-7 show the emission spectra of phosphor $NaEu(MoO_4)_2$:0.08$Sm^{3+}$ under 395 nm and 404 nm excitation, respectively. We find that the emission features are very similar for curves a and b. Both curves show typical $Eu^{3+}$ emission, and the mainline is the $^5D_0\rightarrow{}^7F_2$ transition. The emission from $Sm^{3+}$ was not observed even under direct excitation of the $^6H_{5/2}\rightarrow{}^4K_{11/2}$ transition for $Sm^{3+}$ in the host lattice (curve b). It means that the $Sm^{3+}$ ions can absorb and efficiently transfer energy to the $Eu^{3+}$ ions. A possible four-step energy transfer process from $Sm^{3+}$ to $Eu^{3+}$ is considered as follows. An electron in the ground $^6H_{5/2}$ state of $Sm^{3+}$ ions is promoted to the $^4K_{11/2}$ state upon 404 nm excitation in the first step. Then the $^4G_{5/2}$ level is non-radiatively populated. In the third step, the energy transfer from $Sm^{3+}$ to $Eu^{3+}$ occurs, and the $^5D_0$ state of $Eu^{3+}$ is populated. At last, the radiative transition from the $^5D_0$ to $^7F_2$ level of $Eu^{3+}$ results in the emission.

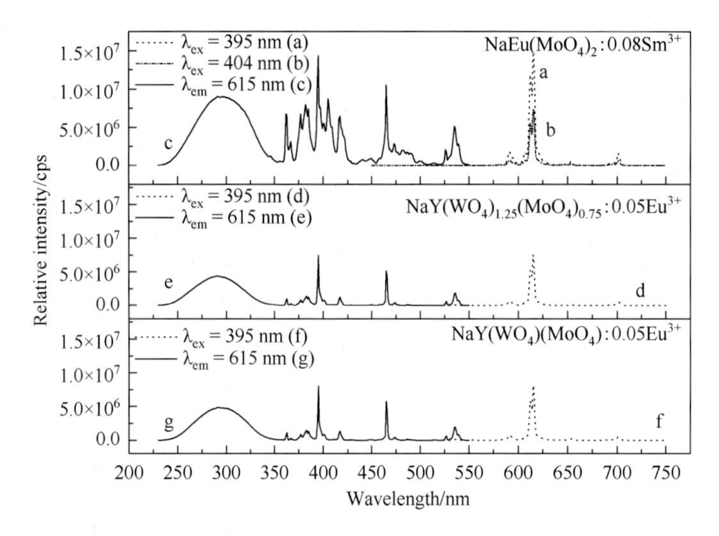


Fig.1-7 The excitation (c, e and g) and emission (a, b, d and f) spectra of $NaEu(MoO_4)_2$:$0.08Sm^{3+}$, as well as the comparison of its luminescence with that of $NaY(WO_4)(MoO_4)$:$0.05Eu^{3+}$ and $NaY(WO_4)_{1.25}(MoO_4)_{0.75}$:$0.05Eu^{3+}$

The CIE chromaticity coordinates of the phosphors $NaEu(MoO_4)_2$:$0.08Sm^{3+}$ are calculated according to the emission curves a and b as listed in Table 1-1. Both the emission spectral feature and the chromaticity coordinates show there is no obvious difference from that of $NaLa(MoO_4)_2$:$xEu^{3+}$. It shows that the chromaticity coordinates are very close to the NTSC standard values for both $NaEu(MoO_4)_2$:$0.08Sm^{3+}$ and $NaLa(MoO_4)_2$:$xEu^{3+}$.

**Table 1-1 The CIE chromaticity coordinates and $^5D_0$-$^7F_2$ relative emission intensity of phosphors**

| Phosphor | Excitation wavelength/nm | CIE chromaticity coordinates* | | $^5D_0$-$^7F_2$ relative intensity |
|---|---|---|---|---|
| | | $x$ | $y$ | |
| $NaEu(MoO_4)_2$ | 395 | 0.66 | 0.34 | 2.0 |
| $NaEu(MoO_4)_2$:$0.08Sm^{3+}$ | 395 | 0.66 | 0.34 | 2.4 |
| | 404 | 0.66 | 0.34 | 1.0 |
| $NaY(WO_4)(MoO_4)$:$0.05Eu^{3+}$ | 395 | 0.65 | 0.34 | 1.1 |
| $NaY(WO_4)_{1.25}(MoO_4)_{0.75}$:$0.05Eu^{3+}$ | 395 | 0.65 | 0.34 | 1.0 |

* The NTSC standard values $x = 0.67$, $y = 0.33$.

More recently, Neeraj et al. reported that the emission intensity of $NaY(WO_4)(MoO_4):0.05Eu^{3+}$ is 7.28 times higher than that of commercial red-emitting phosphor $Y_2OS:Eu^{3+}$ under the excitation at 393 nm. They suggested that $NaY(WO_4)(MoO_4):0.05Eu^{3+}$ and $NaY(WO_4)_{1.25}(MoO_4)_{0.75}:0.05Eu^{3+}$ are probably promising red-emitting LED phosphors. As for comparisons, we prepared these two phosphors according to Ref.8 and measured their luminescent properties. The results are shown in Fig.1-7.

The largest difference among these spectra in Fig.1-7 is the absorption around 400 nm. The absorptions around this wavelength are strengthened and broadened in the spectrum (a). Probably because of the difference in the instrumental response, the $^7F_0 \rightarrow ^5L_6$ transition measured for $NaY(WO_4)(MoO_4):0.05Eu^{3+}$ and $NaY(WO_4)_{1.25}(MoO_4)_{0.75}:0.05Eu^{3+}$ are at 395 nm rather than 393 nm reported by Neeraj et al., which is just the same as that for $NaEu(MoO_4)_2$ and $NaEu(MoO_4)_2:0.08Sm^{3+}$. Therefore, the emission intensities of these phosphors under 395 nm excitation are compared. The comparison of the integrated intensities, as well as the calculated CIE chromaticity coordinates for these phosphors are listed in Table 1-1. It can be found that the emission intensity of $NaEu(MoO_4)_2:0.08Sm^{3+}$ is about 2.4 times stronger than those of $NaY(WO_4)(MoO_4):0.05Eu^{3+}$ and $NaY(WO_4)_{1.25}(MoO_4)_{0.75}:0.05Eu^{3+}$ under 395 nm excitation. Furthermore, it can be observed that even by direct excitation $Sm^{3+}$ upon 404 nm, the $^5D_0$-$^7F_2$ emission intensity of $Eu^{3+}$ in $NaEu(MoO_4)_2:0.08Sm^{3+}$ is also high. From Table 1-1, it might be found that the $^5D_0$-$^7F_2$ emission intensity of $Eu^{3+}$ in $NaEu(MoO_4)_2:0.08Sm^{3+}$ under 404 nm excitation (direct excitation $Sm^{3+}$) is close to those in $NaY(WO_4)(MoO_4):0.05Eu^{3+}$ and $NaY(WO_4)_{1.25}(MoO_4)_{0.75}:0.05Eu^{3+}$ under 395 nm excitation.

#### 1.1.3.3 NaEu $(MoO_4)_2:Bi^{3+},Sm^{3+}$

Although $Eu^{3+}$ ions show the $^7F_0$-$^5L_6$ transition at about 395 nm in most hosts that is near the excitation energy, this transition is parity-forbidden, so it is a narrow line and cannot absorb the excitation energy efficiently. To strengthen and broaden the absorption around 400 nm, one of the important approaches is to introduce a co-activator in the phosphor. We consider that $Sm^{3+}$ and $Bi^{3+}$ are probably two eligible co-activators. On the one hand, $Sm^{3+}$ ions exhibit strong line absorption that belongs to the $^6H_{5/2} \rightarrow ^4K_{11/2}$ transition at about 405 nm in many host lattices, which is close to the emission wavelength (400 nm) of InGaN-based LED, therefore the absorption can be strengthened employing $Sm^{3+}$ and $Eu^{3+}$ co-doping. On the other hand, $Bi^{3+}$ shows a broad band-like absorption, which is also selected. It is well known that the $Bi^{3+}$ ions are very good sensitizers of luminescence in many hosts. It can efficiently absorb the UV-light and transfer the energy to the luminescent center, then the emission intensity of the luminescent center would be strengthened.

The XRD patterns of $NaEu(MoO_4)_2:xBi^{3+}$ ($x$ = 0.02, 0.04, 0.06, 0.08, 0.10, 0.20, 0.30, 0.40, 0.50, 0.60, 0.70, 0.80, 0.90,1.00) were measured. As examples, patterns of the phosphors for $x$ = 0, 0.10, 0.50 and 1.00 are shown in Fig.1-8. Curve a is very consistent with the JCPDS card

No. 51-1508 [$NaBi(MoO_4)_2$], showing the sample has a single phase with the scheelite structure. The other three patterns are very similar to the pattern (a), which means they are of iso-structure with the single-phase. $Eu^{3+}$ occupies the same site as $Bi^{3+}$ in the host. The *d* values are slightly different when varying $Bi^{3+}$ content: it may be since the $Bi^{3+}$ ion radius (131 pm, eight-fold coordination) is a little bigger than that of $Eu^{3+}$ (121 pm, eight-fold coordination), resulting in a $Bi^{3+}$-O distance being larger than the $Eu^{3+}$-O distance.

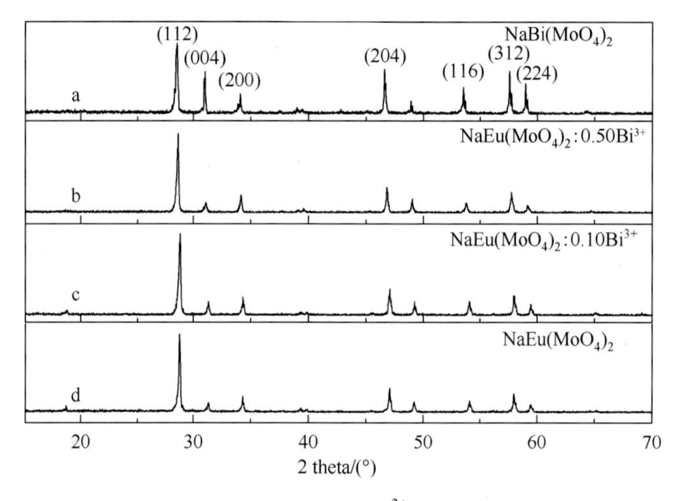


Fig.1-8 The XRD patterns of $NaEu(MoO_4)_2$:$x$$Bi^{3+}$[$x = 0$ (d), 0.10 (c), 0.50 (b), 1.00 (a)]

The PL spectra of $NaEu(MoO_4)_2$:$x$$Bi^{3+}$ ($x = 0$, 0.10, 0.20, 0.30) are shown in Fig.1-9. The excitation spectrum of $NaEu(MoO_4)_2$ (curve a) for monitoring the $^5D_0 \rightarrow {}^7F_2$ emission of $Eu^{3+}$ shows a broad O→Mo CT band with a maximum centered around 270 nm and a band edge at 350 nm. The sharp peaks from 360 nm to 550 nm are ascribed to the intra-configurational 4f-4f transitions of $Eu^{3+}$ in the host lattice, and two of the strongest absorptions are at 395 nm ($^7F_0 \rightarrow {}^5L_6$) and 465 nm ($^7F_0 \rightarrow {}^5D_2$), respectively.

When $Bi^{3+}$ ions are introduced in the phosphors, some changes between curve a and curves c, e and g in the wavelength range below 360 nm can be observed. A rather broad shoulder band appears at the longer wavelength side of the O→Mo CT band (270 nm) in curves c, e and g. Also, it seems that this broad absorption shows somewhat increasing relative to the intensity of the O→Mo CT band when $Bi^{3+}$ content is increased up to 20%. The broad excitation band from 300 nm to 360 nm may be due to the absorption of $Bi^{3+}$. It is well known that the $Bi^{3+}$ ion has an outer $6s^2$ electronic configuration with the ground state of $^1S_0$. The excited states have a 6s6p configuration and are split into the $^3P_0$, $^3P_1$, $^3P_2$ and $^1P_1$ levels in a sequence with the energy increase. If no other configuration is taken into account, the transitions from $^1S_0$ to $^3P_0$ and $^3P_2$ are completely spin forbidden. The two levels, $^3P_1$ and $^1P_1$ are mixed by the spin-orbit coupling. The broad-band absorption from 300 nm to 360 nm is an overlap of the transitions from the ground state $^1S_0$ to $^3P_1$ and $^1P_1$. It can also be observed that the excitation

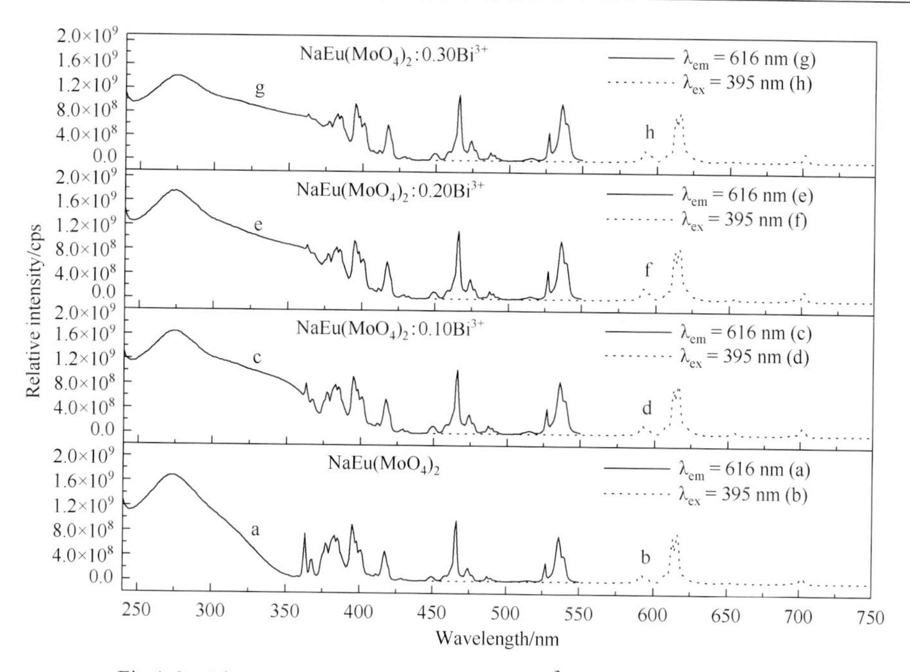


Fig.1-9 The PL spectra of $NaEu(MoO_4)_2$:$xBi^{3+}$($x$ = 0, 0.10, 0.20, 0.30)

intensities of $Eu^{3+}$ at 395 nm show a somewhat increasing relative to the intensity of the O→Mo CT band with the increasing content of $Bi^{3+}$ in the samples. They are higher than that of $Eu^{3+}$ in pure $NaEu(MoO_4)_2$. The excitation intensity around 395 nm in $NaEu(MoO_4)_2$:0.20$Bi^{3+}$ is the highest. When the $Bi^{3+}$ content is more than 20%, the excitation intensities of $Eu^{3+}$ are gradually decreased. Curves b, d, f and h represent the emission spectra of $NaEu(MoO_4)_2$:$xBi^{3+}$ ($x$ = 0, 0.10, 0.20, 0.30) under 395 nm excitation. The four curves are very similar. The strongest peak is located at 616 nm, due to the ${}^5D_0 \rightarrow {}^7F_2$ transition of $Eu^{3+}$. Other f-f transitions (such as ${}^5D_1 \rightarrow {}^7F_J$ transitions) of $Eu^{3+}$ are very weak. The result shows that $Eu^{3+}$ ions occupy the lattice sites without inversion symmetry and the introduction of $Bi^{3+}$ ions does not alter the sub-lattice structure around the luminescent center ions $Eu^{3+}$. This is in good agreement with the result of XRD characterization. The relative emission intensities of the ${}^5D_0 \rightarrow {}^7F_2$ transition and CIE chromaticity coordinates are listed in Table 1-2. These phosphors all show excellent CIE chromaticity coordinates close to the NTSC values. The emission intensities of phosphors doped with $Bi^{3+}$ under 395 nm excitation are higher than that of $NaEu(MoO_4)_2$. Moreover, the emission intensity of $NaEu(MoO_4)_2$:0.20$Bi^{3+}$ is the highest. This result is consistent with their excitation spectra.

**Table 1-2 The CIE chromaticity coordinates and ${}^5D_0$-${}^7F_2$ relative emission intensity of phosphors**

| Phosphor | Excitation wavelength/nm | CIE chromaticity coordinates* | | ${}^5D_0$-${}^7F_2$ relative intensity |
|---|---|---|---|---|
| | | $x$ | $y$ | |
| $NaEu(MoO_4)_2$ | 395 | 0.66 | 0.34 | 1.00 |
| $NaEu(MoO_4)_2$:0.10$Bi^{3+}$ | 395 | 0.66 | 0.34 | 1.13 |

continued

| Phosphor | Excitation wavelength/nm | CIE chromaticity coordinates* | | $^5D_0$-$^7F_2$ relative intensity |
|---|---|---|---|---|
| | | $x$ | $y$ | |
| $NaEu(MoO_4)_2$:$0.20Bi^{3+}$ | 395 | 0.66 | 0.34 | 1.28 |
| $NaEu(MoO_4)_2$:$0.30Bi^{3+}$ | 395 | 0.66 | 0.34 | 1.22 |
| $NaEu(MoO_4)_2$:$0.20Bi^{3+}$, $0.04Sm^{3+}$ | 395 | 0.66 | 0.34 | 1.38 |
| | 405 | 0.66 | 0.34 | 0.76 |

* The NTSC standard values $x = 0.67$, $y = 0.33$.

The decay curves for the $^5D_0 \rightarrow {}^7F_2$ transition (616 nm) of the $Eu^{3+}$ ions in the phosphors $NaEu(MoO_4)_2$: $xBi^{3+}$($x = 0$, 0.10, 0.20, 0.30) are shown in Fig.1-10. They are all well fitted with a single-exponential function, and the lifetime $\tau$ values for $x = 0$, 0.10, 0.20, 0.30 are 0.419 ms, 0.410 ms, 0.398 ms and 0.388 ms respectively. With the increase of the $Bi^{3+}$ content, the samples' lifetime $\tau$ values gradually decrease.

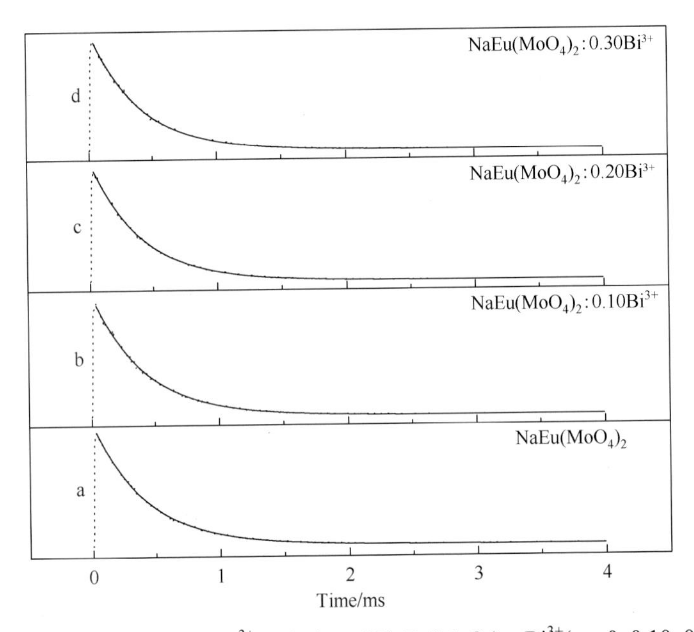


Fig.1-10　Decay curves of the $Eu^{3+}$ emission of $NaEu(MoO_4)_2$:$xBi^{3+}$($x = 0$, 0.10, 0.20, 0.30) ($\lambda_{ex} = 395$ nm; $\lambda_{em} = 616$ nm)

It is well known that $Eu^{3+}$/$Sm^{3+}$ ions exhibit strong absorption at about 395 nm/405 nm. As a consequence, the absorptions around 400 nm are expected to be strengthened and broadened by this co-doping system. As is mentioned above, $NaEu(MoO_4)_2$:$0.20Bi^{3+}$ has the highest emission intensity under 395 nm excitation among the studied $NaEu(MoO_4)_2$:$xBi^{3+}$ samples, so the phosphors $NaEu(MoO_4)_2$:$0.20Bi^{3+}$ doped with $Sm^{3+}$ were prepared. Their XRD patterns are consistent with the pattern of $NaEu(MoO_4)_2$:$0.10Bi^{3+}$. The excitation and emission spectra of $NaEu(MoO_4)_2$:$0.20Bi^{3+}$,$0.04Sm^{3+}$ are shown in Fig.1-11. The 4f-4f transitions of $Sm^{3+}$

appear in the excitation spectrum of $NaEu(MoO_4)_2$:0.20$Bi^{3+}$,0.04$Sm^{3+}$ (curve a) with the introduction of $Sm^{3+}$. Thus, the peak at 405 nm is due to the ${}^6H_{5/2}\rightarrow{}^4K_{11/2}$ transition of $Sm^{3+}$. It results in the broadening of the absorption band around 400 nm for the phosphor when compared with that of $NaEu(MoO_4)_2$:0.20$Bi^{3+}$. Curves b and c show the emission spectra of the phosphor $NaEu(MoO_4)_2$:0.20$Bi^{3+}$, 0.04$Sm^{3+}$ under 395 nm and 405 nm excitation respectively. Both curves show typical $Eu^{3+}$ emission, and the strongest line is at 616 nm (the ${}^5D_0\rightarrow{}^7F_2$ transition of $Eu^{3+}$). The emission of $Sm^{3+}$ cannot be observed even under direct excitation of the ${}^6H_{5/2}\rightarrow{}^4K_{11/2}$ transition of $Sm^{3+}$ in the host lattices (curve c). The emission intensity of $NaEu(MoO_4)_2$:0.20$Bi^{3+}$, 0.04$Sm^{3+}$ under 395 nm excitation is higher than that of $NaEu(MoO_4)_2$:0.20$Bi^{3+}$. It means that the $Sm^{3+}$ ions can absorb energy and efficiently transfer the energy to the $Eu^{3+}$ ions. Its relative emission intensity under 395 nm/405 nm excitation and CIE chromaticity coordinates are listed in Table 1-2. The emission intensity of $NaEu(MoO_4)_2$:0.20$Bi^{3+}$, 0.04$Sm^{3+}$ under 395 nm excitation is 1.38 times stronger than that of $NaEu(MoO_4)_2$:0.20$Bi^{3+}$. Furthermore, the ${}^5D_0\rightarrow{}^7F_2$ relative emission intensity of $NaEu(MoO_4)_2$:0.20$Bi^{3+}$,0.04$Sm^{3+}$ (0.76) under 405 nm (the ${}^6H_{5/2}\rightarrow{}^4K_{11/2}$ transition of $Sm^{3+}$) excitation is a rather significant value, compared with that of $NaEu(MoO_4)_2$:0.20$Bi^{3+}$ (1.28) under 395 nm excitation. The chromaticity coordinates are very close to the NTSC standard values. It means the phosphor $NaEu(MoO_4)_2$:0.20$Bi^{3+}$, 0.04$Sm^{3+}$ share intense and broadened absorption band around 400 nm, as well as strengthened emission with good CIE chromaticity coordinates.

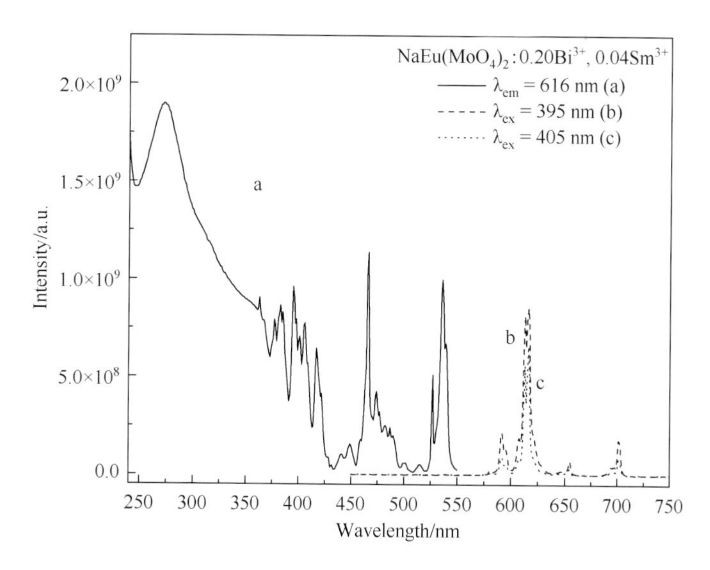


Fig.1-11 The PL spectra of $NaEu(MoO_4)_2$:0.20$Bi^{3+}$,0.04$Sm^{3+}$

Fig.1-12 shows the decay curve for the ${}^5D_0\rightarrow{}^7F_2$ transition (616 nm) of the $Eu^{3+}$ of $NaEu(MoO_4)_2$:0.20$Bi^{3+}$, 0.04$Sm^{3+}$. The curve is also fitted with a single-exponential function, and the lifetime $\tau$ value is 0.362 ms. It is shorter than that of $NaEu(MoO_4)_2$:0.20$Bi^{3+}$.

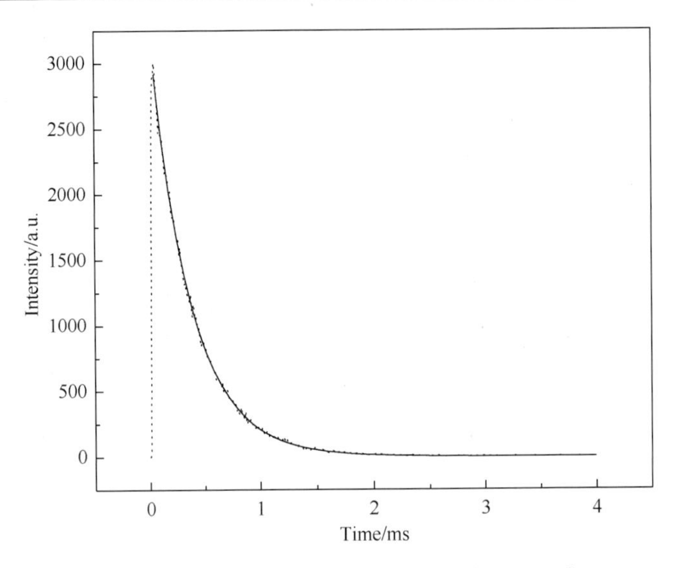


Fig.1-12 Decay curve of the $Eu^{3+}$ emission of $NaEu(MoO_4)_2$:$0.20Bi^{3+}$,$0.04Sm^{3+}$($\lambda_{ex}$ = 395 nm; $\lambda_{em}$ = 616 nm)

#### 1.1.3.4 $AEu(MoO_4)_2$ (A = $Li^+$, $K^+$)

Double molybdates $ALn(MoO_4)_2$ (A = $Li^+$, $Na^+$, $K^+$, $Rb^+$,$Cs^+$; Ln = trivalent rare-earth ions), which share a scheelite-like ($CaMoO_4$) structure, show excellent thermal and hydrolytic stability are considered to be efficient luminescent hosts. As mentioned above, the PL properties of $NaLa(MoO_4)_2$:$xEu^{3+}$ were investigated. These phosphors show efficient red light emission under about 400 nm UV excitation and may be good candidates for the red component of a three-band white LED. In such double molybdates structure, the alkaline ions and rare-earth ions are randomly distributed over the cation sites. The different cations with different radii in the host compound would induce some change in the sub-lattice structure around the luminescent center ions and even change the host structure. Hence doping different ions will lead to different PL properties. Based on these considerations and as a systematic and further work, the spectroscopic properties of $NaLn(MoO_4)_2$:$Eu^{3+}$ (Ln = Y, Gd) were investigated to explore the influence of different rare-earth ions in the host compounds on the luminescence by comparing the results to those of $NaLa(MoO_4)_2$:$Eu^{3+}$ in our previous work. Besides, the phosphors $AEu(MoO_4)_2$ and $A_{0.5}A'_{0.5}Eu$ $(MoO_4)_2$ (A = Li, Na, K) were prepared to find the effect of different alkaline ions in the host compound.

The XRD patterns of $NaLn(MoO_4)_2$:$xEu^{3+}$ (Ln = Y, Gd and La; $x$ = 0, 0.10, 0.20, 0.30, 0.40, 0.50, 0.60, 0.70, 0.80, 0.90, 1.00) were measured. As examples, the XRD patterns of $NaLn(MoO_4)_2$:$0.1Eu^{3+}$ (Ln = Y, Gd) are shown in Fig.1-13. It is obvious that the four curves accord with the JCPDS card No. 25-0828 for $Na_{0.5}Gd_{0.5}MoO_4$. While their $d$ values are slightly different due to the distinct ionic radii between $Gd^{3+}$/$Y^{3+}$ and $Eu^{3+}$.

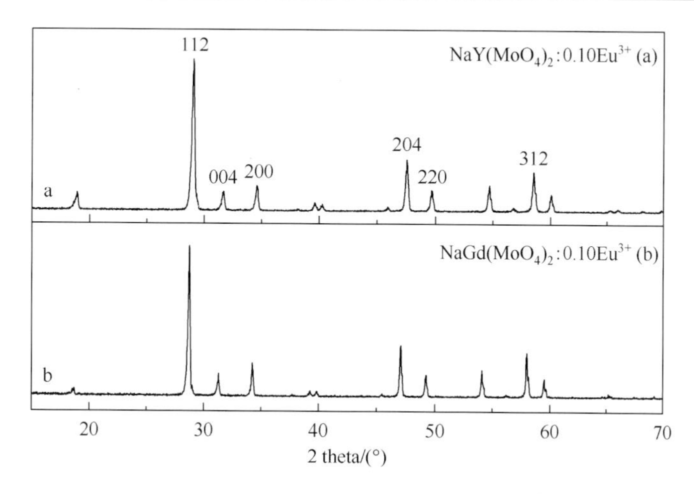


Fig.1-13 The XRD patterns of NaLn($MoO_4$)$_2$:0.10Eu$^{3+}$ (Ln = Y, Gd)

The excitation and emission spectra of NaLn($MoO_4$)$_2$:0.10Eu$^{3+}$ (Ln = Y, Gd, La) and NaEu($MoO_4$)$_2$ are shown in Fig.1-14. The broadband in the range of 250-350 nm can be ascribed to the O→Mo CT transition. The sharp lines in the 360-550 nm range are due to intra-configurational 4f-4f transitions of $Eu^{3+}$ in the host lattice. Two of the strongest excitation peaks are at 395 nm ($^7F_0$→$^5L_6$) and 465 nm ($^7F_0$→$^5D_2$), respectively. The shape of the excitation peaks in the 360-550 nm range of NaLn($MoO_4$)$_2$:0.10Eu$^{3+}$ (Ln = Y, Gd, La) are similar, however, the O→Mo CT bands in spectra (a, c) show a somewhat red-shift compared with that in the spectrum (e). The charge transfer transition from the coordination ions (L) to center metal ions (M) is parity-allowed. The energy is sensitive to the degree of M—L bond covalency. The CT energy decreases with the increase of the M—L bond covalency. For example, it is well known that the covalency of the Eu—S bond is higher than that of the Eu—O bond in general. Therefore, the Eu—S bond shows a lower CT energy than that of the Eu—O band. The electronegativity difference between M and L, the coordination number of M, as well as the M—L bond length, are mainly responsible for the covalency of the M—L bond. The bond covalency increases while the electronegativity difference between M and L decreases. Besides, it is expected that the bond covalency increases with the decrease of the M—L bond length because the overlapping of the electronic cloud of M and L is increased. For example, $MBPO_5$ (M = Sr, Ba) compounds share the stillwellite iso-structure. The M—O average bond length increase from Sr to Ba. It was found in our previous work that the mean covalence of the M—O bond in $MBPO_5$ is calculated to be 5.29% for Sr and 4.77% for Ba, showing a decrease in the M—O bond covalence from Sr to Ba. In NaLn($MoO_4$)$_2$ (Ln = Y, Gd, La) with tetragonal scheelite structure, $Mo^{6+}$ is coordinated by four $O^{2-}$ in a tetrahedral site, $Ln^{3+}$ and $Na^+$ are disordered in the same site and $Ln^{3+}$/$Na^+$ is coordinated by eight $O^{2-}$ of near four $(MoO_4)^{2-}$. Increase in the ionic radii with the relative order of $Y^{3+}$ (89.3 pm) < $Gd^{3+}$ (93.8 pm) < $La^{3+}$ (101.6 pm) and decrease in the electronegativity with the relative order $Y^{3+}$ > $Gd^{3+}$ > $La^{3+}$ result in an increase in the M—O bond covalency from La—O to Gd—O and Y—O. $Ln^{3+}$ is the next-nearest ion of $Mo^{6+}$, the occurrence of $Ln^{3+}$ in the Ln-O-Mo system will weaken the bond ionicity and increase the bond covalency of the Mo—O bond. The bond covalence order of Y—O > Gd—O > La—O results in a Mo—O bond covalence order of $NaY(MoO_4)_2$:0.10$Eu^{3+}$ > $NaGd(MoO_4)_2$:0.10$Eu^{3+}$ > $NaLa(MoO_4)_2$:0.10$Eu^{3+}$, so the O→Mo CT energy decreases with the order $NaY(MoO_4)_2$:0.10$Eu^{3+}$ < $NaGd(MoO_4)_2$:0.10$Eu^{3+}$ < $NaLa(MoO_4)_2$:0.10$Eu^{3+}$. A red-shift of the O→Mo CT excitation band is found with the wavelength order of $NaY(MoO_4)_2$:0.10$Eu^{3+}$ > $NaGd(MoO_4)_2$:0.10$Eu^{3+}$ > $NaLa(MoO_4)_2$:0.10$Eu^{3+}$ in the range from 250 nm to 350 nm, as shown in Fig.1-14.

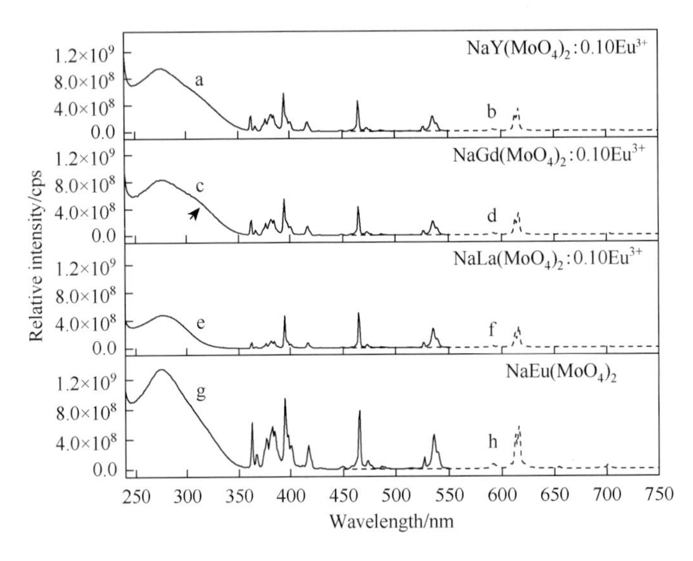


Fig.1-14　The excitation (left, $\lambda_{em}$ = 616 nm) and emission (right, $\lambda_{ex}$ = 395 nm) spectra of $NaLn(MoO_4)_2$:0.10$Eu^{3+}$ (Ln = Y, Gd, La) and $NaEu(MoO_4)_2$

Strong red emission is observed from $NaLn(MoO_4)_2$:0.10$Eu^{3+}$ (Ln = Y, Gd, La) and $NaEu(MoO_4)_2$ samples under 395 nm light excitation, indicating that the compounds are suitable to be excited by a near-UV InGaN chip. The main emission is the $^5D_0 \rightarrow {}^7F_2$ transition of $Eu^{3+}$ at 616 nm. Other f-f transitions of $Eu^{3+}$ are very weak, which is advantageous to obtain a phosphor with good CIE chromaticity coordinates. The emission intensity of $NaEu(MoO_4)_2$ is much higher than that of $NaLn(MoO_4)_2$:0.10$Eu^{3+}$ (Ln = Y, Gd, La). No $Eu^{3+}$ concentration quenching was observed in the series samples of $NaLn(MoO_4)_2$:$xEu^{3+}$ (Ln = Y, Gd). This result is in agreement with that of $NaLa(MoO_4)_2$:$xEu^{3+}$.

The decay curves for the $^5D_0 \rightarrow {}^7F_2$ transition (616 nm) of the $Eu^{3+}$ of these phosphors are shown in Fig.1-15. These decay curves can be well fitted by a single-exponential function as $I = A\exp(-t/\tau)$ and the values of the lifetime of $NaY(MoO_4)_2$:0.10$Eu^{3+}$ and $NaGd(MoO_4)_2$:0.10$Eu^{3+}$ are 0.425 ms and 0.433 ms, respectively.

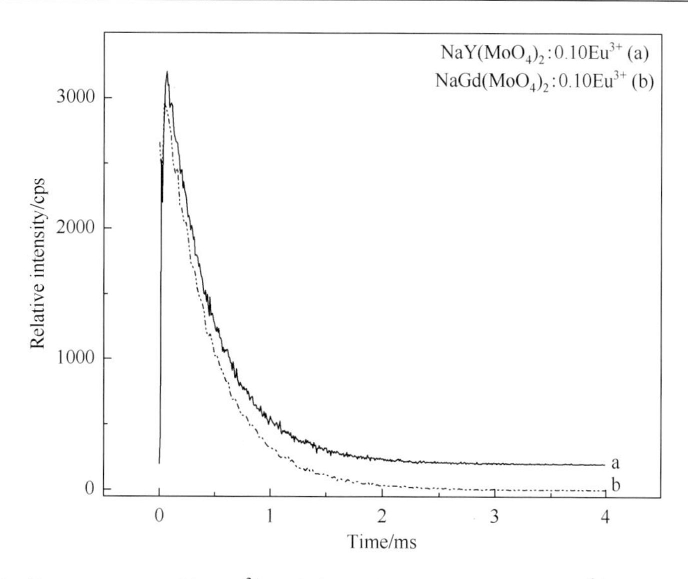


Fig.1-15 Decay curves of the $Eu^{3+}$ emission of $NaLn(MoO_4)_2$:0.10$Eu^{3+}$ (Ln = Y and Gd) ($\lambda_{ex}$ = 395 nm, $\lambda_{em}$ = 616 nm)

As mentioned above, no concentration quenching of $Eu^{3+}$ was observed in the series $NaLn(MoO_4)_2$:$x$$Eu^{3+}$ (Ln = Y, Gd) and Van Vliet et al. reported that the compounds with the scheelite-related crystal structure showed no clear concentration quenching. So a series of $AEu(MoO_4)_2$ (A = Li, K) were prepared and their structures as well as PL properties were investigated and compared with that of $NaEu(MoO_4)_2$.

The XRD patterns of $LiEu(MoO_4)_2$ and $KEu(MoO_4)_2$ are shown in Fig.1-16. The XRD pattern of $LiEu(MoO_4)_2$ reveals that the sample is single-phased and consistent with JCPDS card No. 52-1848[$LiEu(MoO_4)_2$]. The result shows that the phosphor shares the same

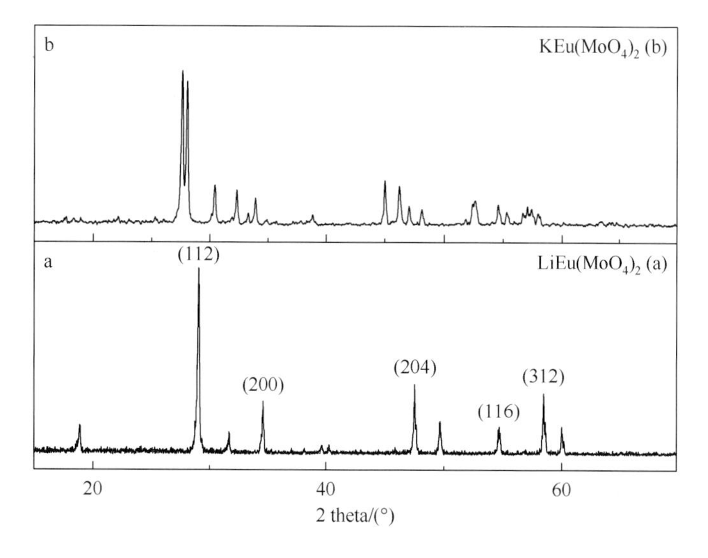


Fig.1-16 The XRD patterns of $AEu(MoO_4)_2$ (A = Li, K)

tetragonal scheelite structure as $NaEu(MoO_4)_2$ with cation disorder. The XRD pattern of $KEu(MoO_4)_2$ is consistent with JCPDS card No. 31-1006 [$KEu(MoO_4)_2$]. It shows that $KEu(MoO_4)_2$ has an ordered triclinic scheelite structure.

The excitation and emission spectra of $AEu(MoO_4)_2$ are shown in Fig.1-17. Comparing the spectra of $LiEu(MoO_4)_2$ and $NaEu(MoO_4)_2$ with that of $KEu(MoO_4)_2$, the O→Mo CT bands of $LiEu(MoO_4)_2$ and $NaEu(MoO_4)_2$ get broader. This result is in agreement with their structure. $LiEu(MoO_4)_2$ and $NaEu(MoO_4)_2$ share a disordered scheelite structure. $Li^+$/$Na^+$ and $Eu^{3+}$ ions are randomly distributed over the cation sites. This accounts for the broadening of the O→Mo CT band. Contrastively, $KEu(MoO_4)_2$ has an ordered triclinic scheelite structure. $Eu^{3+}$ polyhedra are ordered in sheets, forming a two-dimensional sub-lattice. The site symmetry is $C_1$. Therefore, the CT band is broader. $LiEu(MoO_4)_2$ and $NaEu(MoO_4)_2$ share iso-structure. Since the ionic radius of $Na^+$ (97 pm) is bigger than $Li^+$ (68 pm), it is expected that the Li—O bond exhibits more covalency than that of the Na—O bond; this makes the O→Mo CT band for $LiEu(MoO_4)_2$ shift red and broadened as shown in Fig.1-17(a). The excitation bands of $AEu(MoO_4)_2$ at 400 nm are similar. However, their excitation intensities are different. The intensity of $LiEu(MoO_4)_2$ is the strongest among $AEu(MoO_4)_2$ phosphors. This result may be related to the highest covalency and the shortest distance of the Li—O—Mo bond. As discussed above, the $Li^+$ and $Eu^{3+}$ ions are randomly distributed over the cation sites in $LiEu(MoO_4)_2$. Thus the energy transfer from $MoO_4^{2-}$ to $Eu^{3+}$ would be more efficient.

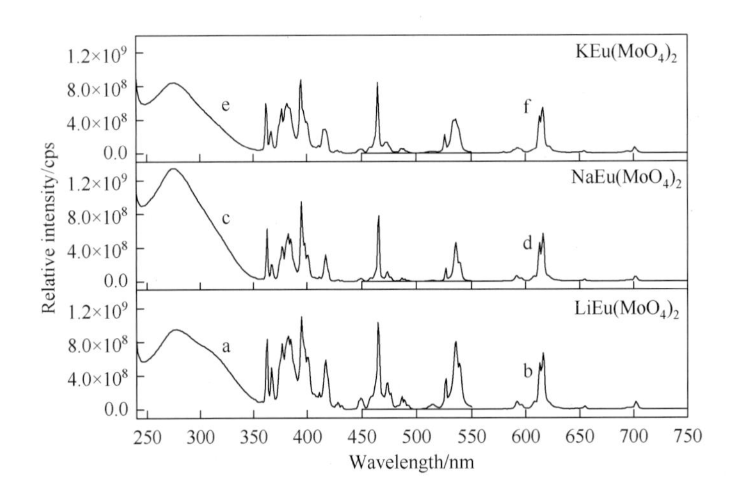


Fig.1-17 Excitation ($\lambda_{em}$ = 616 nm) and emission ($\lambda_{ex}$ = 395 nm) spectra of $AEu(MoO_4)_2$ (A = Li, Na, K)

The emission spectrum of $KEu(MoO_4)_2$ under 395 nm excitation is similar to that of $LiEu(MoO_4)_2$ and $NaEu(MoO_4)_2$. Van Vliet considered that $KEu(MoO_4)_2$ was also slightly disordered, which could be demonstrated by its emission spectrum. Our result also shows that $KEu(MoO_4)_2$ is slightly disordered. The emission intensity and CIE chromaticity coordinates are given in Table 1-3 for all samples studied. Compared with that of $NaEu(MoO_4)_2$ and $KEu(MoO_4)_2$, the result shows $LiEu(MoO_4)_2$ exhibits more intense emission and more appropriate CIE chromaticity coordinates ($x = 0.66$, $y = 0.34$), which are close to the NTSC standard values for red phosphor ($x = 0.67$, $y = 0.33$).

**Table 1-3 The CIE chromaticity coordinates and $^5D_0$-$^7F_2$ relative emission intensity of phosphors**

| Phosphor | Excitation wavelength/nm | CIE chromaticity coordinates* | | $^5D_0$-$^7F_2$ relative intensity |
|---|---|---|---|---|
| | | $x$ | $y$ | |
| $NaEu(MoO_4)_2$ | 395 | 0.66 | 0.34 | 1.00 |
| $LiEu(MoO_4)_2$ | 395 | 0.66 | 0.34 | 1.31 |
| $KEu(MoO_4)_2$ | 395 | 0.66 | 0.34 | 1.09 |
| $Li_{0.5}Na_{0.5}Eu(MoO_4)_2$ | 395 | 0.66 | 0.34 | 1.26 |
| $Na_{0.5}K_{0.5}Eu(MoO_4)_2$ | 395 | 0.66 | 0.34 | 1.19 |

* The NTSC standard values for red are $x = 0.67$, $y = 0.33$.

Their decay curves for the $^5D_0 \rightarrow ^7F_2$ transition (616 nm) of the $Eu^{3+}$ have been measured, and they are also well fitted by a single-exponential function. The values of the lifetime for $LiEu(MoO_4)_2$ and $KEu(MoO_4)_2$ are 0.411 ms and 0.513 ms, respectively. The lifetime values follow the relative order of $LiEu(MoO_4)_2 < NaEu(MoO_4)_2 < KEu(MoO_4)_2$, which is consistent with the radii order of the alkali metal ions $Li^+ < Na^+ < K^+$.

As discussed above, the distortion of the sub-lattice structure results may be in some variation of the excitation band. So phosphors $A_{0.5}A'_{0.5}Eu(MoO_4)_2$ (A, A' = Li, Na, K) were prepared, and their XRD patterns are shown in Fig.1-18. The pattern (b) is similar to the pattern (a). They are consistent with JCPDS card No. 52-1848 [$LiEu(MoO_4)_2$], indicating that $Na_{0.5}K_{0.5}Eu(MoO_4)_2$ and $Li_{0.5}Na_{0.5}Eu(MoO_4)_2$ share the scheelite iso-structure in a single phase. But the $d$ values of $Na_{0.5}K_{0.5}Eu(MoO_4)_2$ are bigger than that of $Li_{0.5}Na_{0.5}Eu(MoO_4)_2$, due to the bigger radius of $K^+$ ion than that of $Li^+$ ion. The patterns (c, d) show the XRD patterns of $LiEu(MoO_4)_2$-$KEu(MoO_4)_2$ fired at different temperatures. Two phases can be found from the related XRD patterns. One phase is in line with $LiEu(MoO_4)_2$, and another is in line with $KEu(MoO_4)_2$. LiEu $(MoO_4)_2$ is of the tetragonal scheelite, while $KEu(MoO_4)_2$ belongs to the triclinic scheelite. The radius of $K^+$ (133 pm) is about 2 times bigger than that of $Li^+$ (68 pm). The different structures between $LiEu(MoO_4)_2$ and $KEu(MoO_4)_2$; the great difference of radii between $K^+$ and $Li^+$ result in the formation of two phases, rather than a single phase of $Li_{0.5}K_{0.5}Eu(MoO_4)_2$.

Fig.1-19 shows the excitation and emission spectra of $Li_{0.5}Na_{0.5}Eu(MoO_4)_2$ and $Na_{0.5}K_{0.5}Eu(MoO_4)_2$. The spectra of $Li_{0.5}Na_{0.5}Eu$ $(MoO_4)_2$ and $Na_{0.5}K_{0.5}Eu(MoO_4)_2$ are similar to that of $LiEu(MoO_4)_2$. This is consistent with their structure. Their relative emission intensity under 395 nm excitation and CIE chromaticity coordinates are also listed in Table 1-3.

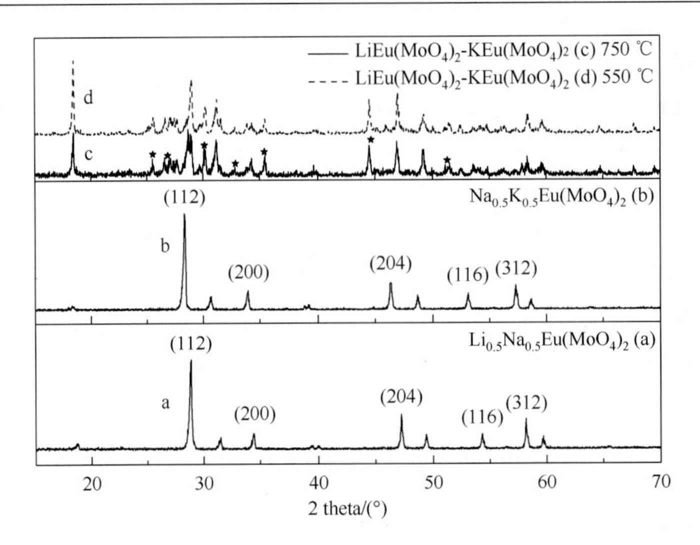


Fig.1-18　The XRD patterns of $Li_{0.5}Na_{0.5}Eu(MoO_4)_2$, $Na_{0.5}K_{0.5}Eu(MoO_4)_2$ and $LiEu(MoO_4)_2$-$KEu(MoO_4)_2$ [★Belong to $KEu(MoO_4)_2$ phase]

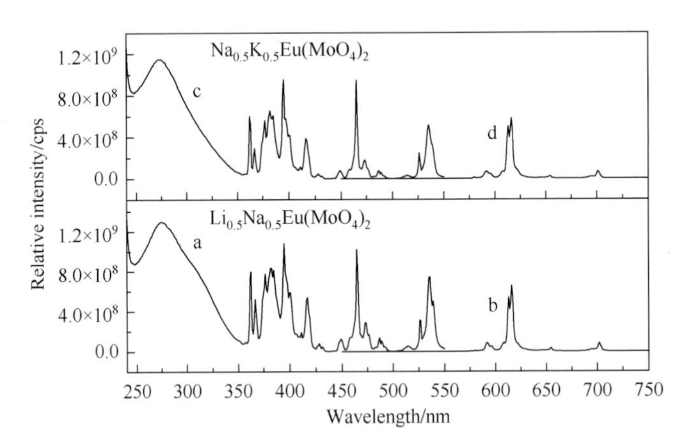


Fig.1-19　The excitation (left, $\lambda_{em} = 616$ nm) and emission (right, $\lambda_{ex} = 395$ nm) spectra of $Li_{0.5}Na_{0.5}Eu(MoO_4)_2$ and $Na_{0.5}K_{0.5}Eu(MoO_4)_2$

The lifetime $\tau$ values of the phosphors $Li_{0.5}Na_{0.5}Eu(MoO_4)_2$ and $Na_{0.5}K_{0.5}Eu(MoO_4)_2$ ($\lambda_{ex} = 395$ nm, $\lambda_{em} = 616$ nm) are 0.439 ms and 0.422 ms respectively. This result is close to those of the phosphors $AEu(MoO_4)_2$ (A = Li, Na, K).

### 1.1.3.5　$(Li_{0.333}Na_{0.334}K_{0.333})Eu(MoO_4)_2$

As discussed above, we reported the luminescence of double molybdates $NaLa(MoO_4)_2{:}xEu^{3+}$ and $NaSm(MoO_4)_2{:}xEu^{3+}$ which were broadened the absorption around 400 nm by the co-doping $Sm^{3+}$-$Eu^{3+}$. These samples show intense red emission, appropriate CIE chromaticity coordinates, and are considered to be a promising red-emitting component in the near-UV LED. Because these double molybdates share the scheelite-like ($CaWO_4$) iso-structure and no concentration quenching of $Eu^{3+}$ was observed in the series samples of

$NaLa(MoO_4)_2:xEu^{3+}$. A systematic and further investigation on the near-UV LED applications of molybdates red-emitting phosphor is necessary. From the viewpoint of the host compound, each spectroscopic line is expected to be narrow when the rare-earth ions enter the lattice sites of a pure host compound in general. Contrastively, if the host compound can form solid solutions by adjusting the cations or anions of this host compound, the sub-lattice structure around the luminescent center ions will be expected to be somewhat diverse, and therefore the spectroscopic lines of rare-earth ions are expected to be broadened. We tried to broaden the absorption around 400 nm by the solid solution $(Li_{0.333}Na_{0.334}K_{0.333})Eu(MoO_4)_2$ (LNKEM). So the luminescence of the sample LNKEM was investigated and compared with that of $Y_2O_2S:Eu^{3+}$.

The XRD patterns of LNKEM and $Y_2O_2S:0.05Eu^{3+}$ are shown in Fig.1-20. Curve a shows that LNKEM is of a single-phase and consistent with the JCPDS card of $Na_{0.5}Gd_{0.5}MoO_4$ (No. 25-0828). It reveals that LNKEM shares a tetragonal scheelite structure. The XRD pattern of $Y_2O_2S:0.05Eu^{3+}$ is consistent with the JCPDS card of $Y_2O_2S$ (No. 24-1424) without the presence of the $Y_2O_3$ phase. It shows that $Y_2O_2S:0.05Eu^{3+}$ has a hexagonal structure with unit cell dimensions of $a = b = 3.7$ Å, $c = 6.5$ Å.

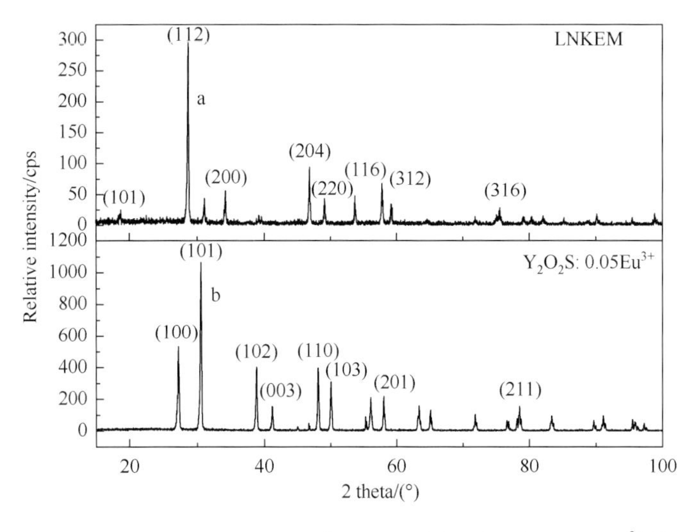


Fig.1-20 The XRD patterns of LNKEM (a) and $Y_2O_2S:0.05Eu^{3+}$(b)

The SEM micrograph of LNKEM is shown in Fig.1-21. These particles reveal highly crystalline with a diameter of about 2 μm, which is very fit to fabricate the solid-state lighting devices.

The excitation spectra of $Y_2O_2S:0.05Eu^{3+}$, LNKEM and $NaEu(MoO_4)_2$ are shown in Fig.1-22. Since the purpose of the present investigation is on the near-UV LED phosphor, only the spectroscopic properties in the range from 300 nm to 500 nm are exhibited in Fig.1-22. In curve a, the band from 300 nm to 390 nm is due to the CT transition of $Eu^{3+}\leftarrow S^{2-}$ in $Y_2O_2S$. In spectra (b, c), the lines in the 360-500 nm range are ascribed to the intra-configurational 4f-4f transitions of $Eu^{3+}$ in the host lattices. Among these peaks, the $^7F_0\rightarrow{}^5L_6$ and

Fig.1-21 The SEM micrograph of LNKEM

$^7F_0 \rightarrow {}^5D_2$ transitions at 395 nm and 465 nm are two of the strongest absorptions. Comparing the excitation spectrum of LNKEM with that of $NaEu(MoO_4)_2$, the f-f transitions of LNKEM were broadened. This may be due to the replacement of partial $Na^+$ ions by $Li^+$ and $K^+$ ions. Alkaline $Li^+$ ($1s^2$). $Na^+$ ($2s^22p^6$) and $K^+$ ($3s^23p^6$) ions are with similar electronic configurations of the noble gases. The ionic radii increase according to the relative order $Li^+ < Na^+ < K^+$, $Na^+$ ionic radius is lying between that of $Li^+$ and $K^+$. As a result, by partial substitution of $Na^+$ ions with smaller $Li^+$ and bigger $K^+$ in appropriate concentrations, it is probably to obtain solid solutions $(Li\text{-}Na\text{-}K)Eu(MoO_4)_2$. The XRD pattern of LNKEM in Fig.1-20, which is in good line with that of $NaEu(MoO_4)_2$, directly show that it is without any detectable impurity phase in the sample and confirm LNKEM is a solid solution. On the other hand, we have prepared the pure single compounds of $AEu(MoO_4)_2$ (A = Li, Na, and K) too, and indeed found that both $LiEu(MoO_4)_2$ and $NaEu(MoO_4)_2$ share tetragonal scheelite, whereas $KEu(MoO_4)_2$ has an order scheelite structure which is a triclinic system. For $AEu(MoO_4)_2$, the alkali metal ions and rare-earth ions are disordered in the same site. Mo (VI) is coordinated by four oxygen atoms in a tetrahedral site and the alkali metal ion/rare-earth ion site is eight coordinates with two sets of rare-oxygen distances. In $KEu(MoO_4)_2$, $Eu^{3+}$ polyhedra are ordered in sheets, forming a two-dimensional sub-lattice, and its site symmetry is $C_1$. In general, if the compound contains below about 5% of some secondary phases, it would be non-detectable on the X-ray patterns. Because $AEu(MoO_4)_2$ are of different structures, confirming that $KEu(MoO_4)_2$ exists as a secondary phase in the sample, it would be about 30% level in the sample. This higher secondary phase is easy to be found in XRD patterns. Hence, we believe LNKEM is a solid solution with a tetragonal scheelite structure and without an impurity phase. $A^+$ ions are the next-near coordination cations of $Eu^{3+}$ ions in the sample. They are with similar electronic configurations, the same electronic charge $Z$ and different ionic radii $r$. Therefore, they show different ionic potentials $Z/r$. The different ionic radii $r$ or ionic potentials $Z/r$ will result in the sub-lattice structure around the luminescent center ions which show diversity. The variation of sub-lattice structure will slightly influence the spin-orbit coupling and the crystal field on $Eu^{3+}$ ions. So, the spectroscopic lines of $Eu^{3+}$ ions are broadened comparing with the single pure compounds.

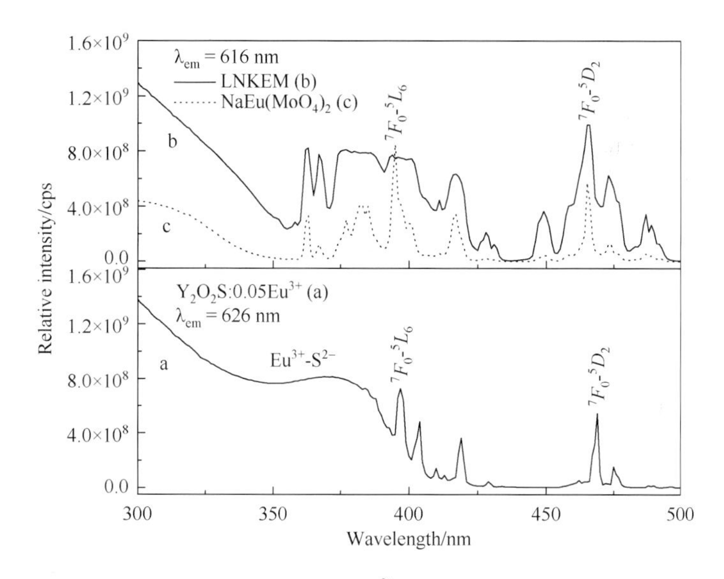


Fig.1-22 The excitation spectra of $Y_2O_2S:0.05Eu^{3+}$ (a) by monitoring 626 nm emission, LNKEM (b) and NaEu $(MoO_4)_2$ (c) by monitoring 616 nm emission

The emission spectra of LNKEM and $Y_2O_2S:0.05Eu^{3+}$ under 395 nm light excitation are shown in Fig.1-23. The main emission line in curve a is the $^5D_0\rightarrow^7F_2$ transition of $Eu^{3+}$ at 616 nm and other transitions from the $^5D_J$ levels to $^7F_J$ ground states are very weak, which is advantageous to obtain good CIE chromaticity coordinates. The results imply that $Eu^{3+}$ ions occupy the lattice sites without inversion symmetry. This is in good agreement with the structural results. $Y_2O_2S:0.05Eu^{3+}$ has a hexagonal structure and the point symmetry of the yttrium site is $C_{3V}$ (3 m). $Eu^{3+}$ was expected to occupy the $Y^{3+}$ site in $Y_2O_2S:0.05Eu^{3+}$. The main emission peaks at 627 nm and 616 nm of $Y_2O_2S:0.05Eu^{3+}$ are ascribed to the transition from the $^5D_0$ to $^7F_2$ and its strongest peak is at 627 nm. Other transitions from the $^5D_J$ ($J=0, 1, 2, 3$) levels to $^7F_J$ ($J=0, 1, 2, 3, 4, 5, 6$) ground states are very weak. Comparing spectrum (a) with the spectrum (b) in Fig.1-23, the following results can be found. Firstly, the emission intensity of LNKEM under 395 nm irradiation is about 5.4 times higher than that of $Y_2O_2S:0.05Eu^{3+}$. Secondly, the CIE chromaticity coordinates are calculated to be $x = 0.65$, $y = 0.35$ for LNKEM, and $x = 0.63$, $y = 0.35$ for $Y_2O_2S:0.05Eu^{3+}$. Compared with NTSC standard values for red, the CIE chromaticity coordinates of LNKEM were closer to the NTSC standard values than that of $Y_2O_2S:0.05Eu^{3+}$. These results imply that the PL properties of LNKEM may be better than those of $Y_2O_2S:0.05Eu^{3+}$ when they are applied in LED.

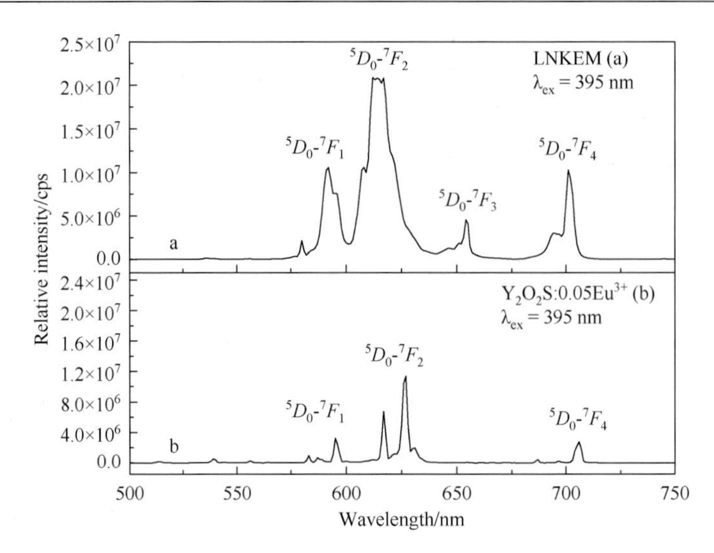


Fig.1-23 The emission spectra of LNKEM (a) and $Y_2O_2S$:0.05$Eu^{3+}$(b) under 395 nm excitation

Our purpose is to obtain a highly efficient red component for LED, so a red LED was fabricated with LNKEM as the red-emitting phosphor. Fig.1-24 shows the emission spectrum of the red light-emitting diode of near-UV InGaN-based on LNKEM. The band at 395 nm is attributed to the emission of the InGaN chip and the sharp peaks at 616 nm and 702 nm are due to the emissions of LNKEM. Brightly red light from the LED is observed by naked eyes. Its CIE chromaticity coordinates are calculated to be $x = 0.56$, $y = 0.27$.

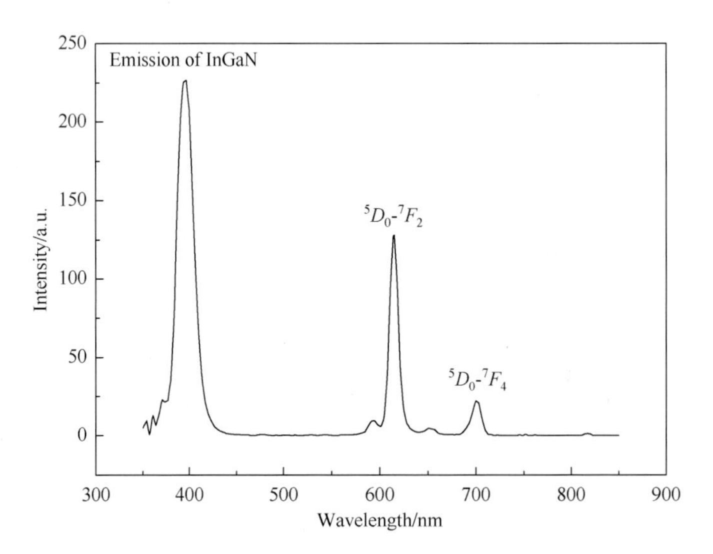


Fig.1-24 The EL spectrum of the red LED-based on LNKEM

The intensive emission of the InGaN chip can still be observed in Fig.1-24. It is advantageous to obtain a w-LED by combining this phosphor with appropriate blue and green phosphors since the most commonly used method is to combine red/green/blue tricolor phosphors with a GaN/InGaN chip. From the standpoint of application, each proper mono-color LED phosphor must meet the following necessary conditions. ① The phosphor must efficiently absorb the 400 nm excitation energy from the InGaN chip. But any mono-color phosphor cannot absorb all this energy; otherwise, another phosphor probably cannot be efficiently excited. ②The phosphor exhibits higher luminescent intensity under 400 nm excitation. ③The chromaticity coordinates of the phosphor are close to the NTSC standard values. Since LNKEM meets all these conditions, it is considered to be a good candidate for the red component of a three-band w-LED.

#### 1.1.3.6 $NaEu(MoO_4)_2{:}\,2xSO_4^{2-}$

In the above section, there were two approaches to broaden the absorption around 400 nm. One method is by co-doping $Sm^{3+}$ and $Eu^{3+}$ ions in the phosphor. Another method is to replace the cations or anions of this host compound, then the sub-lattice structure around the luminescent center ions will be expected to be somewhat diverse, hence the excitation band of the phosphor may be broadened. As further work, we have synthesized $NaEu(MoO_4)_2$ doped with $SO_4^{2-}$ red phosphors and investigated their luminescent properties. The anions $SO_4^{2-}$ and $MoO_4^{2-}$ are equivalent, but they show conspicuous differences in the ionic size and the electronic charge density. It is expected that this distinction will result in larger distortion on the sub-lattice structure and thus lead to the broadening and the intensifying of $Eu^{3+}$ absorption near 400 nm.

The XRD patterns of $NaEu(MoO_4)_2{:}\,2xSO_4^{2-}$ ($x = 0$, 0.10, 0.20, 0.30) are shown in Fig.1-25. Curve a shows that $NaEu(MoO_4)_2$ is of a single-phase and consistent with that given in JCPDS card No. 25-0828 [$Na_{0.5}Gd_{0.5}MoO_4$]. It reveals that $NaEu(MoO_4)_2$ has a scheelite structure with alkali metal ions and rare-earth ions disordered in the same site. $Mo^{6+}$ is coordinated by four $O^{2-}$ in a tetrahedral site. Curves b and c are similar to curve a, which shows the phosphors doped with a little $SO_4^{2-}$ are still of scheelite structure. When the $SO_4^{2-}$ content is over 20%, other phases slowly appear [Fig.1-25(d)]. This result shows that a little $SO_4^{2-}$ can be dissolved in the crystal lattice of $NaEu(MoO_4)_2$ to form a solid-state solution. However, due to the difference in the structure between $SO_4^{2-}$ and $MoO_4^{2-}$, the other phases will appear with the higher content $SO_4^{2-}$.

Fig.1-26 shows the excitation and emission spectra of $NaEu(MoO_4)_2{:}\,2xSO_4^{2-}$ ($x = 0$, 0.10). Curves a and b are the excitation spectra of $NaEu(MoO_4)_2$ and $NaEu(MoO_4)_2{:}0.20\,SO_4^{2-}$ by monitoring emission at 616 nm. The broadband from 250 nm to 350 nm is ascribed to the CT band of the O→Mo. The sharp lines in the 360-550 nm range are intra-configurational 4f-4f transitions of $Eu^{3+}$ in the host lattices, and two of the strongest absorptions are at 395 nm and 465 nm, which are attributable to the $^7F_0\rightarrow{}^5L_6$ and $^7F_0\rightarrow{}^5D_2$ transitions of $Eu^{3+}$, respectively. Comparing the spectra b with a, it can be found that the charge transfer band in $NaEu(MoO_4)_2{:}0.20\,SO_4^{2-}$ shows a little red-shift. The excitation intensity of the 4f-4f transitions

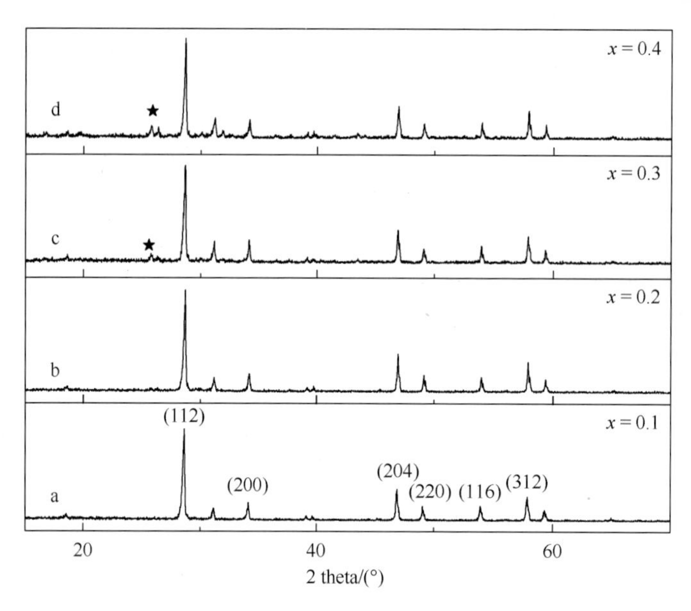


Fig.1-25　The XRD patterns of $NaEu(MoO_4)_2$: $2xSO_4^{2-}$ ($x$ = 0, 0.10, 0.20, 0.30)

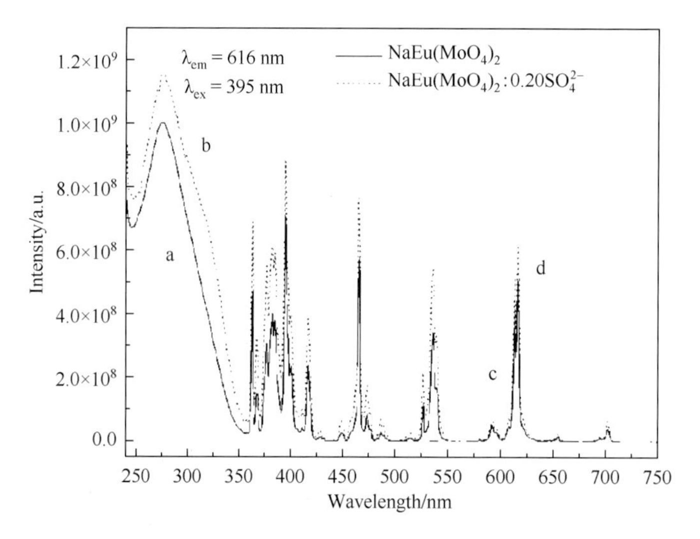


Fig.1-26　Excitation ($\lambda_{em}$ = 616 nm) and emission ($\lambda_{ex}$ = 395 nm) spectra of $NaEu(MoO_4)_2$: $2xSO_4^{2-}$ ($x$ = 0, 0.10)

of $Eu^{3+}$ is strengthened. In our previous works, we discussed that the charge transfer transition from the coordination ions (L) to center ions (M) is parity-allowed, whose energy is sensitive to the degree of the M—L bond covalency. The CT energy decreases with the increase of the M—L bond covalency. And the M—L bond covalency is influenced by the electronegative difference between M and L, the ligand number, and the M—L bond length. The bond covalency increases with the decrease of the electronegative difference of M and L, and the decrease of the M—L bond length. It is well known that the electronegativity of $S^{6+}$ (2.6) is bigger than that of $Mo^{6+}$ (1.6). The doping $SO_4^{2-}$ will change the overlapping magnitude of the electronic cloud for the Mo—O bond, resulting in the covalency of the Mo—O bond increases. Besides, the doping $SO_4^{2-}$ will result in larger distortion on the sub-lattice structure, which will probably decrease the M—L bond length. Thus, the covalency of the Mo—O bond increases too. With the increasing covalency of the Mo—O bond, the O→Mo CT band of $NaEu(MoO_4)_2$:0.20$SO_4^{2-}$ shows a red-shift, and the excitation intensity of the 4f-4f transitions of $Eu^{3+}$ is strengthened compared with that of $NaEu(MoO_4)_2$.

Curves c and d are the emission spectra of $NaEu(MoO_4)_2$ and $NaEu(MoO_4)_2$:0.20$SO_4^{2-}$ under 395 nm light excitation. The strongest emission peak is the $^5D_0 \rightarrow ^7F_2$ transition of $Eu^{3+}$ at 616 nm and other f-f transitions of $Eu^{3+}$ are very weak, which is advantageous to obtain a good phosphor with good CIE chromaticity coordinates. The emission intensity of $NaEu(MoO_4)_2$:0.20$SO_4^{2-}$ is higher than that of $NaEu(MoO_4)_2$. The CIE chromaticity coordinates for $NaEu(MoO_4)_2$:0.20$SO_4^{2-}$ are calculated to be $x = 0.66$, $y = 0.34$, which are close to the NTSC standard values. The concentration quenching curve is shown in Fig.1-27. When the $SO_4^{2-}$ content is up to 10%, the emission intensity is the strongest.

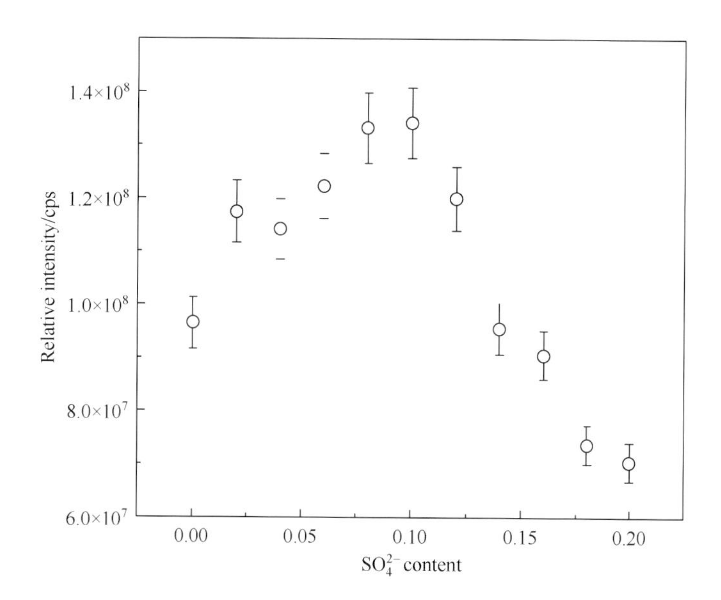


Fig.1-27 Concentration dependence of the relative emission intensity of $Eu^{3+}$ $^5D_0 \rightarrow ^7F_2$ transition in $NaEu(MoO_4)_2$: 2$x$$SO_4^{2-}$ ($\lambda_{ex}$ = 395 nm)

The decay curve for the $^5D_0 \rightarrow ^7F_2$ transition (616 nm) of the $Eu^{3+}$ in $NaEu(MoO_4)_2$:0.20$SO_4^{2-}$ has been measured, as shown in Fig.1-28. The lifetime value of $NaEu(MoO_4)_2$:0.20$SO_4^{2-}$ is 0.344 ms.

Fig.1-29 is the concentration dependence of the relative decay time of $Eu^{3+}$ $^5D_0 \rightarrow ^7F_2$ transition in $NaEu(MoO_4)_2$: 2$x$$SO_4^{2-}$. With the increase of the $SO_4^{2-}$ content, the value of decay time exhibits a down trend, meaning the increase of the non-radiative transition process among $Eu^{3+}$ ions with the increasing content of $SO_4^{2-}$.

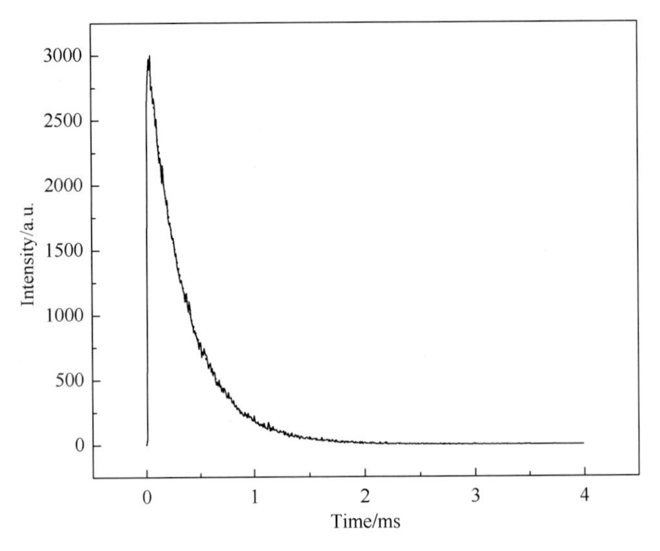


Fig.1-28 Decay curve of the $Eu^{3+}$ emission of $NaEu(MoO_4)_2$:0.20 $SO_4^{2-}$ ($\lambda_{ex}$ = 395 nm; $\lambda_{em}$ = 616 nm)

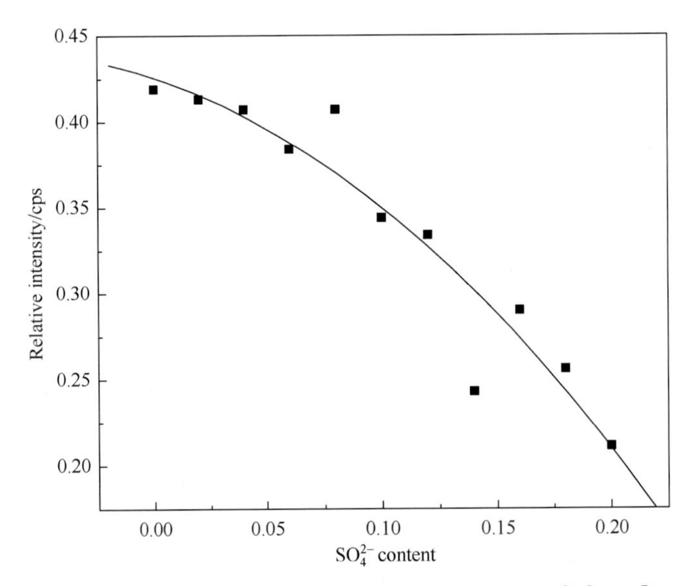


Fig.1-29 Concentration dependence of the relative decay time of $Eu^{3+}$ $^5D_0 \rightarrow {}^7F_2$ transition in $NaEu(MoO_4)_2$: $2xSO_4^{2-}$ ($\lambda_{ex}$ = 395 nm, $\lambda_{em}$ = 616 nm)

Our target is to search novel red phosphors for near-UV LED chips, then two single red LEDs were fabricated with $NaEu(MoO_4)_2$:0.20 $SO_4^{2-}$ and $NaEu(MoO_4)_2$. Fig.1-30 shows the emission spectra of the red LEDs of near-UV InGaN-based on $NaEu(MoO_4)_2$:0.20 $SO_4^{2-}$ and $NaEu(MoO_4)_2$. The band at 395 nm is attributed to the emission of the InGaN chip and red-emitting peaks at 616 nm and 702 nm are due to the emissions of the red phosphor. The red light emission intensity of the red LED-based $NaEu(MoO_4)_2$:0.20 $SO_4^{2-}$ is higher than that of the red LED-based on $NaEu(MoO_4)_2$. This result shows that $NaEu(MoO_4)_2$:0.20 $SO_4^{2-}$ is an excellent red-emitting component for w-LEDs.

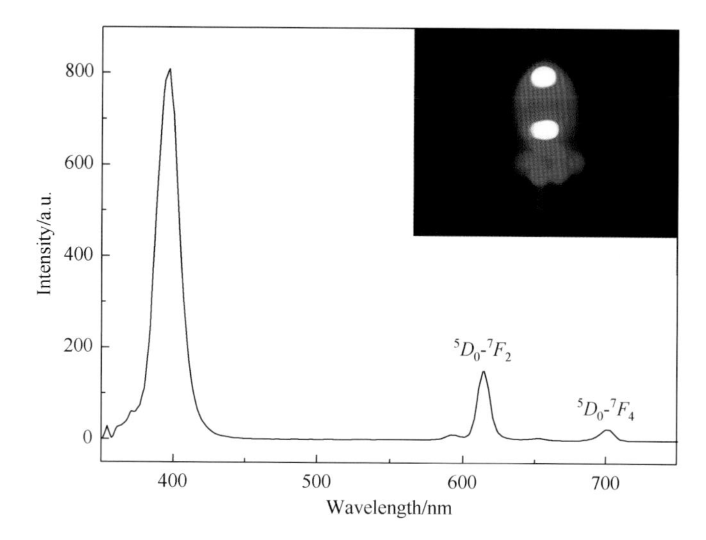


Fig.1-30 The EL spectrum of the $NaEu(MoO_4)_2$:0.20 $SO_4^{2-}$ -InGaN-based LED under 20 mA current excitation. The inserted figure is the lighting photograph of the red LED

1.1.3.7 $NaEu(MoO_4)_2$: $2xB_4O_7^{2-}$

Boron oxide is not only a good flux but also able to replace some compositions of the phosphors and enhance the luminescent properties. In the present work, we have synthesized $NaEu(MoO_4)_2$ doped with boron oxide red phosphor and investigated their luminescent properties.

The XRD patterns of $NaEu(MoO_4)_2$ doped with different boron oxide contents are shown in Fig.1-31. When the content of $H_3BO_3$ is less than 40%, the as-obtained samples show a pure phase with that of $NaEu(MoO_4)_2$. This result shows that the phosphors are of the same crystal structure as that of $NaEu(MoO_4)_2$ with a tetragonal scheelite structure. In the crystal structure of $NaEu(MoO_4)_2$, each $Mo^{6+}$ is coordinated by four $O^{2-}$ in a tetrahedral site, $Eu^{3+}$ and $Na^+$ are disordered in the same site and $Eu^{3+}/Na^+$ is coordinated by eight $O^{2-}$ of near four $MoO_4^{2-}$. The proper boron oxide can perfectly replace the $MoO_4^{2-}$ group in the tetragonal scheelite structure. When the content of $H_3BO_3$ is up to 40%, other phases slowly appear [Fig.1-31(d)]. Hence, the optimized content of $H_3BO_3$ is 30% in this case.

The excitation and emission spectra of the phosphors $NaEu(MoO_4)_2$: $2xB_4O_7^{2-}$ ($x = 0, 0.12$) are shown in Fig.1-32. The curves a and c are the excitation spectra of $NaEu(MoO_4)_2$ and $NaEu(MoO_4)_2$:0.24 $B_4O_7^{2-}$ by monitoring emission at 616 nm. The broadband from 250 nm to 350 nm is the CT band of the O→Mo. The intra-configurational 4f-4f transitions of $Eu^{3+}$ appear in the 350-550 nm range, and two of the strongest absorptions are located at 395 nm and 465 nm, which are attributable to the $^7F_0 \rightarrow ^5L_6$ and $^7F_0 \rightarrow ^5D_2$ transitions of $Eu^{3+}$, respectively. With the introduction of $H_3BO_3$, the CT band of the O→Mo shows a little red-shift, and the excitation intensity of the 4f-4f transitions of $Eu^{3+}$ is strengthened.

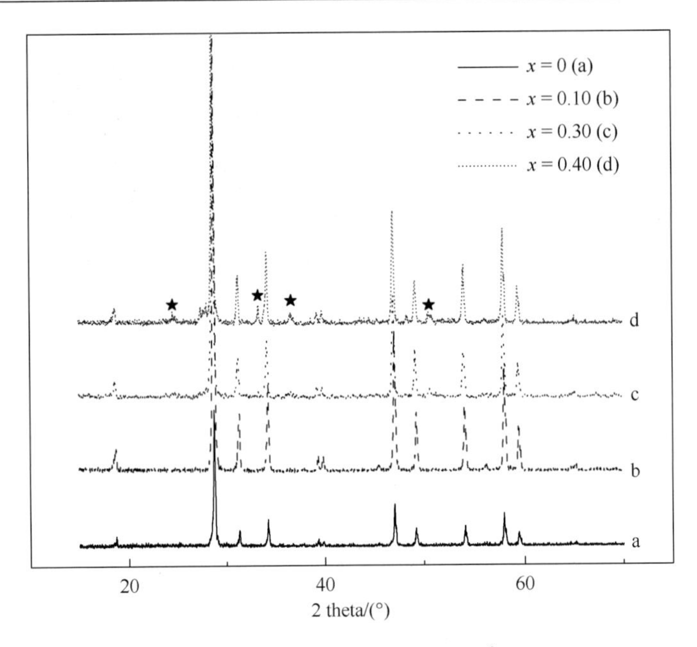


Fig.1-31 The XRD patterns of $NaEu(MoO_4)_2$: $2xB_4O_7^{2-}$ ($x$ = 0, 0.10, 0.30, 0.40)

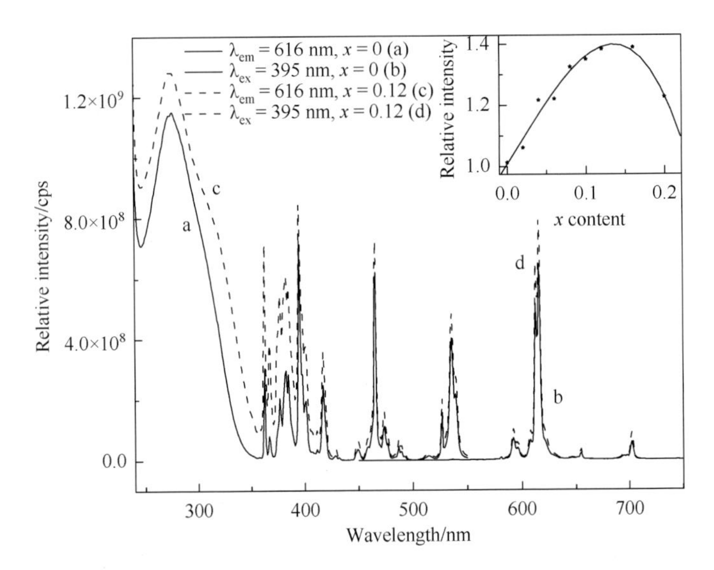


Fig.1-32 The excitation ($\lambda_{em}$ = 616 nm) and emission ($\lambda_{ex}$ = 395 nm) spectra of $NaEu(MoO_4)_2$: $2xB_4O_7^{2-}$ ($x$ = 0, 0.12). The inseted figure is the concentration dependence of the relative emission intensity of $Eu^{3+}$ $^5D_0 \rightarrow {}^7F_2$ transition in NaEu $(MoO_4)_2$: $2xB_4O_7^{2-}$ ($\lambda_{ex}$ = 395 nm)

In our previous work, we discussed that the charge transfer transition from the coordination ions (L) to center ions (M) is parity-allowed, whose energy is sensitive to the degree of the M—L bond covalency. The CT energy decreases with the increase of the M—L bond covalency. And the M—L bond covalency is influenced by the electronegative difference between M and L, the ligand number, and the M—L bond length. The bond covalency increases with the decreasing electronegative difference of M and L, and the decreasing of the M—L bond length. The electronegativity of $B^{3+}$ (2.0) is bigger than that of $Mo^{6+}$ (1.6). The introduction of $H_3BO_3$ will improve the overlapping magnitude of the electronic cloud for the Mo—O bond, and then Mo—O bond covalency will be increased. Moreover, the introduction of $H_3BO_3$ will result in some distortions on the sub-lattice structure, which will probably decrease the M—L bond length. Thus, the Mo—O bond covalency may also be increased. With the increase of the Mo—O bond covalency, the O→Mo CT band of $NaEu(MoO_4)_2$:0.24 $B_4O_7^{2-}$ shows a red-shift, and the excitation intensity of the 4f-4f transitions of $Eu^{3+}$ is strengthened compared with that of $NaEu(MoO_4)_2$.

The strongest emission peak (Fig.1-32) is the $^5D_0 \rightarrow {}^7F_2$ transition of $Eu^{3+}$ at 616 nm, which is advantageous to obtain a good phosphor with good CIE chromaticity coordinates. The emission intensity of $NaEu(MoO_4)_2$:0.24 $B_4O_7^{2-}$ is 1.37 times higher than that of $NaEu(MoO_4)_2$. The CIE chromaticity coordinates for $NaEu(MoO_4)_2$:0.24 $B_4O_7^{2-}$ are calculated to be $x = 0.66$, $y = 0.33$, which are close to the NTSC standard values. The emission intensity for phosphors $NaEu(MoO_4)_2$ doped with different content boron oxide was investigated as a function of the dopant concentrations $x$ as exhibited in insert of Fig.1-32; it can be observed that phosphors $NaEu(MoO_4)_2$:0.24 $B_4O_7^{2-}$ show the strongest emission under 395 nm excitation.

Our target is to search novel red phosphors for near-UV LED chips, then the single red light-emitting LED has been fabricated with the red phosphor $NaEu(MoO_4)_2$ doped with boron oxide. Fig.1-33 shows the emission spectrum of the red light-emitting diodes of near-UV InGaN-based $NaEu(MoO_4)_2$:0.24 $B_4O_7^{2-}$ under 20 mA current excitation. The band at 395 nm is attributed to the emission of the InGaN chip, and the emission peaks at 616 nm and 702 nm are due to the emissions of the red phosphors.

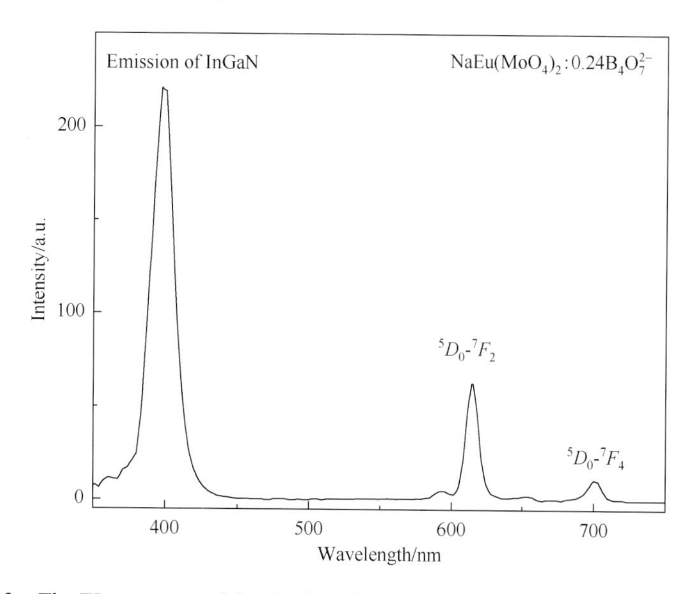


Fig.1-33 The EL spectrum of the single red LEDs-based on $NaEu(MoO_4)_2$:0.24 $B_4O_7^{2-}$

Fig.1-34 is the CIE diagram of $NaEu(MoO_4)_2$:0.24 $B_4O_7^{2-}$ -InGaN LED. Its CIE chromaticity coordinates are calculated to be $x$ = 0.49, $y$ = 0.23. The bright red light from the LED-based on $NaEu(MoO_4)_2$:0.24 $B_4O_7^{2-}$ can be observed by naked eyes when it is excited with 20 mA current.

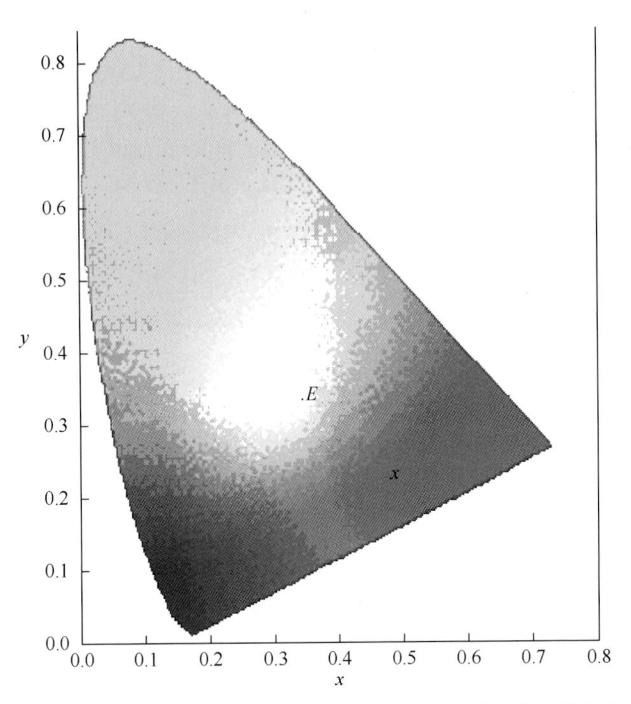


Fig.1-34　CIE diagram of the red LED-based on $NaEu(MoO_4)_2$:0.24 $B_4O_7^{2-}$

1.1.3.8　$NaEu(MoO_4)_2$:0.04$Sm^{3+}$ prepared by the Pechini process

In consideration of the application, the phosphor needs an appropriate particle shape and size. To the best of our knowledge, the double molybdates $AB(MoO_4)_2$ (A = $Li^+$, $Na^+$, $K^+$, $Rb^+$, $Cs^+$; B = trivalent rare-earth ions) particles were mainly prepared by the solid-state reaction technique. The Pechini method has the advantages of good homogeneity through mixing the starting materials at the molecular level in solution, a lower calcining temperature, and a shorter firing time. Hence, a phosphor with narrower particle size distribution is expected to be obtained by this method. In the present work, the phosphor $NaEu(MoO_4)_2$:0.04$Sm^{3+}$ was prepared by the Pechini method as a further and systematic work.

For $NaEu(MoO_4)_2$:0.04$Sm^{3+}$ prepared by the Pechini method fired at different temperatures, the formation of the crystalline phase can be seen in the XRD patterns as shown in Fig.1-35. The crystalline phase of the scheelite structure appears at 500 ℃ and the XRD patterns of the samples are consistent with that given in the JCPDS card No. 25-0828. The diffraction intensities are increased with the increase of the firing temperature due to the increased crystallinity. The XRD patterns of the samples prepared by the solid-state reaction are also in agreement with the standard card.

The morphology and grain size show an important influence on its PL properties when it is coated on a LED chip. Then the SEM micrographs of the phosphors were investigated. A series of SEM micrographs (Fig.1-36) reveal the morphology and grain size changes for the samples prepared by the Pechini method under the different firing temperatures. The particles fired under 500 ℃ have a spherical shape with an average diameter of 60 nm. With the increase of the firing temperature, the particle sizes increase. The sizes fired at 600 ℃ and 700 ℃ are about 300 nm and 400 nm, respectively. When the firing temperature is 800 ℃, the particles aggregate and show irregular shapes with a diameter of about 2 μm. Fig.1-36(e) shows the irregular morphology and size of the particles prepared by the solid-state reaction method. These particles show the heavy agglomerate phenomenon with a size of about 3μm. The result indicates that the uniform and superfine phosphor can be prepared by the Pechini method at lower temperatures.

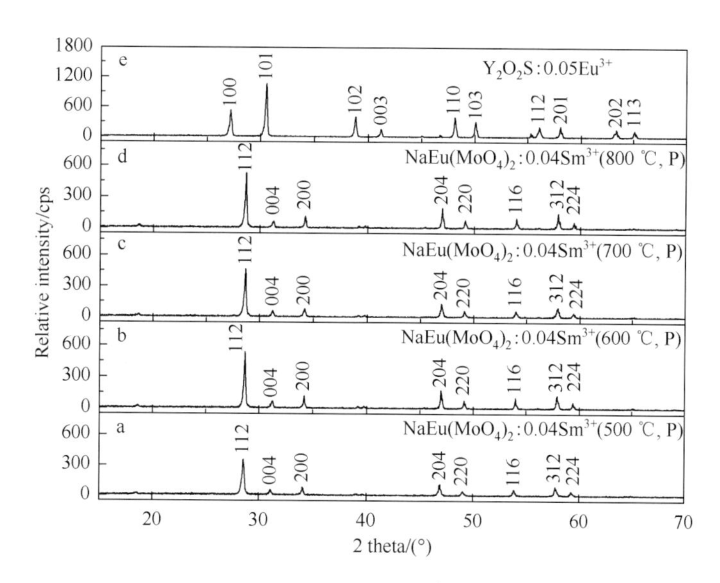


Fig.1-35 The XRD patterns of $NaEu(MoO_4)_2:0.04Sm^{3+}$ fired at 500 ℃ (a), 600 ℃ (b), 700 ℃ (c), 800 ℃ (d), and the XRD pattern of $Y_2O_2S:0.05Eu^{3+}$ (e)

$Eu^{3+}/Sm^{3+}$ ions incorporated in the molybdate host lattice exhibit strong absorption at about 395 nm/405 nm, which is close to the emission wavelength (400 nm) of InGaN-based near-UV LED. Hence, it is possible to obtain a phosphor with broadened absorption around 400 nm and intensive red emission by the $Eu^{3+}$-$Sm^{3+}$ co-doped system. In our previous work, it was found that the sample $NaEu(MoO_4)_2:0.04Sm^{3+}$ prepared by the solid-state reaction technique exhibits the strongest emission. So we prepared $NaEu(MoO_4)_2:0.04Sm^{3+}$ with the Pechini method here and investigated its luminescent properties.

Fig.1-37 shows the excitation spectra of $NaEu(MoO_4)_2:0.04Sm^{3+}$ prepared by the Pechini method (labeled as P in the figure) with the different firing temperatures of 500 ℃, 600 ℃, 700 ℃

and 800 ℃ by monitoring the emission at 616 nm. Four curves exhibit a similar spectroscopic feature, the broadband in the range of 250-350 nm is assignable to the O→Mo CT transition while the lines in the range from 360 nm to 550 nm belong to f-f transitions of $Eu^{3+}$ and $Sm^{3+}$ ions in the host lattices. The ${}^7F_0 \to {}^5L_6$ and ${}^7F_0 \to {}^5D_2$ transitions of $Eu^{3+}$ at 395 nm and 465 nm are two of the strongest absorptions. The line at 405 nm is due to the ${}^6H_{5/2} \to {}^4K_{11/2}$ transition of $Sm^{3+}$. Some differences also can be found in these spectra. The CT band and the lines in the range from 360 nm to 550 nm in the spectrum (d) are broader than those in spectra (a, b and c).

Fig.1-36 The SEM micrographs of $NaEu(MoO_4)_2$:0.04$Sm^{3+}$ fired at different temperatures: 500 ℃ (a), 600 ℃ (b), 700 ℃ (c), 800 ℃ (d) and that prepared by the solid-state reaction technique (e)

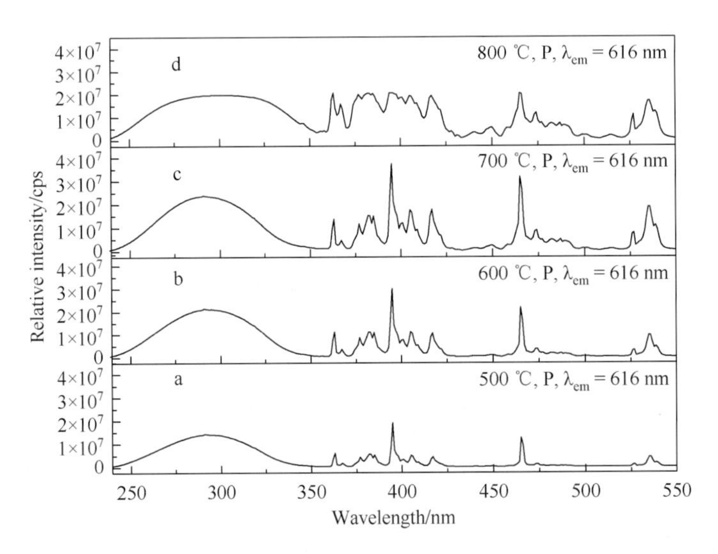


Fig.1-37 The excitation spectra of $NaEu(MoO_4)_2$:0.04$Sm^{3+}$ prepared by the Pechini method with the different firing temperatures: 500 ℃ (a), 600 ℃ (b), 700 ℃ (c), 800 ℃ (d)

Consulting the XRD patterns and the SEM micrographs, it can be seen that the phosphor prepared at 800 ℃ shows the best crystallinity, but with obvious aggregation, irregular shape, and the largest particle size, suggesting the morphologies of phosphors show some influences on the spectroscopic properties. Moreover, the center of the CT band is located at about 290 nm in curve a while it is at about 300 nm in the curve d, showing a slight shorter-wavelength shifting in nano-scale phosphor prepared by the Pechini method with the firing temperature of 500 ℃. This may be due to the quantum size effect of the phosphor nanoparticles. The reason is that the nano size of the phosphor increases the kinetic energy of the electrons and results in a larger bandgap. So, higher energy is required to excite the phosphor.

Fig.1-38 represents the emission spectra upon direct excitation of the $^7F_0 \rightarrow {}^5L_6$ transition of $Eu^{3+}$ at 395 nm or the $^6H_{5/2} \rightarrow {}^4K_{11/2}$ transition of $Sm^{3+}$ at 405 nm in the host lattice for $NaEu(MoO_4)_2$:$0.04Sm^{3+}$ prepared by the Pechini method at different annealing temperatures. The main emission is the $^5D_0 \rightarrow {}^7F_2$ transition of $Eu^{3+}$ at 616 nm. Other f-f transitions of $Eu^{3+}$ are very weak, which is advantageous to obtain a phosphor with good CIE chromaticity coordinates.

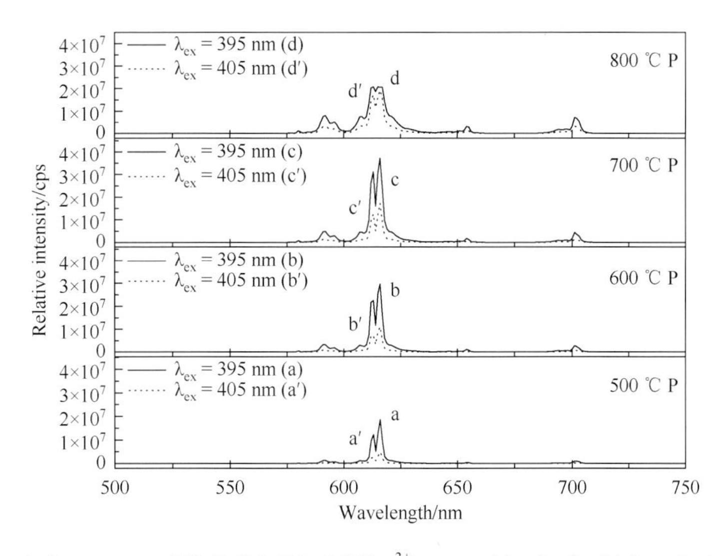


Fig.1-38 The emission spectra of $NaEu(MoO_4)_2$:$0.04Sm^{3+}$ prepared by the Pechini method with the different firing temperatures under 395 nm/405 nm excitation

The CIE chromaticity coordinates of the phosphors $NaEu(MoO_4)_2$:$0.04Sm^{3+}$ are calculated according to these curves, as listed in Table 1-4. It shows that the chromaticity coordinates are very close to the NTSC standard values. Moreover, it can be seen that the $^5D_0 \rightarrow {}^7F_2$ transitions show bad resolution in curves d and d′, which are in line with the excitation spectrum [Fig.1-37(d)]. The emission of $Sm^{3+}$ can not be observed in the emission spectra even under direct excitation of the $^6H_{5/2} \rightarrow {}^4K_{11/2}$ transition of $Sm^{3+}$ in the host lattice (Fig.1-38 a′, b′, c′, d′), indicating that $Sm^{3+}$ ions can absorb and efficiently transfer the excitation energy to $Eu^{3+}$ ions, the energy transfer process has been depicted in our previous work. The emission peaks of $Eu^{3+}$

got broader in the curve d. The reason may be as follows. Firstly, the irregular morphologies of phosphors prepared by the Pechini method maybe show some influences on the spectroscopic properties. Secondly, the organic component perhaps still exists in the phosphor, which may result in bad resolution phenomena.

**Table 1-4 The CIE chromaticity coordinates and $^5D_0$-$^7F_2$ relative emission intensity of phosphors under 395 nm and 405 nm excitation**

| Phosphor | Preparation condition | Excitation wavelength/nm | $^5D_0$-$^7F_2$ relative intensity | Chromaticity coordinates* | |
|---|---|---|---|---|---|
| | | | | $x$ | $y$ |
| $Y_2O_2S$:0.05$Eu^{3+}$ | solid-state reaction | 395 | 1.0 | 0.63 | 0.35 |
| | | 405 | 0.7 | 0.58 | 0.36 |
| $NaEu(MoO_4)_2$:0.04$Sm^{3+}$ | sol-gel method (500 ℃) | 395 | 1.1 | 0.66 | 0.33 |
| | | 405 | 0.4 | 0.59 | 0.35 |
| | sol-gel method (600 ℃) | 395 | 2.3 | 0.66 | 0.34 |
| | | 405 | 0.8 | 0.64 | 0.34 |
| | sol-gel method (700 ℃) | 395 | 3.2 | 0.66 | 0.34 |
| | | 405 | 1.2 | 0.65 | 0.34 |
| | sol-gel method (800 ℃) | 395 | 4.0 | 0.65 | 035 |
| | | 405 | 2.3 | 0.64 | 034 |
| | solid-state reaction | 395 | 4.2 | 0.66 | 0.34 |
| | | 405 | 2.0 | 0.65 | 0.34 |

* The NTSC standard values for red: $x$ = 0.67, $y$ = 0.33.

The phosphors $NaEu(MoO_4)_2$:0.04$Sm^{3+}$ with different firing temperatures show similar excitation and emission spectra, but their intensities are different. Therefore, the emission intensities of these phosphors under 395 nm/405 nm excitation have been compared, and the comparison of the integrated intensities for these phosphors is listed in Table 1-4. It can be found that the emission intensity under 395 nm/405 nm excitation of $NaEu(MoO_4)_2$:0.04$Sm^{3+}$ fired at 500 ℃ is the weakest and the intensity of $NaEu(MoO_4)_2$:0.04$Sm^{3+}$ increases with the increase of the firing temperature. In general, the two factors, either higher UV absorption or higher quantum efficiency will result in higher emission intensity. Comparing the excitation spectra in Fig.1-37 with the emission spectra in Fig.1-38, the change of emission intensity may be mainly the result of altering UV absorption around 400 nm in this case. ①A little organic component probably still exists at the lower temperature that reduces the absorption and emission intensity; ②the nanoparticles yield more defects and crystallographic distortions, which results in the weakest emission intensity of phosphor nanoparticles. With the increase of the firing temperature, the number of defects and distortion in the inner phosphors decreases, and the organic component disappears. So, the emission intensity increases.

Taking a comprehensive factor of the firing temperature into account, the luminescent intensities, the chromaticity coordinates, the particles morphology and the grain size, 700 ℃ is chosen as an optimum firing temperature to prepare the phosphor by the Pechini method. Hence the luminescent properties of the phosphor prepared by the Pechini method at 700 ℃ are compared with that prepared by solid-state reaction at 800 ℃. All the emission spectra are shown in Fig.1-39 and the spectroscopic data are listed in Table 1-4 for comparison. It can be found that the emission intensities of $NaEu(MoO_4)_2$:$0.04Sm^{3+}$ prepared by the Pechini method at 700 ℃ under 395 nm/405 nm excitation are close to that by solid-state reaction at 800 ℃ while the former has smaller particle size and narrower size distribution [Fig.1-36(c) and Fig.1-36(e)], which is favorable to the fabrication of LEDs.

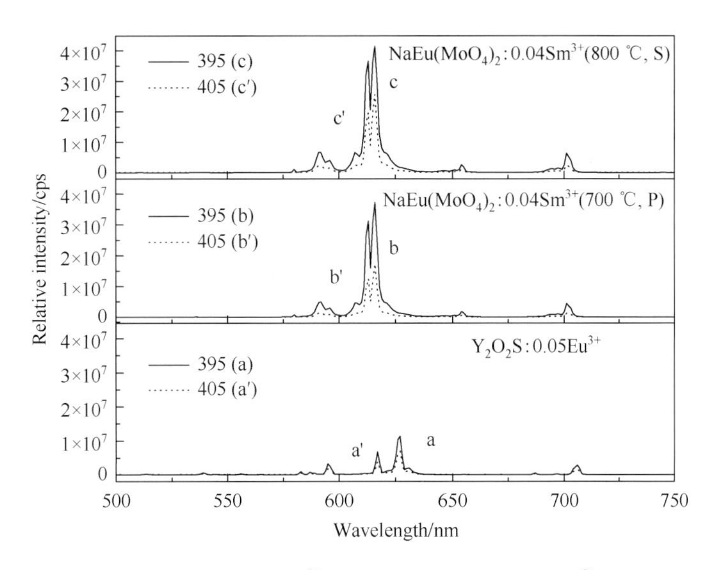


Fig.1-39 The emission spectra of $Y_2O_2S$:$0.05Eu^{3+}$(a/a′), $NaEu(MoO_4)_2$:$0.04Sm^{3+}$(b/b′) prepared by the Pechini method with the firing temperature at 700 ℃(labeled as P) and $NaEu(MoO_4)_2$:$0.04Sm^{3+}$ (c/c′) prepared by the solid-state reaction technique at 800 ℃ (labeled as S) under 395 nm/405 nm excitation

$Y_2O_2S$:$0.05Eu^{3+}$, the red phosphor currently used in the near-UV LED was also prepared. Its emission spectra under 395 nm/405 nm excitation are shown in Fig.1-39(a, a′). The main emission peaks at 627 nm and 616 nm of $Y_2O_2S$:$0.05Eu^{3+}$ are ascribed to the $^5D_0$ to $^7F_2$ transition of $Eu^{3+}$ with the strongest peak at 627 nm, whose results in its CIE chromaticity coordinates are not as good as that of $NaEu(MoO_4)_2$:$0.04Sm^{3+}$. Its emission intensities with the calculated CIE chromaticity coordinates under 395 nm/405 nm excitation were compared with that of $NaEu(MoO_4)_2$:$0.04Sm^{3+}$ (Table 1-4).The results show that the emission intensities of $NaEu(MoO_4)_2$:$0.04Sm^{3+}$ prepared by the Pechini method with the firing temperature of 700 ℃ are much stronger than that of $Y_2O_2S$: $0.05Eu^{3+}$, and its CIE chromaticity coordinates are closer to the NTSC standard values than that of $Y_2O_2S$:$0.05Eu^{3+}$.

As discussed above, $NaEu(MoO_4)_2$:$0.04Sm^{3+}$ may be a candidate to substitute conventional

$Y_2O_2S$:0.05$Eu^{3+}$ for the near-UV LED chips. We fabricated the red LEDs with $NaEu(MoO_4)_2$: 0.04$Sm^{3+}$ prepared by two different methods for comparison. Fig.1-40 shows the EL spectra of the red LEDs-based on $NaEu(MoO_4)_2$:0.04$Sm^{3+}$ prepared by the Pechini method with the firing temperature at 700 ℃ and by the solid-state reaction technique at 800 ℃ The near-UV emission at about 400 nm in curve a is ascribed to the emission of the InGaN chip, and the sharp peaks at 615 nm and 702 nm are due to the emissions of $NaEu(MoO_4)_2$:0.04$Sm^{3+}$ phosphor. Intense red light emission from the LED can be observed by naked eyes, and its CIE chromaticity coordinates are calculated to be $x = 0.47$, $y = 0.23$. Curve b shows a similar feature with curve a. The ratio of the emission intensity of the InGaN chip to that of $NaEu(MoO_4)_2$:0.04$Sm^{3+}$ is calculated to be 3.9 and 3.6 for the phosphors prepared by the Pechini method at 700 ℃ and the solid-state reaction at 800 ℃, respectively. It shows the emission intensities of the phosphor prepared by different methods are very close. Mono-phosphor $NaEu(MoO_4)_2$:0.04$Sm^{3+}$ can not absorb the whole 400 nm emission from the LED chip.

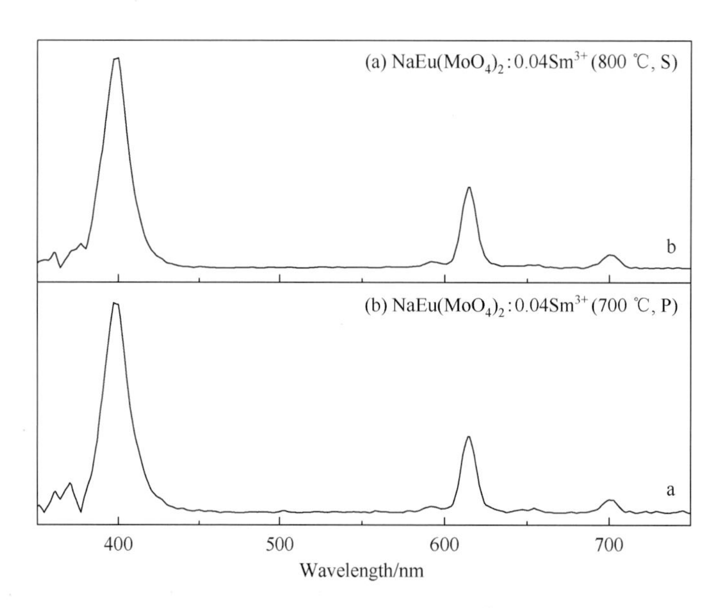


Fig.1-40 The EL spectra of red LEDs-based on $Eu(MoO_4)_2$:0.04$Sm^{3+}$ prepared by the Pechini method with the firing temperature at 700 ℃ (a), prepared by solid-state reaction technique at 800 ℃ (b)

A strong 400 nm emission band is still observed as shown in Fig.1-40. This is another characteristic of phosphor $NaEu(MoO_4)_2$:0.04$Sm^{3+}$. We think this feature will probably make it favorable to fabricate a w-LED by combining this phosphor with appropriate blue and green phosphors. The residual 400 nm emission of the LED chip might be used to excite green and blue phosphors in devices. The phosphor $NaEu(MoO_4)_2$:0.04$Sm^{3+}$ prepared by the Pechini method with the firing temperature at 700 ℃ shares suitable particle size and intense red light emission, so it may be a good candidate for the red component of a three-band w-LED.

#### 1.1.3.9 $NaEu(MoO_4)_2$ prepared by the combustion method

As we know, the particle size of the phosphor prepared by the conventional solid-state reaction is big with irregular morphology, which is not suitable for the application. Therefore, the wet-chemical methods seem to be an attractive alternative to the classical approach. The combustion process to prepare the phosphor is very facile, and it only takes a few minutes, which has been extensively applied to the preparation of various nano-scale inorganic phosphors. So $NaEu(MoO_4)_2$ was prepared by the combustion method, and its PL properties were investigated.

The XRD patterns of $NaEu(MoO_4)_2$ prepared by the combustion method fired at 600 ℃, 700 ℃, 800 ℃ and prepared by the solid-state reaction at 800 ℃ are shown in Fig.1-41. The single crystalline phase of a scheelite structure has been formed at 600 ℃, and the XRD patterns of the samples are consistent with that given in JCPDS card No. 25-0828 ($Na_{0.5}Gd_{0.5}MoO_4$). The result shows that these phosphors are of a single phase with the scheelite structure.

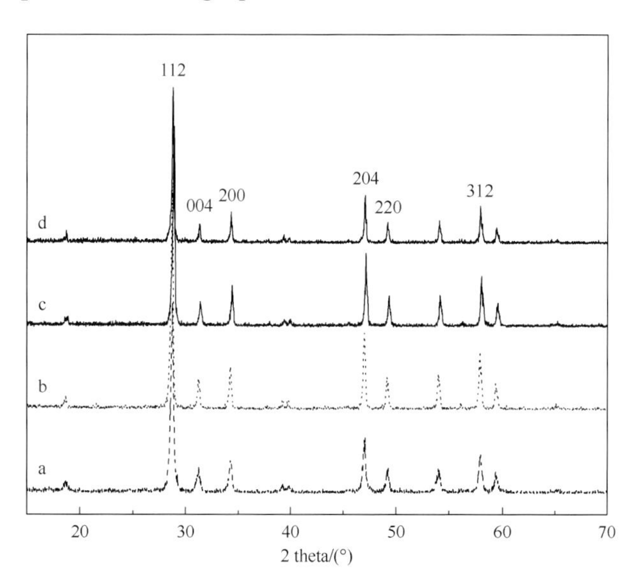


Fig.1-41 The XRD patterns of $NaEu(MoO_4)_2$ by the combustion method fired at 600 ℃ (a), 700 ℃ (b), 800 ℃ (c) and by the solid-state reaction at 800 ℃ (d)

The SEM micrographs of $NaEu(MoO_4)_2$ prepared by the combustion method under various firing temperatures are shown in Fig.1-42. The particles fired at 600 ℃ have a spherical shape with an average diameter of about 200 nm. With the increase of the firing temperature, the particle sizes increase. The particle sizes fired at 700 ℃ and 800 ℃ are about 300 nm and 500 nm respectively. The morphology is regular and uniform. The SEM micrograph of the sample prepared by the solid-state reaction at 800 ℃ is shown in Fig.1-42(d). These particles show a heavy agglomerate phenomenon with irregular morphology. The result indicates that the uniform and superfine phosphor can be obtained by the combustion method.

Fig.1-42 The SEM micrographs of $NaEu(MoO_4)_2$ fired at different temperatures: (a) 600 ℃, (b) 700 ℃, (c) 800 ℃ and (d) that prepared by the solid-state reaction technique

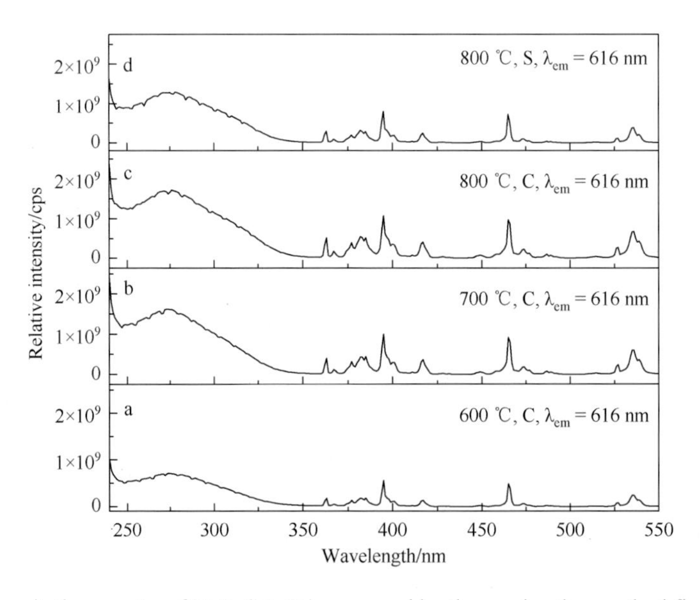


Fig.1-43 The excitation spectra of $NaEu(MoO_4)_2$ prepared by the combustion method fired at (a) 600 ℃, (b) 700 ℃, (c) 800 ℃ and (d) that prepared by the solid-state reaction technique

The excitation spectra of $NaEu(MoO_4)_2$ prepared by the combustion with the various temperatures and by monitoring the emission at 616 nm are shown in Fig.1-43. These four curves are of similar features. The broadband between 250 nm to 350 nm is assignable to the O→Mo CT transition, the sharp peaks in the 360-550 nm range belong to f-f transitions of $Eu^{3+}$ in the host lattices. The strongest absorption peak is at 395 nm, which is due to the $^7F_0 \rightarrow {}^5L_6$ transitions of $Eu^{3+}$. With the increase of the firing temperature, the excitation intensities of the phosphors also increase. The intensity of the sample prepared by the combustion method fired at 800 ℃ is higher than that prepared by the solid-state reaction.

Fig.1-44 represents the emission spectra of the phosphors under 395 nm excitation. The main emission is the $^5D_0 \rightarrow {}^7F_2$ transition of $Eu^{3+}$ at 616 nm. Other f-f transitions of $Eu^{3+}$ are very weak, which is advantageous to obtain a phosphor with good CIE chromaticity coordinates.

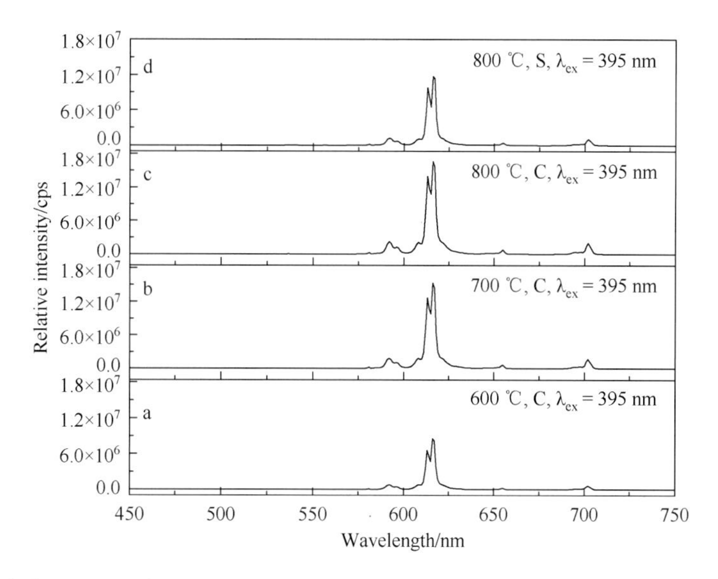


Fig.1-44 The emission spectra of $NaEu(MoO_4)_2$ prepared by the combustion method fired at (a) 600 ℃, (b) 700 ℃, (c) 800 ℃ and (d) that prepared by the solid-state reaction technique under 395 nm excitation

The CIE chromaticity coordinates of the phosphors $NaEu(MoO_4)_2$ are calculated to be $x = 0.66$, $y = 0.33$ from these curves. The emission intensities of the phosphors prepared by different methods are compared. The emission intensity of the phosphor prepared by the combustion method with the firing temperature at 800 ℃ is about 1.6 times stronger than that prepared by the solid-state reaction fired at 800 ℃. With the increasing temperature, the grains size increases. Meanwhile, many defects and distortions in the inner phosphors decrease, and the organic component disappears. So the emission intensity increases. The sample prepared by the solid-state reaction shows the heavy agglomerate phenomenon, which influences the absorption of the phosphor in the UV light region. Then the emission intensity is lower than that prepared by the combustion method fired at 800 ℃.

## 1.2 $Eu^{3+}$-activated double tungstates red phosphors for the near-UV LED chips

### 1.2.1 Introduction

Double alkaline rare-earth tungstates $AB(WO_4)_2$ ($A = Li^+$, $Na^+$; B = trivalent rare-earth ions) share the tetragonal scheelite-like ($CaWO_4$) iso-structure. The alkali-metal ions and rare-earth ions are disordered in the same site. W (VI) is coordinated by four oxygen atoms in a tetrahedral site and the alkali metal ion/rare-earth ion site is eight coordinated with two sets of rare-oxygen distances. In the past, the investigations on the double tungstates $AB(WO_4)_2$ focused on their single crystals to take advantage of their laser properties. Recently, the powder samples of these compounds were investigated for they might be promising candidates as phosphors for solid-state lighting.

In this section, the red phosphors double tungstates doped or co-doped with $Eu^{3+}$, $Bi^{3+}$ and $Sm^{3+}$ were prepared by the solid-state method, and their PL properties were investigated.

### 1.2.2 Preparation method

The relative red phosphors have been prepared by the solid-state reaction technique at high temperature. The starting stoichiometric mixtures are $WO_3$ (A. R. grade), $NaHCO_3$ (A.R. grade), (A.R. grade), $La_2O_3$(99.99% purity), $Sm_2O_3$ (99.99% purity) and $Eu_2O_3$ (99.99% purity). They were first ground and pre-fired at 500 ℃ for 4 h, then heated at 800 ℃ for 4 h.

### 1.2.3 Structures and PL properties

#### 1.2.3.1 $NaLa(WO_4)_2{:}xEu^{3+}$

The XRD patterns of $NaLa(WO_4)_2{:}xEu^{3+}$ ($x = 0$, 0.05, 1.0) are shown in Fig.1-45. The XRD patterns of the sample $NaLa(WO_4)_2{:}xEu^{3+}$ ($x = 0$, 0.05, 1.0) are of similar shapes, which are consistent with the JCPDS card No. 79-1118 [$NaLa(WO_4)_2$]. It shows that these compounds are of the single-phase with the tetragonal scheelite-like ($CaWO_4$) isostructure. $Eu^{3+}$ ions occupy the sites of the $La^{3+}$ ions in the crystal structure.

The PL spectra of $NaLa(WO_4)_2{:}xEu^{3+}$ samples ($x = 0$, 0.10, 0.20, 0.30, 0.40, 0.50, 0.60, 0.70, 0.80, 0.90, 1.00) are measured. The spectra of $NaLa(WO_4)_2{:}xEu^{3+}$ for $x = 0.10$, 0.50, and 1.00 are shown in Fig.1-46. Curves a, c and e are the excitation spectra of the $NaLa(WO_4)_2{:}xEu^{3+}$ ($x = 0.10$, 0.50, and 1.00) monitored for the $Eu^{3+}$ emission (616 nm). The broad excitation band from 250 nm to 300 nm in the spectrum (a) is attributable to the O→W CT transition. The

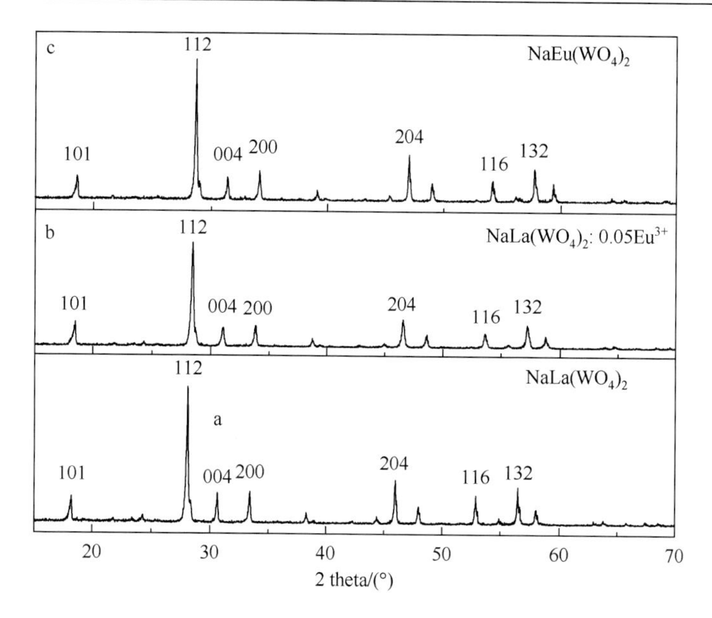


Fig.1-45 The XRD patterns of $NaLa(WO_4)_2$:$xEu^{3+}$ ($x = 0, 0.05, 1.0$)

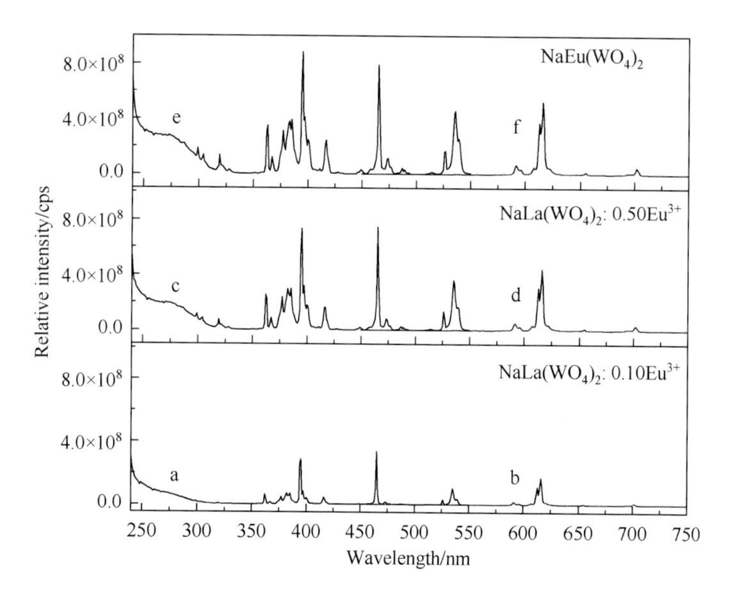


Fig.1-46 PL spectra of $NaLa(WO_4)_2$:$xEu^{3+}$ ($x = 0.1, 0.5, 1.0$)

sharp lines in the 350-550 nm range are intra-configurational 4f-4f transitions of $Eu^{3+}$ in the host lattices. The $^7F_0\rightarrow^5L_6$ and $^7F_0\rightarrow^5D_2$ transitions at 395 nm and 465 nm are two of the strongest absorptions among these absorption peaks. With the increase of the $Eu^{3+}$ content, the excitation intensity is also increasing. Some sharp peaks of $Eu^{3+}$ in the range from 290 nm to 350 nm appear in the excitation spectra. They are due to the $^7F_0\rightarrow^5F_{2,4}$ and $^7F_0\rightarrow^5H_3$ transitions of $Eu^{3+}$ respectively. Curves b, d and f represent the emission spectra under excitation at 395 nm. The main emission line is the $^5D_0\rightarrow^7F_2$ transition of $Eu^{3+}$ at 616 nm. Other f-f transitions (such as $^5D_0 \rightarrow {}^7F_{1,3,4}$ transition, $^5D_0 \rightarrow {}^7F_J$ transitions) are very weak. The intensity of the $^5D_0 \rightarrow {}^7F_2$ transition (magnetic dipole transition) is about 6.9 times higher than that of the $^5D_0 \rightarrow {}^7F_1$ transition (electric dipole transition) in this case. This result shows the magnetic dipole transitions of $Eu^{3+}$ are allowed in these phosphors. Hence, $Eu^{3+}$ ions mainly occupy the lattice sites without inversion symmetry, which is in good agreement with the results of the crystal structure.

With the increase of the $Eu^{3+}$ content, the emission intensity of the phosphors is also increasing (Fig.1-47). The emission intensity of $NaEu(WO_4)_2$ is about 3.4 times higher than that of $NaLa(WO_4)_2$:0.10$Eu^{3+}$. No concentration quenching of $Eu^{3+}$ is observed in these samples of $NaLa(WO_4)_2$:$xEu^{3+}$.This phenomenon also can be observed in the series of phosphors $NaLa(MoO_4)_2$:$xEu^{3+}$. Hence, $NaLa(WO_4)_2$ is a good host for doping $Eu^{3+}$.

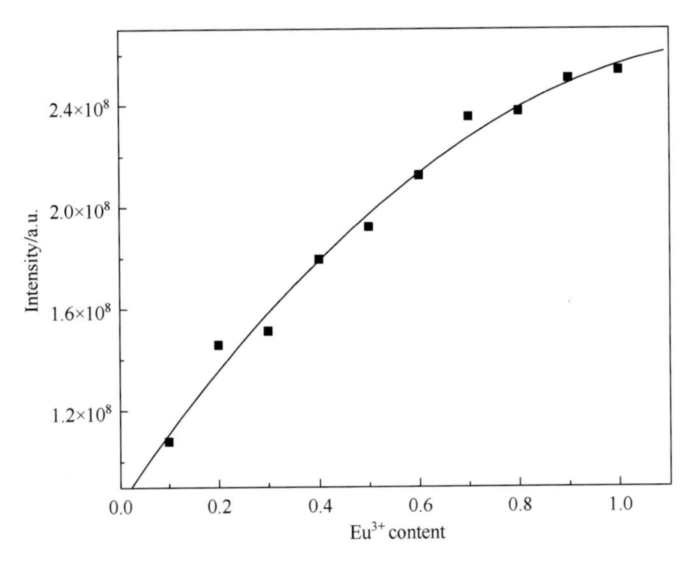


Fig.1-47 Concentration dependence of the relative emission intensity of $Eu^{3+}$ $^5D_0 \rightarrow {}^7F_2$ transition in $NaLa(WO_4)_2$:$xEu^{3+}$

#### 1.2.3.2 $NaEu(WO_4)_2$:$xBi^{3+}$

Fig.1-48 shows the XRD patterns of $NaEu(WO_4)_2$:$xBi^{3+}$ ($x = 0.05$, 0.50). These two curves are consistent with the JCPDS card No. 79-1118 [$NaLa(WO_4)_2$]. This shows that these phosphors doped with the different $Bi^{3+}$ ions content are still of the single-phase with the tetragonal scheelite-like ($CaWO_4$) isostructure. It is expected that $Bi^{3+}$ occupies the same site as $Eu^{3+}$ in the as-obtained phosphor. However, the *d* (interplanar distance) values are slightly different when varying $Bi^{3+}$ content. It may be because the $Bi^{3+}$ ion radius (131 pm, eight-fold coordination) is a little bigger than that of $Eu^{3+}$ (121 pm, eight-fold coordination), which results in a $Bi^{3+}$-O distance being larger than the $Eu^{3+}$-O distance. So the *d* values are different.

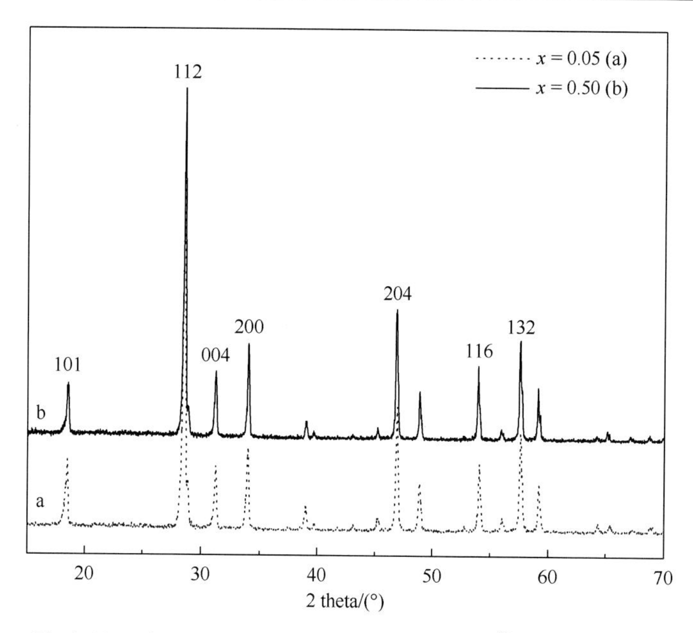


Fig.1-48 The XRD patterns of $NaEu(WO_4)_2$:$xBi^{3+}$ ($x$ = 0.05, 0.50)

Fig.1-49 shows the excitation and emission spectra of $NaEu(WO_4)_2$:$xBi^{3+}$ ($x$ = 0, 0.05, 0.30). The broad excitation band (curve a) from 250 nm to 290 nm is due to the O→W CT band, and the sharp peaks from 290 nm to 550 nm are ascribed to the intra-configurational 4f-4f transitions of $Eu^{3+}$ in the host lattice. With the introduction of $Bi^{3+}$ ions into the phosphors, some changes can be observed in the excitation band from 290 nm to 350 nm. A rather broad shoulder band appears at the longer wavelength side of the O→W CT band in curves c and e. With the increasing of $Bi^{3+}$ ions content further, the broad shoulder band overlaps a portion of the 4f-4f transitions (such as $^7F_0 \to ^5F_{2,\ 4}$, and $^7F_0 \to ^5H_3$ transitions) of $Eu^{3+}$. Besides, the intensity of the O→W CT band and the 4f-4f transitions of $Eu^{3+}$ are also increasing with the introduction of the $Bi^{3+}$ ions. The broad excitation band from 300 nm to 350 nm may be due to the absorption of $Bi^{3+}$. It is well known that the $Bi^{3+}$ ion has an outer $6s^2$ electronic configuration with the ground state of $^1S_0$. The excited states have a 6s6p configuration and split into the $^3P_0$, $^3P_1$, $^3P_2$ and $^1P_1$ levels in a sequence with the energy increase. The transitions from $^1S_0$ to $^3P_0$ and $^3P_2$ are completely spin forbidden if no other configurations are taken into account. The two levels, $^3P_1$ and $^1P_1$, are mixed by spin-orbit coupling. The broad-band absorption from 300 nm to 350 nm is an overlap of the transitions from the ground state $^1S_0$ to $^3P_1$ and $^1P_1$. The excitation intensity around 395 nm in $NaEu(WO_4)_2$:$0.30Bi^{3+}$ is the highest. When the $Bi^{3+}$ content is over 30%, the excitation intensities of $Eu^{3+}$ are gradually decreased. Curves b, d and f represent the emission spectra of $NaEu(WO_4)_2$:$xBi^{3+}$ ($x$ = 0, 0.10, 0.20, 0.30) under 395 nm excitation. These curves are of similar shapes. The strongest peak is at 616 nm, due to the $^5D_0 \to ^7F_2$ transition of $Eu^{3+}$. Other f-f transitions of $Eu^{3+}$ are very weak. The results are constant with those of emission spectra of $NaLa(WO_4)_2$:$xEu^{3+}$ samples. So the introduction of $Bi^{3+}$ ions does not alter the sub-lattice structure around the luminescent center ions $Eu^{3+}$. This is in good agreement with the result of XRD characterization.

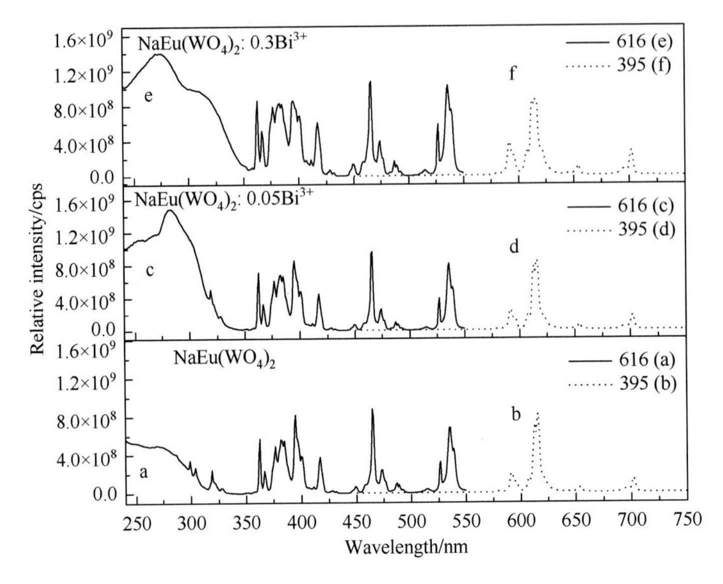


Fig.1-49　The excitation and emission spectra of $NaEu(WO_4)_2$:$xBi^{3+}$ ($x$ = 0, 0.05, 0.30)

The relative emission intensities of the $^5D_0\rightarrow^7F_2$ transition and CIE chromaticity coordinates are listed in Table 1-5. These phosphors show excellent CIE chromaticity coordinates are close to NTSC standard values. The emission intensities of phosphors doped with $Bi^{3+}$ under 395 nm excitation are higher than that of $NaEu(WO_4)_2$. The emission intensity of $NaEu(WO_4)_2$:$0.30Bi^{3+}$ is 1.6 times stronger than that of $NaEu(WO_4)_2$.

**Table 1-5　The $^5D_0$-$^7F_2$ relative emission intensity and the CIE values of the phosphors**

| Phosphor | Excitation wavelength/nm | $^5D_0$-$^7F_2$ relative intensity | CIE chromaticity coordinates* | |
|---|---|---|---|---|
| | | | $x$ | $y$ |
| $NaEu(WO_4)_2$ | 395 | 1.00 | 0.66 | 0.34 |
| $NaEu(WO_4)_2$:$0.05Bi^{3+}$ | 395 | 1.17 | 0.66 | 0.34 |
| $NaEu(WO_4)_2$:$0.30Bi^{3+}$ | 395 | 1.60 | 0.66 | 0.34 |

* The NTSC standard values for red are $x$ = 0.67, $y$ = 0.33.

### 1.2.3.3　$NaEu(WO_4)_2$:$xSm^{3+}$

As discussed above, $Sm^{3+}$ is an excellent luminescent center and a sensitizer with the main absorption around 405 nm in most host lattices. Therefore, we investigated the $Sm^{3+}$-$Eu^{3+}$ co-doping system in $NaLa(WO_4)_2$.

Fig.1-50 is the excitation and emission spectra of $NaLa(WO_4)_2$:$0.04Sm^{3+}$. These sharp lines in the 340-550 nm range are due to the f-f transitions of $Sm^{3+}$ with the strongest absorption of the ${}^6H_{5/2}\rightarrow{}^4K_{11/2}$ transition located at 404 nm. Curve b represents the emissions under 404 nm excitation. The spectra show typical $Sm^{3+}$ f-f transitions with three main peaks at 563 nm, 597 nm and 644 nm, which are corresponding to the ${}^4G_{5/2}\rightarrow{}^6H_{5/2}$, ${}^4G_{5/2}\rightarrow{}^6H_{7/2}$ and ${}^4G_{5/2}\rightarrow{}^6H_{9/2}$ transitions, respectively.

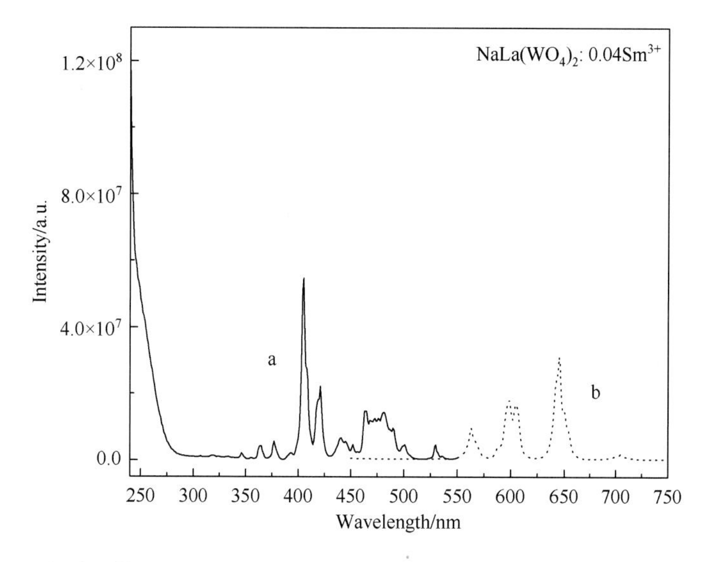


Fig.1-50 The excitation ($\lambda_{em}$ = 645 nm) and emission ($\lambda_{ex}$ = 405 nm) spectra of $NaLa(WO_4)_2$:$0.04Sm^{3+}$

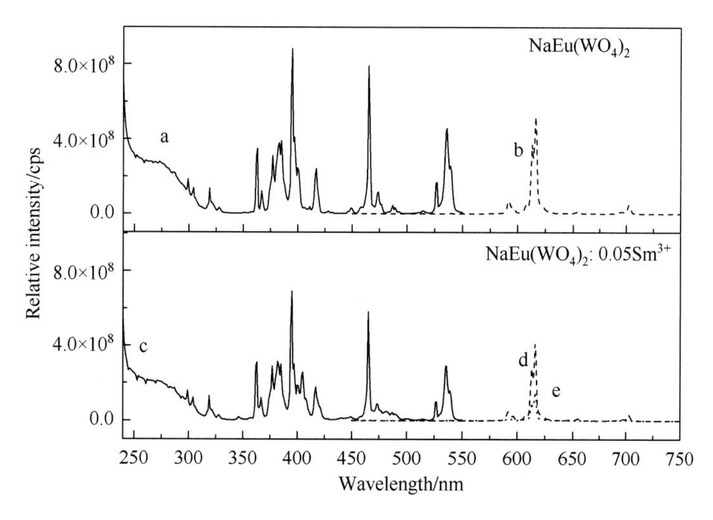


Fig.1-51 The photo-luminescent spectra of $NaEu(WO_4)_2$:$xSm^{3+}$ ($x$ = 0, 0.05)
(a), (c): $\lambda_{em}$ = 616 nm; (b), (d): $\lambda_{ex}$ = 395 nm; (e): $\lambda_{ex}$ = 405 nm

Since no concentration quenching is observed in the series samples of $NaLa(WO_4)_2$:$xEu^{3+}$, the PL properties of $NaEu(WO_4)_2$ doped with $Sm^{3+}$ are studied. Fig.1-51 exhibits the excitation and emission spectra of $NaEu(WO_4)_2$:$xEu^{3+}$ ($x$ = 0, 0.05). Compared with the excitation spectrum of $NaEu(WO_4)_2$ (curve a), the excitation spectrum of $NaEu(WO_4)_2$:$0.05Sm^{3+}$ presents the typical $Sm^{3+}$

f-f transition at 405 nm, which broadens the excitation intensity of this phosphor in the blue light range. Moreover, only emission peaks of $Eu^{3+}$ can be found in the emission spectrum of $NaEu(WO_4)_2$:0.05$Sm^{3+}$ under 405 nm light excitation. No emission of $Sm^{3+}$ can be observed. This means that $Sm^{3+}$ can transfer energy efficiently to $Eu^{3+}$.

Fig.1-52 is the decay curves of $NaEu(WO_4)_2$:$x$$Sm^{3+}$ ($x$ = 0, 0.05). The values of lifetime are 514 μs and 412 μs, respectively.

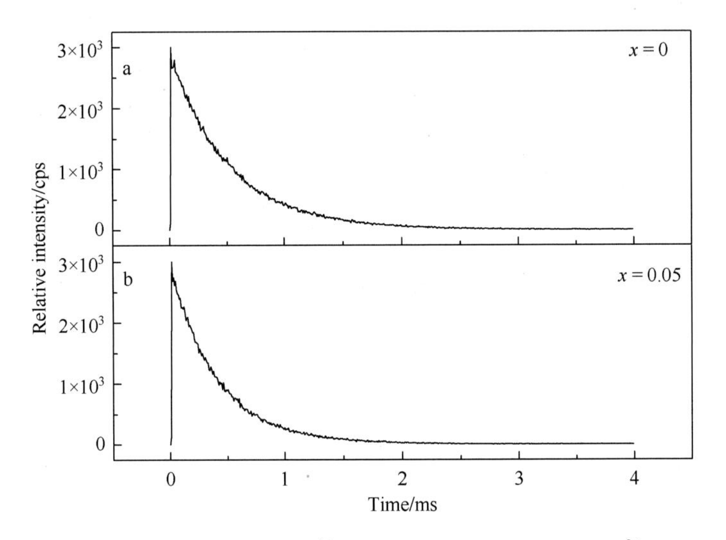


Fig.1-52 The decay curves of the $Eu^{3+}$ emission of $NaEu(WO_4)_2$:$x$$Sm^{3+}$ ($x$ = 0, 0.05) ($\lambda_{ex}$ = 395 nm; $\lambda_{em}$ = 616 nm)

## 1.3 $Eu^{3+}$-activated tetra-molybdates/tungstates red phosphors for the near-UV LED chip

### 1.3.1 Introduction

Being similar with double molybdates/tungstates, the compounds $Na_5RE(MO_4)_4$ ($RE^{3+} = Y^{3+}$, $La^{3+}$; $M^{6+} = Mo^{6+}$, $W^{6+}$) share the scheelite-related structure with $RE^{3+}$ occupying the eight-coordinate sites. Tetra-molybdates/tungstates may be used as excellent hosts for doping rare-earth ions. Hence, $Na_5RE(MO_4)_4$ doped with different contents of $Eu^{3+}$ have been prepared and their structures, PL properties, and application are studied in detail.

### 1.3.2 Preparation method

The stoichiometric mixtures of starting materials $(NH_4)_6Mo_7O_{24}·4H_2O$, $WO_3$, $NaHCO_3$ and $RE_2O_3$ (RE = Eu, Sm) were ground in an agate mortar. The mixtures were fired at 600 ℃

for 8 h in the atmosphere. The samples were slowly cooled down to room temperature. Then the pink phosphors were obtained.

## 1.3.3 Structures, morphologies, PL properties and applications

### 1.3.3.1 $Na_5La(MoO_4)_4$:$x$$Eu^{3+}$

The XRD patterns of $Na_5La(MoO_4)_4$:$x$$Eu^{3+}$ ($x$ = 0, 0.5, 1) are shown in Fig.1-53. All patterns are consistent with the JCPDS card No. 72-2158 [$Na_5La(MoO_4)_4$, space group $I41/a$]. This result reveals that the as-prepared phosphors share a pure phase with the scheelite-related isostructure as $Na_5La(MoO_4)_4$. The introduction of $Eu^{3+}$ does not change the lattice of $Na_5La(MoO_4)_4$.

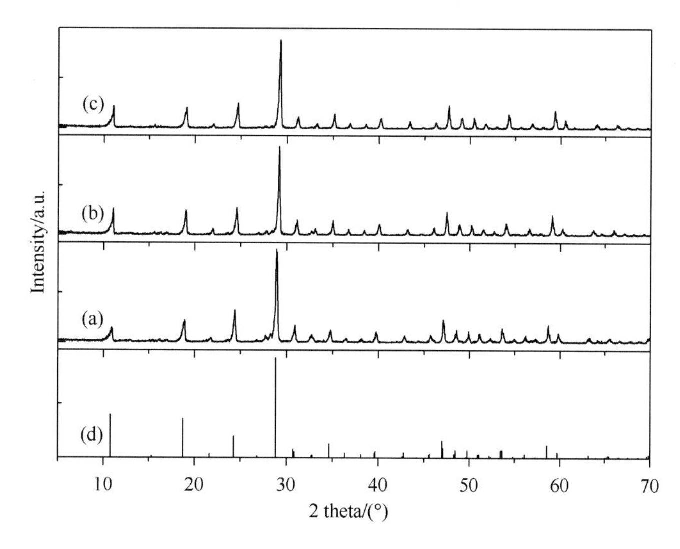


Fig.1-53 XRD patterns of $Na_5La(MoO_4)_4$:$x$$Eu^{3+}$ [$x$ = 0(a), 0.5(b), 1(c)] and JCPDS card for $Na_5La(MoO_4)_4$(d)

Fig.1-54 is the crystal structure of $Na_5La(MoO_4)_4$ with the tetragonal structure ($I41/a$, $a = b$ = 11.574 Å, $c$ = 11.620 Å, $V$ = 1556.59 $Å^3$). In this structure, each $Mo^{6+}$ is coordinated by four $O^{2-}$ to form a tetrahedron. $Na^+$ and $La^{3+}$ share the same crystal site. Due to the similar radius and the same charge between $La^{3+}$ and $Eu^{3+}$, $Eu^{3+}$ ions prefer to occupy the crystal sites of $La^{3+}$ ions.

The PL spectra of $Na_5La(MoO_4)_4$:$x$$Eu^{3+}$ ($x$ = 0.1, 0.5, 1.0) are shown in Fig.1-55. The broad bands from 250 nm to 310 nm in curves a, c and e are attributable to the O→Mo CT transition. The sharp lines in the 360-550 nm range are intra-configurational 4f-4f transitions of $Eu^{3+}$ ions in the host lattices. The $^7F_0 \rightarrow {^5L_6}$ and $^7F_0 \rightarrow {^5D_2}$ transitions at 395 nm and 465 nm are two of the strongest absorptions. Other f-f transitions of $Eu^{3+}$are weak. With the increase of the $Eu^{3+}$ concentration, the maxima of the O→Mo CT band of $Na_5La(MoO_4)_4$:$x$$Eu^{3+}$ phosphors

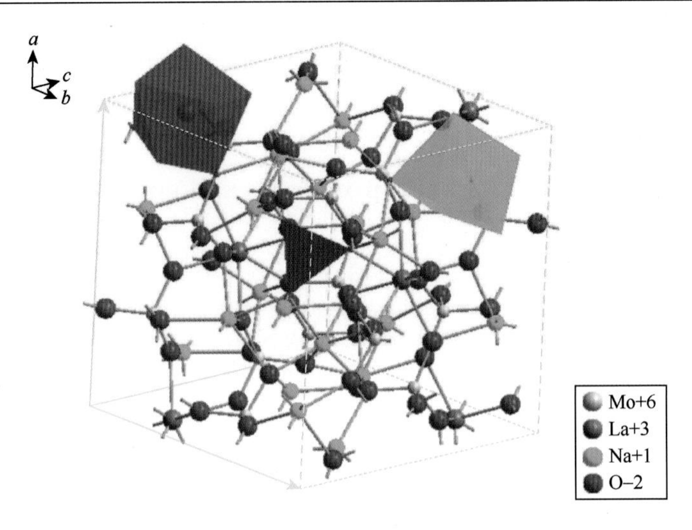


Fig.1-54 The crystal structure of $Na_5La(MoO_4)_4$

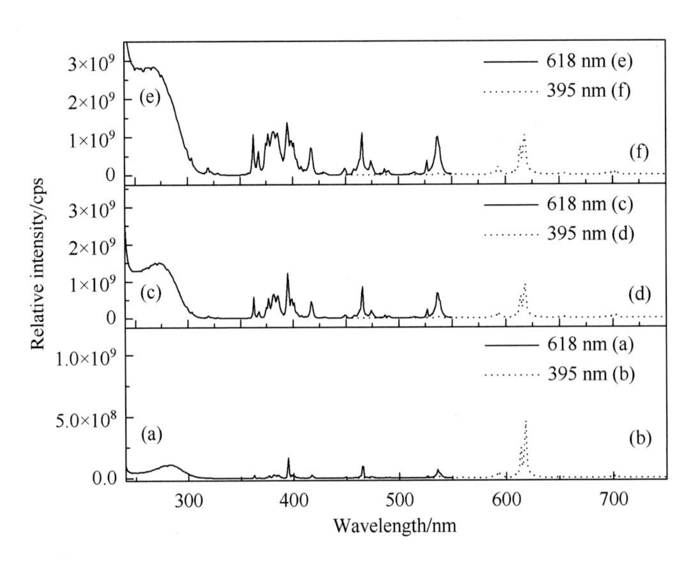


Fig.1-55 The excitation ($\lambda_{em}$ = 618 nm) and emission ($\lambda_{ex}$ = 395 nm) spectra of $Na_5La(MoO_4)_4$: $xEu^{3+}$ [$x$ = 0.10 (a, b), 0.50 (c, d) and 1.0 (e, f)]

exhibit a somewhat blue-shift. The intensities of the CT absorption are also increased, showing the influence of $Eu^{3+}$ concentrations on the O→Mo CT band. The shapes of emission spectra for $Na_5La(MoO_4)_4$:$xEu^{3+}$ with different $Eu^{3+}$-doping concentrations are similar. The strongest emission peak is at 618 nm, which is due to the $^5D_0 \rightarrow {}^7F_2$ transition of $Eu^{3+}$. The results imply that $Eu^{3+}$ ions occupy the lattice sites without inversion symmetry, which is in good agreement with the results of its crystal structure. Other transitions from the $^5D_0$ excited level to $^7F_J$ ground states are very weak, which is advantageous to obtain good CIE coordinates for phosphors. The CIE values for the phosphor $Na_5Eu(MoO_4)_4$ are calculated to be $x$ = 0.65, $y$ = 0.34 in terms of its emission spectrum, which is close to the NTSC standard values for red ($x = 0.67$, $y = 0.33$). The $^5D_0 \rightarrow {}^7F_2$ emission intensity of $Na_5Eu(MoO_4)_4$ is about 3 times higher than that of $Na_5La(MoO_4)_4$:$0.10Eu^{3+}$. No concentration quenching of $Eu^{3+}$is also observed in the series samples of $Na_5La(MoO_4)_4$:$xEu^{3+}$. Pan et al. reported that $Na_5Eu(WO_4)_4$ (with cell parameters: $a = 11.507$ Å, $c = 11.406$ Å; space group $I41/a$) had no concentration quenching due to the special structure of Eu-O-W-O-Eu. The bond-angles of Eu—O—W and O—W—O are 100° and 105°, respectively. So, the energy transfer between $Eu^{3+}$ ions is difficult. $Na_5Eu(MoO_4)_4$ also shows a similar XRD pattern and cell parameters ($a = 11.439$ Å, $c = 11.406$ Å; space group $I41/a$) with those of $Na_5Eu(WO_4)_4$. No concentration quenching is found for $Na_5Eu(MoO_4)_4$ too. This may be due to a similar special structure of Eu-O-Mo-O-Eu in $Na_5Eu(MoO_4)_4$.

Since both $Na_5Eu(MoO_4)_4$ and $NaEu(MoO_4)_2$ share the scheelite-like ($CaWO_4$) isostructure as well as exhibit strong red emission, their PL spectra are compared in Fig.1-56. Some differences can be found between the excitation spectra. The maximum of the O→Mo CT band for $NaEu(MoO_4)_2$ is located at a longer wavelength compared with that of $Na_5Eu(MoO_4)_4$. The strongest emission peak of $NaEu(MoO_4)_2$ under 395 nm excitation is at 616 nm, which is due to the $^5D_0 \rightarrow {}^7F_2$ transition of $Eu^{3+}$. The CIE values for the phosphor $NaEu(MoO_4)_2$ are calculated to be $x = 0.66$, $y = 0.33$. The emission intensity of $Na_5Eu(MoO_4)_4$ is about two times as strong as that of $NaEu(MoO_4)_2$. $Na_5Eu(MoO_4)_4$ with a 1 ∶ 4 molar ratio of Eu to Mo can save half of the $Eu^{3+}$ consumption, compared with $NaEu(MoO_4)_2$ with a 1 ∶ 2 molar ratio.

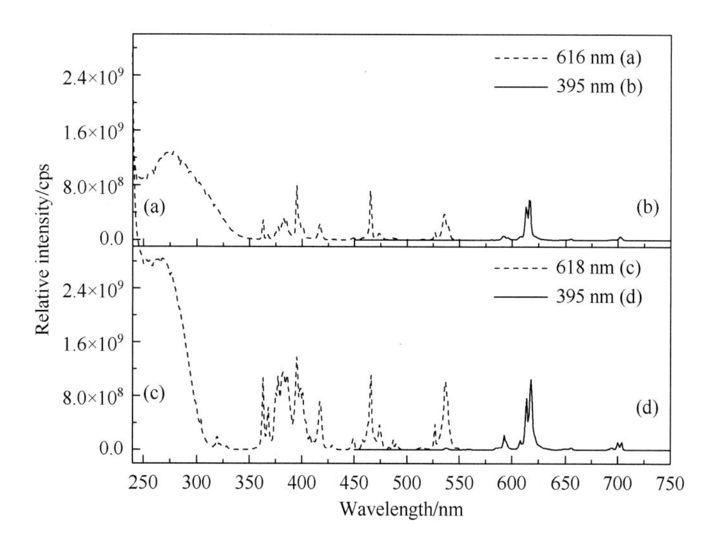


Fig.1-56 The excitation ($\lambda_{em} = 616$ nm or 618 nm) and emission ($\lambda_{ex} = 395$ nm) spectra of $NaEu(MoO_4)_2$ (a, b) and $Na_5Eu(MoO_4)_4$ (c, d)

Since the phosphor $Na_5Eu(MoO_4)_4$ shows stronger absorption around 400 nm, more intense red emission with suitable CIE chromaticity coordinates, and lower $Eu^{3+}$ consumption, it may be applied as a cheap and excellent red component for the fabrication of the near-UV

InGaN based w-LEDs. Hence, single red LEDs are fabricated by the combination of a near-UV InGaN chip with phosphors $Na_5Eu(MoO_4)_4$ and $NaEu(MoO_4)_2$, respectively. Their emission spectra under a forward bias of 20 mA are shown in Fig.1-57. The emission bands at 400 nm in curves a and b are attributed to the emission of the InGaN chip. The sharp peak at 616 nm is due to the $Eu^{3+}$ emission of $Na_5Eu\ (MoO_4)_4/NaEu\ (MoO_4)_2$.

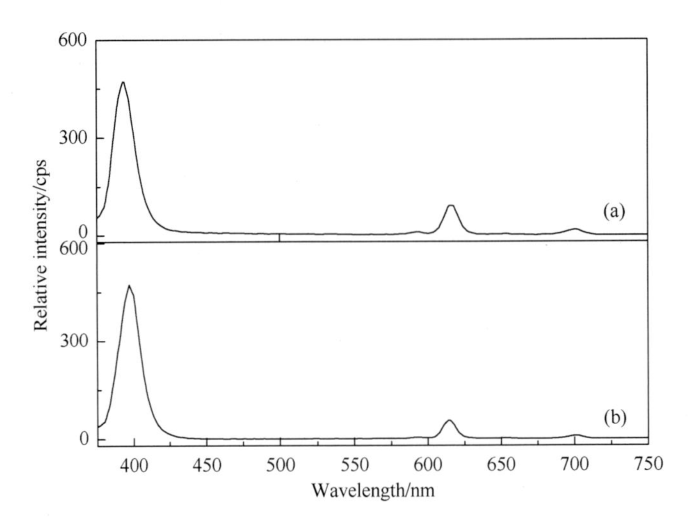


Fig.1-57 The EL spectra of the red LED-based $Na_5Eu\ (MoO_4)_4$ (a) and $NaEu\ (MoO_4)_2$ (b) under a forward bias of 20 mA

Photographes of the lighting red LEDs with $Na_5Eu(MoO_4)_4/NaEu(MoO_4)_2$ is shown in Fig.1-58. Intense red light can be observed by naked eyes, and the LED with $Na_5Eu(MoO_4)_4$ phosphor shows stronger red-emitting than that of $NaEu(MoO_4)_2$, which is in agreement with Fig.1-57.

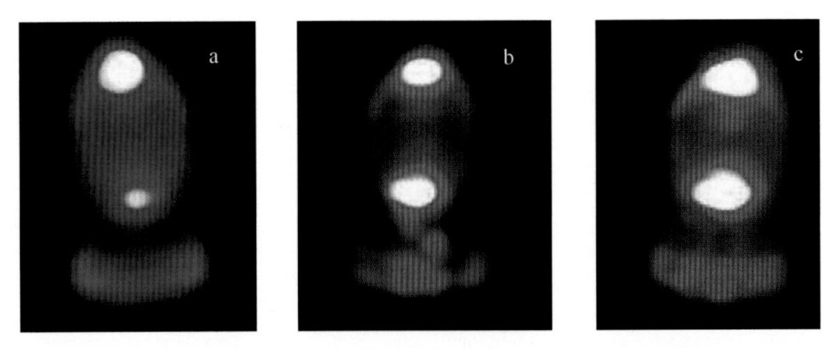


Fig.1-58 The photographs of the LEDs [(a) InGaN chip, (b) InGaN chip+ phosphor $Na_5Eu\ (MoO_4)_4$, (c) InGaN chip+phosphor $NaEu\ (MoO_4)_2$]

### 1.3.3.2 $Na_5Eu(MoO_4)_4{:}xSm^{3+}$

Since $Na_5Eu(MoO_4)_4$ exhibits excellent PL properties, the next work is to optimize this red phosphor. Then we prepared $Na_5La(MoO_4)_4{:}xSm^{3+}$ phosphors and investigated their PL spectra.

The XRD patterns of $Na_5La(MoO_4)_4$:$x$$Sm^{3+}$ ($x$ = 0.02, 0.08) are shown in Fig.1-59. These patterns are consistent with the JCPDS card No. 72-2158 [$Na_5La$ $(MoO_4)_4$, space group $I41/a$]. The result indicates these phosphors are of a single-phase with the scheelite structure. $Sm^{3+}$ ions occupy the $La^{3+}$ sites in these compounds. The $d$ value difference between $Sm^{3+}$ ion and $La^{3+}$ ion is due to their distinct ionic radii.

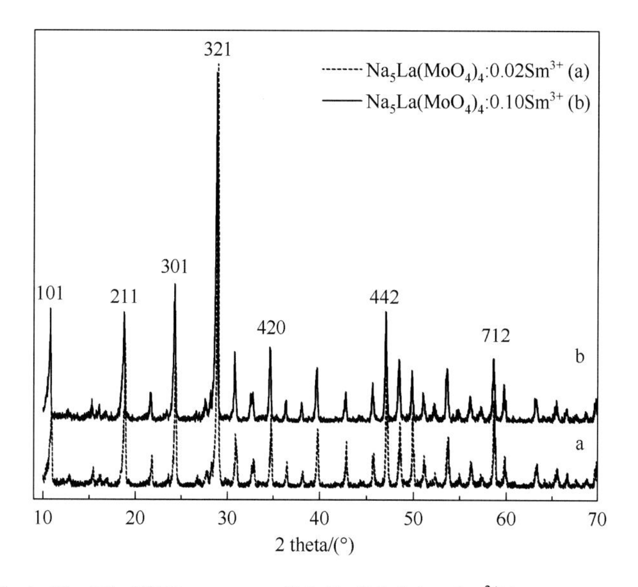


Fig.1-59 The XRD patterns of $Na_5La(MoO_4)_4$:$x$$Sm^{3+}$ ($x$ = 0.02, 0.08)

The excitation and emission spectra of $Na_5La(MoO_4)_4$:$x$$Sm^{3+}$ ($x$ = 0.02, 0.08) are shown in Fig.1-60. The broad excitation bands from 240 nm to 300 nm in curves a and c are attributable to the O→Mo CT transition. The sharp peaks in the 300-550 nm range are intra-configurational 4f-4f

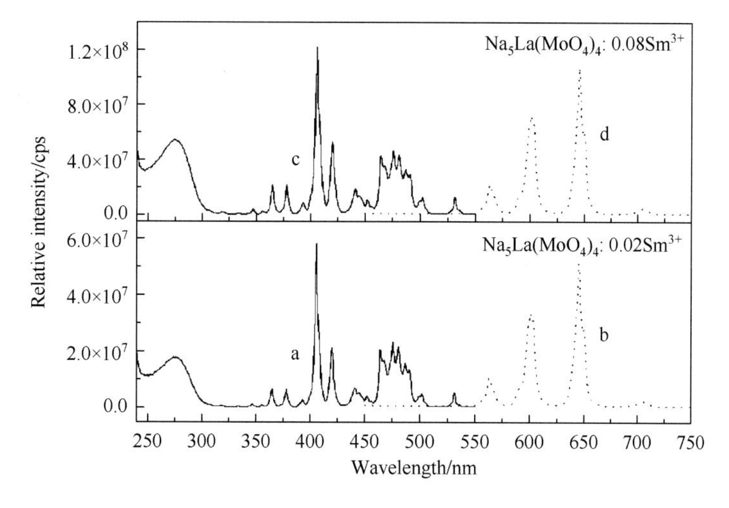


Fig.1-60 The excitation (a, c; $\lambda_{em}$ = 645 nm) and emission (b, d; $\lambda_{ex}$ = 405 nm) spectra of $Na_5La(MoO_4)_4$:$x$$Sm^{3+}$ ($x$ = 0.02, 0.08)

transitions of $Sm^{3+}$ in the host lattices. The strongest excitation peak is at 405 nm, which is due to the $^6H_{5/2}\rightarrow{}^4K_{11/2}$ transition of $Sm^{3+}$. When the $Sm^{3+}$ content is 8%, the excitation intensity is the strongest. Three main emission peaks at 563 nm, 597 nm and 645 nm are corresponding to the $^4G_{5/2}\rightarrow{}^6H_{5/2}$, $^4G_{5/2}\rightarrow{}^6H_{7/2}$ and $^4G_{5/2}\rightarrow{}^6H_{9/2}$ transitions, respectively.

The photo-luminescence of red phosphor $Na_5Eu(MoO_4)_4$ was investigated in our previous work, which may be found in the application on the near-UV white LED with intense red emission. As mentioned above, $Na_5La(MoO_4)_4$:$x$$Sm^{3+}$ phosphors exhibit a strong absorption line at 405 nm. Therefore, a phosphor with broadened and strengthened absorptions around 400 nm, which fits the emission wavelength of the near-UV LEDs, will probably be obtained by an appropriate $Sm^{3+}$-$Eu^{3+}$ co-doping. In view of this consideration, the phosphors $Na_5Eu(MoO_4)_4$:$x$$Sm^{3+}$ ($x = 0$, 0.02, 0.04, 0.06, 0.08, 0.10, 0.12) were prepared, and their structure as well as PL properties were investigated.

The XRD patterns of $Na_5Eu(MoO_4)_4$:$x$$Sm^{3+}$ ($x = 0.02$, 0.10) are shown in Fig.1-61. They are also consistent with the JCPDS card of $Na_5La(MoO_4)_4$ (No. 72-2158). The result indicates these phosphors are of similar scheelite-related structures.

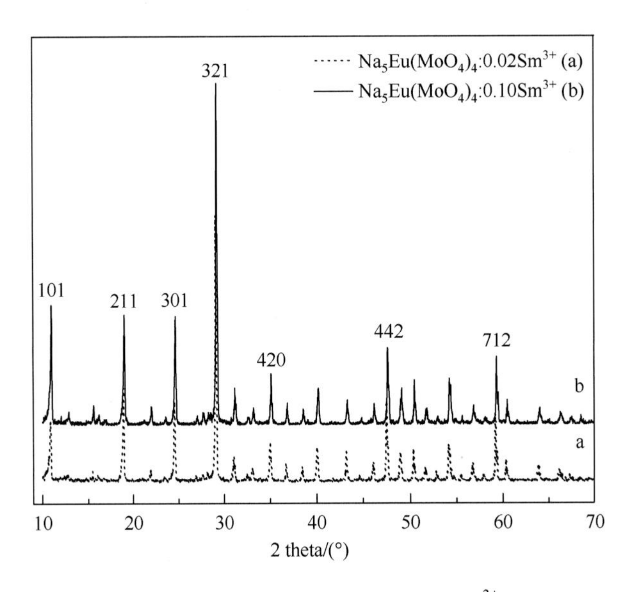


Fig.1-61 The XRD patterns of $Na_5Eu(MoO_4)_4$:$x$$Sm^{3+}$ ($x = 0.02$, 0.10)

Fig.1-62 is the excitation and emission spectra of $Na_5Eu(MoO_4)_4$:$x$$Sm^{3+}$ ($x = 0$, 0.10). The excitation band of $Na_5Eu(MoO_4)_4$:0.10$Sm^{3+}$ is broader than that of $Na_5Eu(MoO_4)_4$. This indicated the phosphor $Na_5Eu(MoO_4)_4$:0.10$Sm^{3+}$ shares stronger absorption than that of $Na_5Eu(MoO_4)_4$. These two emission spectra are of similar shapes. They show typical $Eu^{3+}$ emission, and the mainline is the $^5D_0\rightarrow{}^7F_2$ transition of $Eu^{3+}$ at 618 nm. The emission of $Sm^{3+}$ is not observed even under direct excitation of the $^6H_{5/2}\rightarrow{}^4K_{11/2}$ transition of $Sm^{3+}$ in the host lattice (curve e). This means that $Sm^{3+}$ ions can absorb and efficiently transfer energy to the

$Eu^{3+}$ ions. A possible four-step energy transfer process from $Sm^{3+}$ to $Eu^{3+}$ is considered as follows. Firstly, an electron in the ground $^6H_{5/2}$ state of $Sm^{3+}$ ions is promoted to the $^4K_{11/2}$ state upon 405 nm excitation. Then, the $^4G_{5/2}$ level is non-radiatively populated. In the next step, the energy transfer from $Sm^{3+}$ to $Eu^{3+}$ occurs and the $^5D_0$ state of $Eu^{3+}$ is populated. In the last step, mainly radiative transition from $^5D_0$ to $^7F_2$ level of $Eu^{3+}$ results in the red emission. This energy transfer process from $Sm^{3+}$ to $Eu^{3+}$ is in agreement with the result reported by Lee et al. Curves b and c in Fig.1-62 show the emission spectra of phosphor $Na_5Eu(MoO_4)_4$ under 395 nm and 405 nm light excitation, respectively. The emission intensity under 405 nm light excitation is very weak, compared with that under 395 nm light excitation.

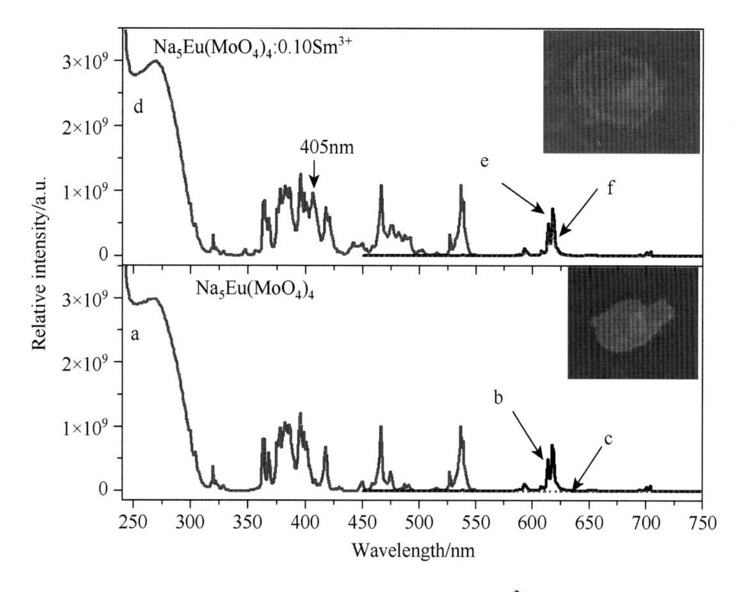


Fig.1-62 PL spectra of $Na_5Eu(MoO_4)_4$:$x$$Sm^{3+}$ ($x$ = 0, 0.10).
The inseted figures are the photos of the phosphors excited by 365 nm light

The intense red light can be observed from the phosphors excited by the UV light (Fig.1-62). The emission intensities of the phosphors according to the emission spectra are compared. The comparison of the integrated intensities is listed in Table 1-6. The emission intensity of $Na_5Eu(MoO_4)_4$:0.10$Sm^{3+}$ under 395 nm excitation is close to that of $Na_5Eu(MoO_4)_4$. But the emission intensity of $Na_5Eu(MoO_4)_4$:0.10$Sm^{3+}$ under 405 nm excitation (direct excitation $Sm^{3+}$) is stronger than that of $Na_5Eu(MoO_4)_4$ under 405 nm light excitation, and close to that of $Na_5Eu(MoO_4)_4$ under 395 nm excitation. This result shows $Na_5Eu(MoO_4)_4$:0.10$Sm^{3+}$'s exhibition strengthened red emission under 400 nm light excitation. The CIE chromaticity coordinates of the phosphors are calculated, and the results are also listed in Table 1-6. The chromaticity coordinates are very close to NTSC standard values for red ($x$ = 0.67, $y$ = 0.33).

**Table 1-6　The $^5D_0$-$^7F_2$ relative emission intensity and the CIE chromaticity coordinates of the phosphors**

| Phosphors | Excitation wavelength/nm | Relative intensity | CIE coordinates* | |
|---|---|---|---|---|
| | | | $x$ | $y$ |
| $Na_5Eu(MoO_4)_4$ | 395 | 1 | 0.66 | 0.34 |
| | 405 | 0.004 | 0.60 | 0.35 |
| $Na_5Eu(MoO_4)_4$:0.10$Sm^{3+}$ | 395 | 1.03 | 0.66 | 0.34 |
| | 405 | 0.81 | 0.66 | 0.34 |

* NTSC standard values for red are $x = 0.67$, $y = 0.33$.

Fig.1-63 shows the decay curve for the $^5D_0 \rightarrow ^7F_2$ transition (618 nm) of the $Eu^{3+}$ in $Na_5Eu(MoO_4)_4$:$x$$Sm^{3+}$ ($x = 0$, 0.02, 0.10) phosphors under 395 nm light excitation. These curves are fitted into a single-exponential function, and the lifetime $\tau$ values are calculated to be 0.558 ms, 0.550 ms, 0.541 ms, respectively.

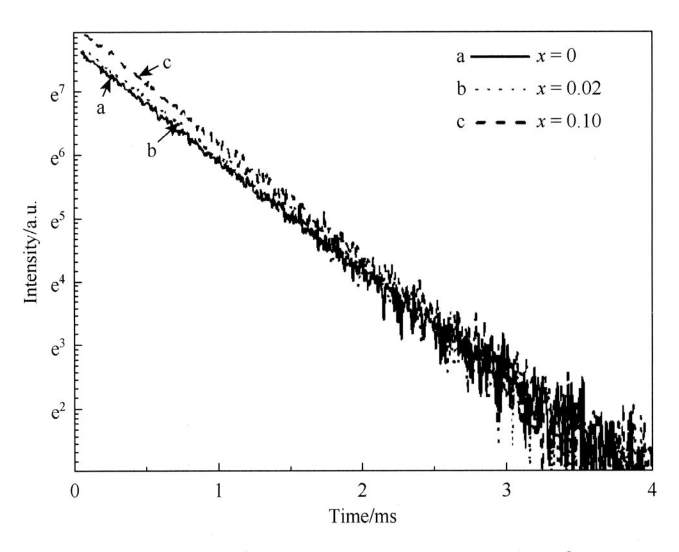


Fig.1-63　Decay curves of the $Eu^{3+}$ emission of $Na_5Eu(MoO_4)_4$:$x$$Sm^{3+}$ ($x = 0$, 0.02, 0.10) ($\lambda_{ex} = 395$ nm; $\lambda_{em} = 618$ nm)

Our target is to obtain a highly efficient red component for the near-UV LEDs. So the single red LEDs have been fabricated by combining these red phosphors with InGaN chips. Fig.1-64 shows the EL spectra of the red light-emitting diodes of near-UV InGaN-based $Na_5Eu(MoO_4)_4$:$x$$Sm^{3+}$ ($x = 0$, 0.10) under 20 mA current excitation. Curve a is the emission of the InGaN chip with the main emission peak at 395 nm. The sharp peaks at 616 nm and 702 nm are due to the emissions of red phosphor. Comparing with the red LED-based $Na_5Eu(MoO_4)_4$, bright red light is observed from the red LED-based $Na_5Eu(MoO_4)_4$:0.10$Sm^{3+}$ by naked eyes.

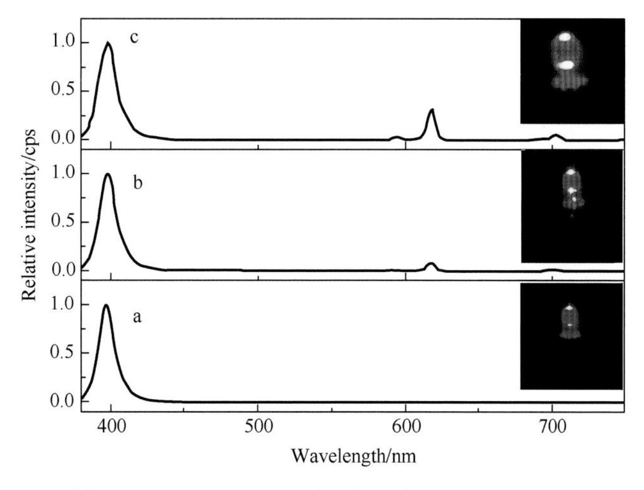


Fig.1-64 The EL spectra of the near-UV InGaN LED (a), the red LED-based $Na_5Eu(MoO_4)_4$(b), $Na_5Eu(MoO_4)_4$: $0.10Sm^{3+}$ (c) under 20 mA current excitation. The inserted figures are the lighting photographs of three LEDs

### 1.3.3.3 $Na_5Eu(MoO_4)_4$ doped with boron oxide

In the above section, the luminescence of $Na_5La(MoO_4)_4$:$x$$Eu^{3+}$ was reported. The phosphor $Na_5Eu(MoO_4)_4$ showed intense red emission, which could find an application on near-UV w-LEDs. Boron oxide was not only a good flux but also able to replace some competent phosphors and enhance the luminescent properties. In this section, red phosphors $Na_5Eu(MoO_4)_4$ doped with different contents of boron oxide were prepared, and their PL properties were investigated. At last, a single-red LED was fabricated by combining the red phosphor with a 400 nm emitting InGaN chip.

The XRD patterns of $Na_5Eu(MoO_4)_4$ doped with different contents of boron oxide are shown in Fig.1-65. The XRD patterns are similar to the JCPDS card No. 72-2158 [$Na_5La(MoO_4)_4$,

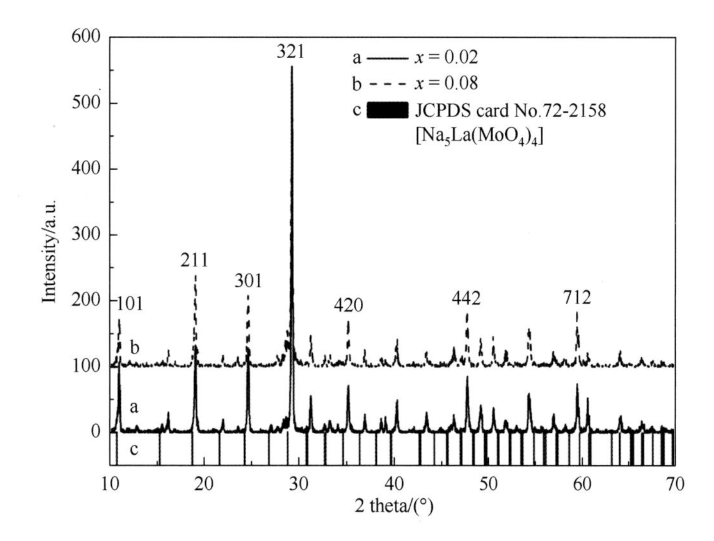


Fig.1-65 The XRD patterns of EMB0, NEMB2 and NEMB4

space group *I*41/*a*]. This reveals that the phosphors share the scheelite-related isostructure as $Na_5La(MoO_4)_4$. The diffraction peaks of the phosphors show a slight shift compared with the JCPDS card. This is due to the distinct ionic radii between $Eu^{3+}$ and $La^{3+}$.

The PL spectra of NEMB0, NEMB2 and NEMB4 are shown in Fig.1-66. The broadband from 200 nm to 310 nm is attributable to the O→Mo CT transition. The sharp lines in the 310-550 nm range are intra-configurational 4f-4f transitions of $Eu^{3+}$ ions in the host lattices. The strongest excitation peak is at 395 nm, which is due to the $^7F_0 \rightarrow ^5L_6$ transitions of $Eu^{3+}$ ion. Curves b, d and f are the emission spectra under 395 nm excitation. The strongest emission peak is at 617 nm, which is due to the $^5D_0 \rightarrow ^7F_2$ transition of $Eu^{3+}$. This result implies that $Eu^{3+}$ ions occupy the lattice sites without inversion symmetry, which is in good agreement with the results of its crystal structure. Other transitions from the $^5D_0$ excited level to $^7F_J$ ground states are very weak, which is advantageous to obtain good CIE chromaticity coordinates for phosphors. The sample NEMB2 shows the strongest emission, and the CIE values for the phosphor are calculated to be $x = 0.673$, $y = 0.327$ in terms of its emission spectrum, which are very close to the NTSC standard values for red.

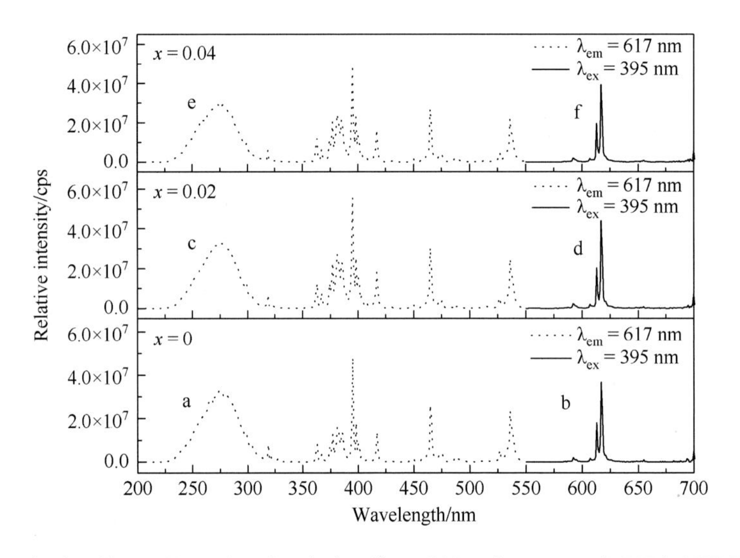


Fig.1-66 The excitation ($\lambda_{em}$ = 617 nm) and emission ($\lambda_{ex}$ = 395 nm) spectra of EMB0, NEMB2 and NEMB4

The decay curve for the $^5D_0 \rightarrow ^7F_2$ transition (616 nm) of the $Eu^{3+}$ of NEMB2 is shown in Fig.1-67. The decay curve can be well fitted by a single-exponential function as $I = A \exp(-t/\tau)$. The lifetime value of this phosphor is 0.549 ms.

Fig.1-68 shows the emission spectrum of the red light-emitting diode of one near-UV InGaN-based the red phosphor NEMB2. The emission band at 395 nm is attributed to the emission of the InGaN chip. The emission peaks at 617 nm and 702 nm are due to the emissions of red phosphor. Bright red light from the LED is observed by naked eyes. Its CIE chromaticity coordinates are calculated to be $x = 0.465$, $y = 0.222$.

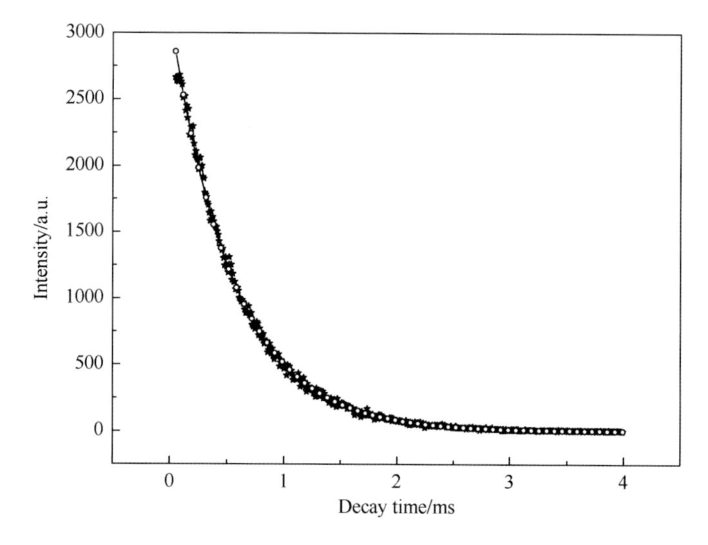


Fig.1-67 Decay curve of the $Eu^{3+}$ emission of NEMB2 ($\lambda_{ex}$ = 395 nm, $\lambda_{em}$ = 617 nm)

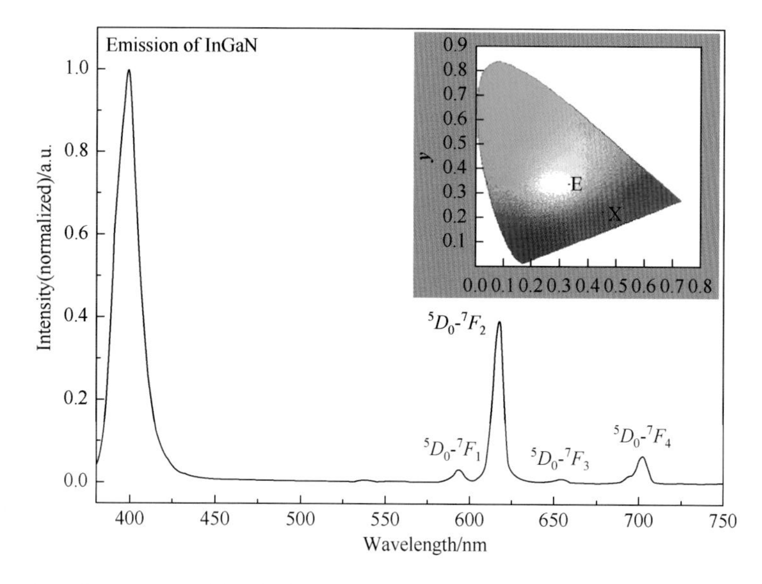


Fig.1-68 The EL spectrum of the single-red LEDs-based NEMB2 under 20 mA current excitation. The inserted figure is the chromaticity coordinate diagram of the red LED

The emission of the InGaN chip (395 nm) can still be observed in Fig.1-67. This is advantageous to obtain a white-emitting LED by combining this phosphor with appropriate blue and green phosphors. We think that each proper mono-color LED phosphor must meet the following necessary conditions from the standpoint of application. ① The phosphor must efficiently be excited in the near-UV range. But any mono-color phosphor cannot absorb all this energy. Otherwise, other phosphors probably cannot be efficiently excited. ② The phosphor exhibits higher luminescent intensity under 400 nm excitation. ③ The chromaticity coordinates of the phosphor are close to the NTSC standard values.

### 1.3.3.4 $Na_5Eu(WO_4)_4$:$x$$Sm^{3+}$

The tungstates $Na_5Ln(WO_4)_4$(Ln = La, Y) which share the scheelite-like (double molybdates/tungstates) iso-structure, were considered to be efficient luminescent hosts. But to the best of our knowledge, little research of these compounds for the potential application in LEDs was reported. Moreover, it is well known that $Eu^{3+}$/$Sm^{3+}$ ions exhibit strong absorption at about 395 nm/405 nm in many host lattices, which is close to the emission wavelength (400 nm) of the near-UV LED. Hence, it is possible to obtain a phosphor with broadened absorption around 400 nm for the $Eu^{3+}$-$Sm^{3+}$ co-doped system. In this section, we discussed the PL properties of $Na_5La(WO_4)_4$:$x$$R^{3+}$ (R = Eu, Sm) and $Na_5Eu(WO_4)_4$:$x$$Sm^{3+}$.

The XRD patterns of $Na_5La(WO_4)_4$:$x$$Sm^{3+}$ ($x$ = 0, 0.20, 0.50, 1.0) are shown in Fig.1-69. These patterns are consistent with the JCPDS card No. 85-1096, indicating these phosphors are of a single-phase with the scheelite structure. $Eu^{3+}$ ions occupy the $La^{3+}$ sites in these compounds. The $d$ values are slightly different due to the distinct ionic radii between $La^{3+}$ and $Eu^{3+}$.

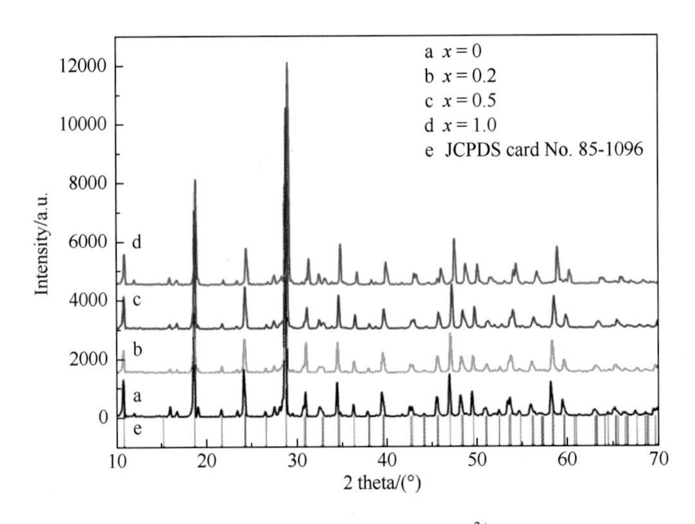


Fig.1-69 The XRD patterns of $Na_5La(WO_4)_4$:$x$$Eu^{3+}$ ($x$ = 0, 0.20, 0.50, 1.0)

The excitation and emission spectra of $Na_5La(WO_4)_4$:$x$$Eu^{3+}$ ($x$ = 0.1, 0.5, 1.0) are shown in Fig.1-70. The broad excitation band from 240 nm to 290 nm is attributable to the O→W CT transition. The sharp peaks in the 290-550 nm range are intra-configurational 4f-4f transitions of $Eu^{3+}$ in the host lattices, and two of the strongest peaks at 396 nm and 465 nm are due to the $^7F_0$→$^5L_6$ and $^7F_0$→$^5D_2$ transitions of $Eu^{3+}$. With the increase of the $Eu^{3+}$ content, the excitation intensity also increases.

Curves b, d and f represent the emission spectra under excitation at 396 nm. The strongest emission is at 618 nm, which is due to the $^5D_0$→$^7F_2$ transition of $Eu^{3+}$. Other transitions from the $^5D_J$ levels to $^7F_J$ ground states, such as $^5D_0$→$^7F_J$ ($J$ = 1, 3, 4) lines and $^5D_1$→$^7F_J$ transitions ($J$ = 1, 2) are very weak, which is advantageous to obtain a phosphor with good CIE

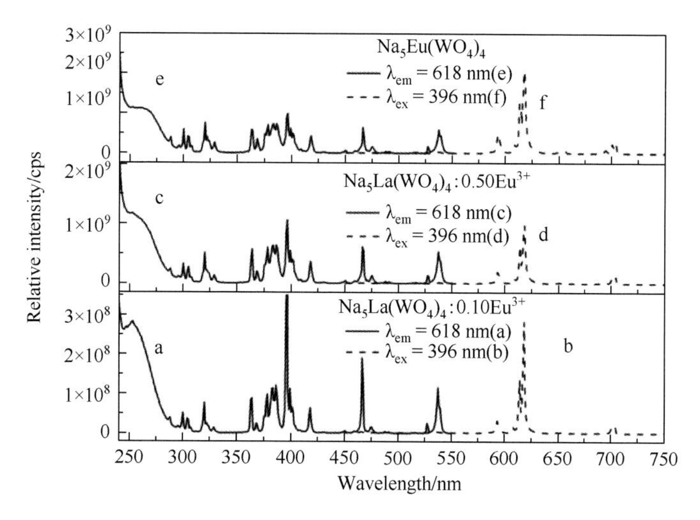


Fig.1-70 The excitation (a, c, d) and emission (b, d, f) spectra of $Na_5La(WO_4)_4$:$x$$Eu^{3+}$ ($x$ = 0.1, 0.5, 1.0)

chromaticity coordinates. The intensity of the $^5D_0{\rightarrow}^7F_2$ transition (electric dipole transition) is much higher than that of the $^5D_0{\rightarrow}^7F_1$ transition (magnetic dipole transition). This result implies that the electric dipole transitions of $Eu^{3+}$ are allowed in these phosphors. With the increase of the $Eu^{3+}$ content, the emission intensity of the phosphors also increases. No concentration quenching is observed in the series samples of $Na_5La(WO_4)_4$:$x$$Eu^{3+}$. The CIE chromaticity coordinates for $Na_5Eu(WO_4)_4$ are calculated to be $x$ = 0.65, $y$ = 0.34 (Table 1-7), which is close to the NTSC standard values ($x$ = 0.67, $y$ = 0.33).

**Table 1-7 The CIE chromaticity coordinates and the $^5D_0$-$^7F_2$ relative emission intensity of phosphors**

| Phosphors | Excitation wavelength/nm | CIE chromaticity coordinates* | | $^5D_0$-$^7F_2$ relative intensity |
|---|---|---|---|---|
| | | $x$ | $y$ | |
| $Na_5Eu(WO_4)_4$ | 396 | 0.65 | 0.34 | 1.0 |
| $Na_5Eu(WO_4)_4$:0.06$Sm^{3+}$ | 396 | 0.65 | 0.34 | 1.80 |
| | 405 | 0.64 | 0.34 | 0.95 |

* The NTSC standard values $x$ = 0.67, $y$ = 0.33.

Fig.1-71 is the XRD patterns of $Na_5La(WO_4)_4$:0.04$Sm^{3+}$ and $Na_5Eu(WO_4)_4$:0.06$Sm^{3+}$. These two patterns are also consistent with the JCPDS card No. 85-1096, indicating that these two phosphors are of the scheelite structure. The ionic radius of $La^{3+}$ is bigger than that of $Eu^{3+}$. Then, the corresponding $d$ values in XRD patterns are small. Hence, pattern (a) shows a red-shift comparing with pattern (b).

The excitation and emission spectra of $Na_5La(WO_4)_4$:0.04$Sm^{3+}$ are shown in Fig.1-72. The broad excitation band from 240 nm to 300 nm is assignable to the O→W CT absorption as described above. The sharp lines in the 300-550 nm range are due to f-f transitions of $Sm^{3+}$ in the host lattice, and the main absorption is due to the $^6H_{5/2}{\rightarrow}^4K_{11/2}$ transition at 405 nm. Curve

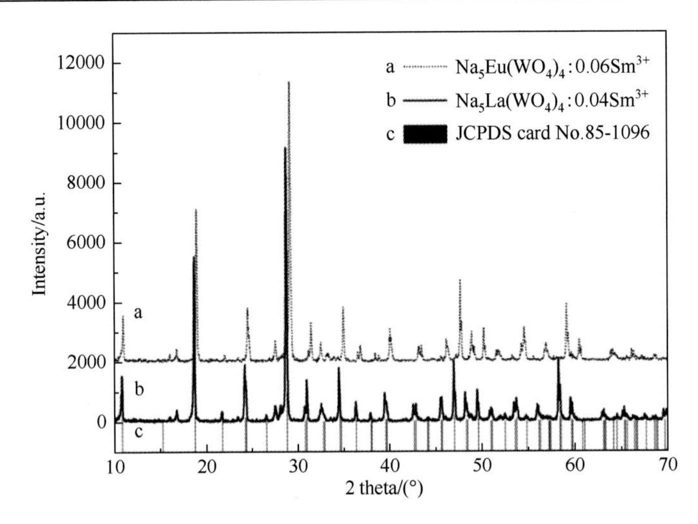


Fig.1-71 The XRD patterns of $Na_5La(WO_4)_4$:0.04$Sm^{3+}$ and $Na_5Eu(WO_4)_4$:0.06$Sm^{3+}$

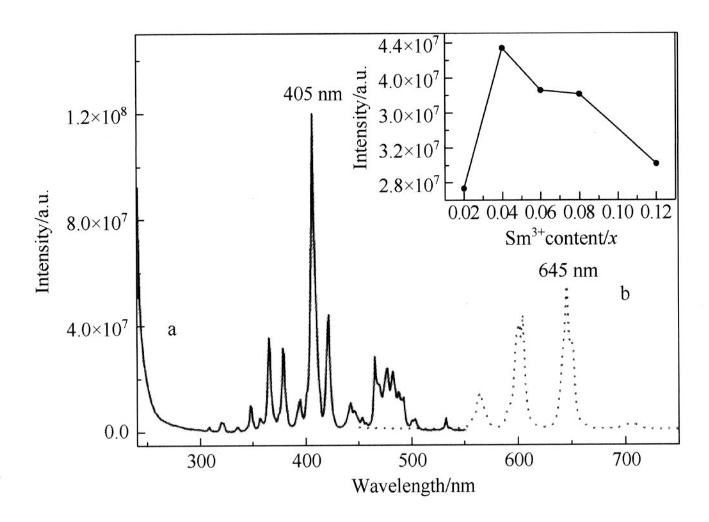


Fig.1-72 The excitation and emission spectra of $Na_5La(WO_4)_4$:0.04$Sm^{3+}$. The inseted figure is the concentration dependence of the relative emission intensity of $Sm^{3+}$ $^4G_{5/2}\rightarrow{}^6H_{9/2}$ transition in $Na_5La(WO_4)_4$:$x$$Sm^{3+}$ ($\lambda_{ex}$ = 405 nm)

b represents the emission under 405 nm excitation. The spectrum shows typical red emission of $Sm^{3+}$. Three main peaks at 563 nm, 597 nm and 645 nm are corresponding to the $^4G_{5/2}\rightarrow{}^6H_{5/2}$, $^4G_{5/2}\rightarrow{}^6H_{7/2}$ and $^4G_{5/2}\rightarrow{}^6H_{9/2}$ transitions, respectively. The relative emission intensity of $Sm^{3+}$ $^4G_{5/2}\rightarrow{}^6H_{9/2}$ transition in $Na_5La(WO_4)_4$:$x$$Sm^{3+}$ is shown in the inset of Fig.1-72. It can be observed that the sample $Na_5La(WO_4)_4$:0.04$Sm^{3+}$ shows the strongest emission under 405 nm excitation.

As mentioned above, the phosphors $Na_5La(WO_4)_4$:$x$$Eu^{3+}$ exhibit a strong absorption line at 396 nm, and $Na_5La(WO_4)_4$:$x$$Sm^{3+}$ exhibit a strong absorption line at 405 nm. Therefore a phosphor with broadened and strengthened absorptions around 400 nm, which fits the emission wavelength of the near-UV LEDs. The phosphor will probably be obtained by an appropriate

$Sm^{3+}$-$Eu^{3+}$ co-doping. In general, because the trivalent rare-earth ions share similar ionic radii, it was assumed that they entered the same lattice sites when different rare-earth ions were doped in a specific host lattice. In view of this consideration, the phosphors $Na_5Eu(WO_4)_4$:$x$$Sm^{3+}$ ($x = 0$, 0.02, 0.04, 0.06, 0.08, 0.10, 0.12) were prepared, and their PL properties were investigated.

Fig.1-73 is the excitation and emission spectra of $Na_5Eu(WO_4)_4$ and $Na_5Eu(WO_4)_4$:0.06$Sm^{3+}$. Comparing the spectrum c with a, it can be found that both $Eu^{3+}$ and $Sm^{3+}$ f-f transitions absorptions are observed in the spectrum c. The excitation band in the near-UV range of Fig.1-73c is broader than that in Fig.1-73(a).

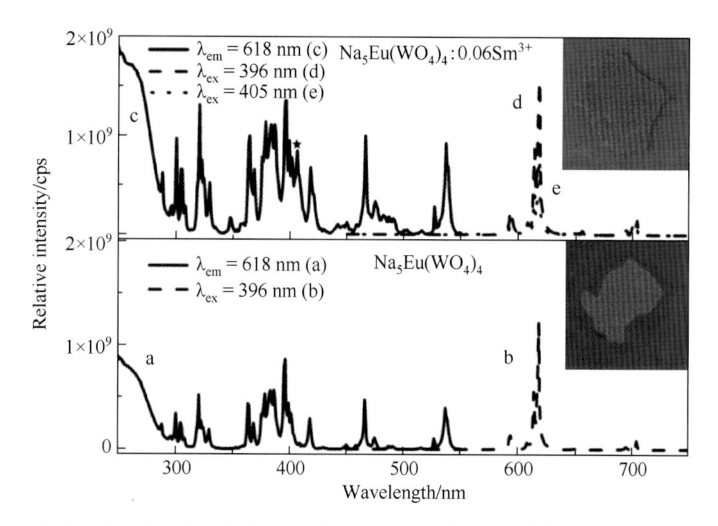


Fig.1-73 The excitation (a, c) and emission (b, d, e) spectra of $Na_5Eu(WO_4)_4$ and $Na_5Eu(WO_4)_4$:0.06$Sm^{3+}$ (The inseted figures are the photos of the phosphors excited by 365 nm light)

Curves d and e in Fig.1-73 show the emission spectra of phosphor $Na_5Eu(WO_4)_4$:0.06$Sm^{3+}$ under 396 nm and 405 nm excitation, respectively. The spectra are similar. They show the typical $Eu^{3+}$ emission with the mainline at 618 nm, which is ascribed to the $^5D_0 \rightarrow ^7F_2$ transition of $Eu^{3+}$. The emission of $Sm^{3+}$ is not found even under the direct excitation of the $^6H_{5/2} \rightarrow ^4K_{11/2}$ transition of $Sm^{3+}$ in the host lattice (curve e). It means that the $Sm^{3+}$ ions can absorb and efficiently transfer energy to the $Eu^{3+}$ ions. A possible four-step energy transfer process from $Sm^{3+}$ to $Eu^{3+}$ is considered as Fig.1-74. An electron in the ground $^6H_{5/2}$ state of $Sm^{3+}$ ions will be promoted to $^4K_{11/2}$ state upon 405 nm excitation in the first step. Then the $^4G_{5/2}$ level is non-radiatively populated, followed this step the energy transfer from $Sm^{3+}$ to $Eu^{3+}$ occurs and the $^5D_0$ state of $Eu^{3+}$ is populated. In the last step, mainly radiative transition from the $^5D_0$ to $^7F_2$ level of $Eu^{3+}$ results in the red emission. And these energy transfer processes are in agreement with the works reported by Biju et al. and Lee et al.

These phosphors show intense red emission (Fig.1-73), and the emission intensities of the phosphors according to the spectra are compared (Table 1-7). It can be found that the emission intensity of $Na_5Eu(WO_4)_4$:0.06$Sm^{3+}$ under 395 nm excitation is about 1.8 times

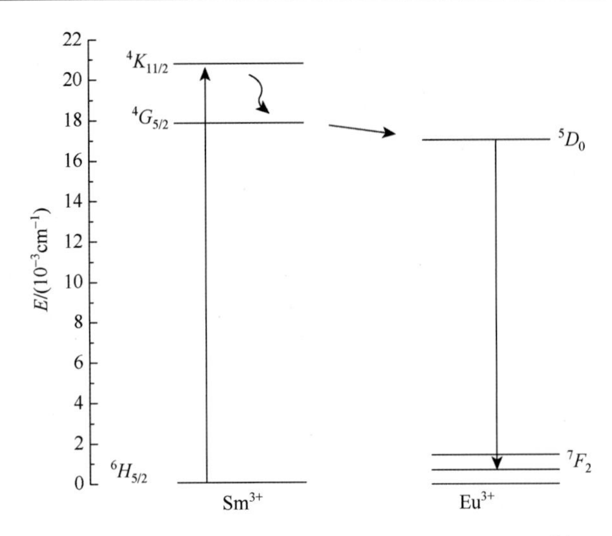


Fig.1-74　The diagram of energy transfer processes from $Sm^{3+}$ to $Eu^{3+}$

stronger than that of $Na_5Eu(WO_4)_4$. The emission intensity of $Na_5Eu(WO_4)_4$:$0.06Sm^{3+}$ under 405 nm excitation (direct excitation $Sm^{3+}$) is close to that of $Na_5Eu(WO_4)_4$ under 395 nm excitation. The CIE chromaticity coordinates of the phosphors are calculated, and the results are also listed in Table 1-7. The chromaticity coordinates are very close to the NTSC standard values.

Fig.1-75 is the EL spectra of the red LEDs of the near-UV InGaN-based on $Na_5Eu(WO_4)_4$: $xSm^{3+}$ ($x = 0, 0.06$) under 20 mA current excitation. Curve a is the emission spectrum of the InGaN chip, and the main emission peak is at about 395 nm. The sharp peaks at 616 nm and 702 nm are due to the emissions of red phosphor. Comparing with the red LED-based on $Na_5Eu(WO_4)_4$, the bright

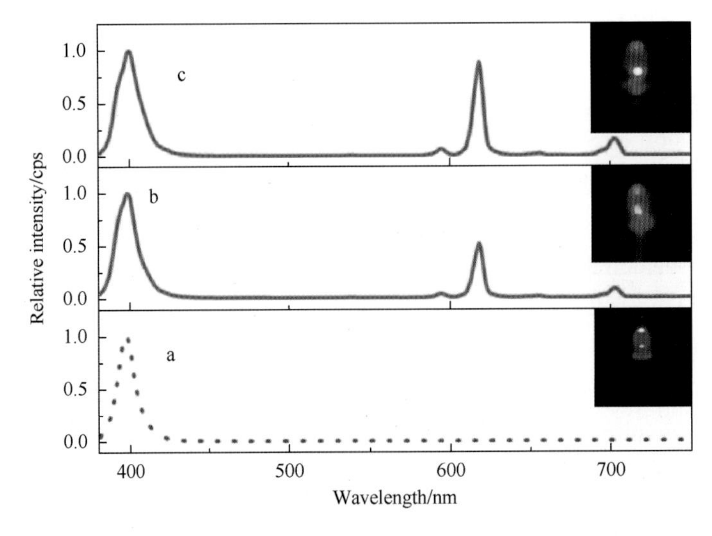


Fig.1-75　The EL spectra of the near-UV InGaN LED (a), the red LED-based $Na_5Eu(WO_4)_4$ (b), $Na_5Eu(WO_4)_4$:$0.06Sm^{3+}$ (c) under 20 mA current excitation (The inseted figures are the photos of their LEDs)

red light can be observed from the red LED-based $Na_5Eu(WO_4)_4$:$0.06Sm^{3+}$ by naked eyes. The CIE chromaticity coordinates of the red LED-based on $Na_5Eu(WO_4)_4$:$0.06Sm^{3+}$ are $x = 0.534$, $y = 0.256$. The intensive emission of the InGaN chip can still be observed in Fig.1-75. It is advantageous to obtain a w-LED by combining this phosphor with appropriate blue and green phosphors.

## 1.4 $Eu^{3+}$-activated other inorganic hosts

### 1.4.1 $LaBSiO_5$ co-doped with $Eu^{3+}$, $Al^{3+}$

The compound $LaBSiO_5$ shows excellent thermal as well as hydrolytic stability and is considered to be an efficient luminescent host. Recently, Xue et al. reported the PL properties of $LaBSiO_5$:$Eu^{3+}$ red-emitting phosphors. The red emission intensity of $LaBSiO_5$:$Eu^{3+}$ could be greatly enhanced by co-doping $Bi^{3+}$ and $Sm^{3+}$ ions.

The appropriate red phosphors for the near-UV LEDs must show strong and broad absorption around 400 nm firstly. There are two approaches to broaden the absorption in this range. One is to introduce some sensitizers, such as co-doping $Bi^{3+}$/$Sm^{3+}$ and $Eu^{3+}$ ions. On the other hand, from the viewpoint of the host compound, each spectroscopic line is expected to be narrow when the rare-earth ions (such as $Eu^{3+}$) enter the lattice sites of a pure host compound in general. Contrastively, if the host compound can form solid solutions by adjusting the cations or anions of this host compound, the sub-lattice structure around the luminescent center ions will be expected to be somewhat diverse. Therefore, the spectroscopic lines of rare-earth ions are expected to be broadened. $Al^{3+}$ ($2s^22p^6$) ion shares the similar electronic configurations of $B^{3+}$ ($1s^2$). But their ionic radii and electronegativities are different. The introduction of $Al^{3+}$ changes the sub-lattice structure, and the PL properties of the phosphors are expected to be optimized.

In this part, $LaBSiO_5$:$Eu^{3+}$ and $LaBSiO_5$:$Eu^{3+}$,$Al^{3+}$ are synthesized by the solid-state reaction, and their structure and PL properties were investigated. At last, the EL properties of red LEDs based on $LaBSiO_5$:$0.30Eu^{3+}$,$0.25Al^{3+}$ were investigated.

Fig.1-76 presents the XRD patterns of $LaBSiO_5$:$0.05Eu^{3+}$, $LaBSiO_5$:$0.30Eu^{3+}$ and $LaBSiO_5$:$0.30Eu^{3+}$,$0.25Al^{3+}$. The XRD pattern of $LaBSiO_5$:$0.05Eu^{3+}$ fired at 1100 ℃ is corresponding to the standard JCPDS card of $LaBSiO_5$ (a trigonal crystal structure with a space group of *P*31). In this crystal structure, the $La^{3+}$ ion is in six coordination with $O^{2-}$ ions. Both $BO_4$ tetrahedron and $SiO_4$ tetrahedron are interlinked by corner-sharing with each other to form a six-membered ring system. A little $Eu^{3+}$ does not cause significant changes in the structure of the host, and $Eu^{3+}$ ions occupy the sites of $La^{3+}$. Curves b and c are also similar to curve a, indicating that the phosphor $LaBSiO_5$:$0.30Eu^{3+}$,$0.25Al^{3+}$ shows the single-phase as same as $LaBSiO_5$. $Al^{3+}$ ions occupy the sites of $B^{3+}$ ions in this solid-state solution.

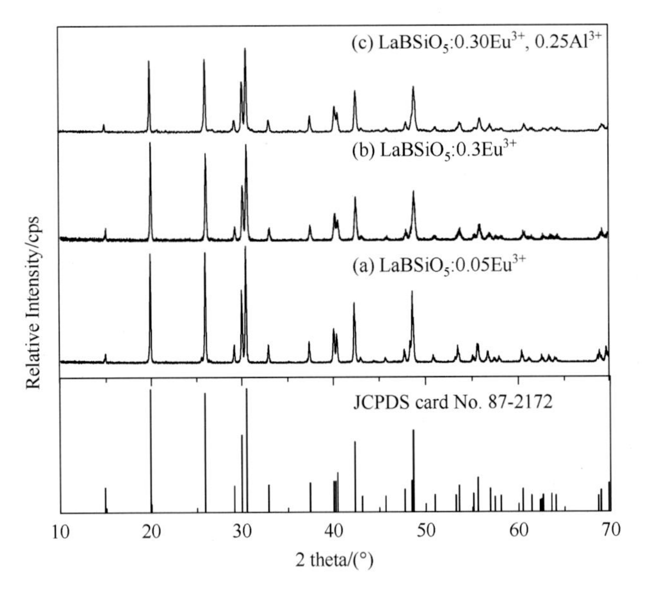


Fig.1-76　XRD patterns of $LaBSiO_5:0.05Eu^{3+}$ (a), $LaBSiO_5:0.30Eu^{3+}$ (b) and $LaBSiO_5:0.30Eu^{3+},0.25Al^{3+}$ (c)

Fig.1-77 is the excitation and emission spectra of $LaBSiO_5:0.30Eu^{3+}$. The excitation spectrum (curve a) is recorded by monitoring emission at 619 nm. The broadband from 200 nm to 300 nm is attributed to the charge transfer (CT) from $O^{2-}$ to $Eu^{3+}$. Many sharp peaks in the range from 300 nm to 500 nm are ascribed to the 4f-4f transition of $Eu^{3+}$ ions. The strongest excitation peak is located at 395 nm, which is due to the transition of $^7F_0$-$^5L_6$. The emission spectrum of $LaBSiO_5:0.30Eu^{3+}$ (curve b) is obtained under the excitation of 395 nm. These

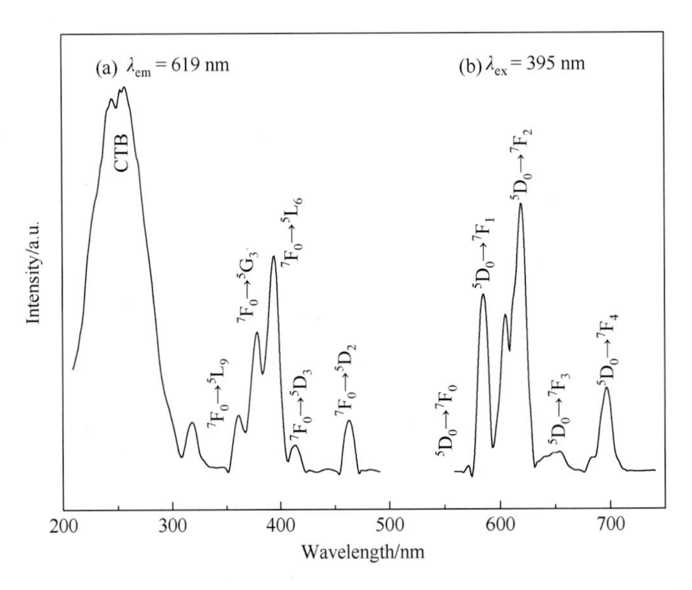


Fig.1-77　The excitation (a) and emission (b) spectra of $LaBSiO_5:0.30Eu^{3+}$

emission peaks are caused by the 4f-4f transitions of $Eu^{3+}$ ions, including the $^5D_0$-$^7F_1$ transition (585 nm), the $^5D_0$-$^7F_2$ transition (605 nm, 619 nm), the $^5D_0$-$^7F_3$ transition (654 nm) and the $^5D_0$-$^7F_4$ transition (697 nm). There are two splitting peaks of the $^5D_0$-$^7F_2$ emission, which probably results from the energy-level splitting of the $^7F_2$ state. Among all emission peaks, the emission peak with the largest intensity is located at around 619 nm. The CIE chromaticity coordinates for the phosphor $LaBSiO_5$:0.30$Eu^{3+}$ are $x = 0.631$, $y = 0.368$, which are calculated in terms of the emission spectrum.

Fig.1-78 exhibits the excitation and emission spectra of $LaBSiO_5$:$xEu^{3+}$ ($x = 0.10$, 0.20, 0.30, 0.40, 0.50). These excitation spectra are of similar shapes. With the increasing content of $Eu^{3+}$, the excitation intensity is enhanced up to the critical concentration of 30%. The influence of $Eu^{3+}$ concentration on the emission intensity of $LaBSiO_5$:$xEu^{3+}$ is shown in Fig.1-78(b). These phosphors share intense red emission under 395 nm excitation. The luminescence intensity increases linearly as the $Eu^{3+}$ concentration rises until it reaches 30%. This result is in agreement with that reported by Xue et al. The concentration dependence of the relative emission integral intensity of $LaBSiO_5$:$xEu^{3+}$ ($x = 0.10$, 0.20, 0.30, 0.40, 0.50) is shown in inserted figure of Fig.1-78(b).

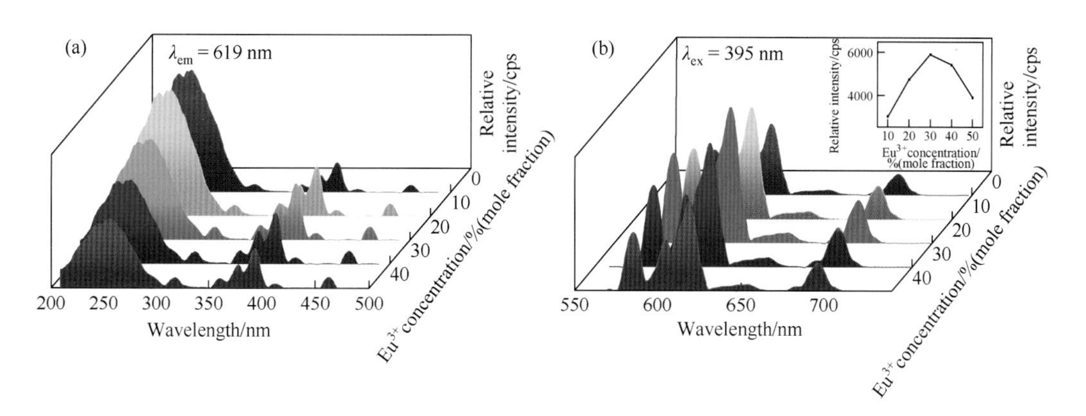


Fig.1-78. The excitation (a) and emission (b) spectra of $LaBSiO_5$:$xEu^{3+}$ ($x = 0.10$, 0.20, 0.30, 0.40, 0.50). The inserted figure is the concentration dependence of the relative integral emission intensity of $LaBSiO_5$:$xEu^{3+}$

To furtherly enhance the luminescent property of $LaBSiO_5$:0.3$Eu^{3+}$, $Al^{3+}$ ions were doped into $LaBSiO_5$:0.30$Eu^{3+}$. The excitation and emission spectra of $LaBSiO_5$:0.30$Eu^{3+}$,$yAl^{3+}$ ($y = 0$, 0.05, 0.15, 0.35) are shown in Fig.1-79. The excitation intensities of these phosphors are enhanced with the introduction of $Al^{3+}$. When the content of $Al^{3+}$ is 25%, the excitation intensity is the strongest. Fig.1-79(b) shows the emission spectra of these phosphors under 395 nm light excitation. With the doping of $Al^{3+}$, the intensity of the splitting peak at 605 nm declines. When the content of $Al^{3+}$ is about 15%, only one emission of $^5D_0$-$^7F_2$ transitions can be observed, which is located at 616 nm. In the structure of $LaBSiO_5$:0.30$Eu^{3+}$,$yAl^{3+}$, the $Eu^{3+}$ ion is coordinated by six $O^{2-}$. Since $Al^{3+}$ has a larger ionic radius than that of $B^{3+}$, the doped

$Al^{3+}$ will enlarge the volume of the crystal cell, and result in the Eu-O distance longer. So, the intensity of the crystal field around $Eu^{3+}$ declines and is accompanied by the diminution of the emission splitting.

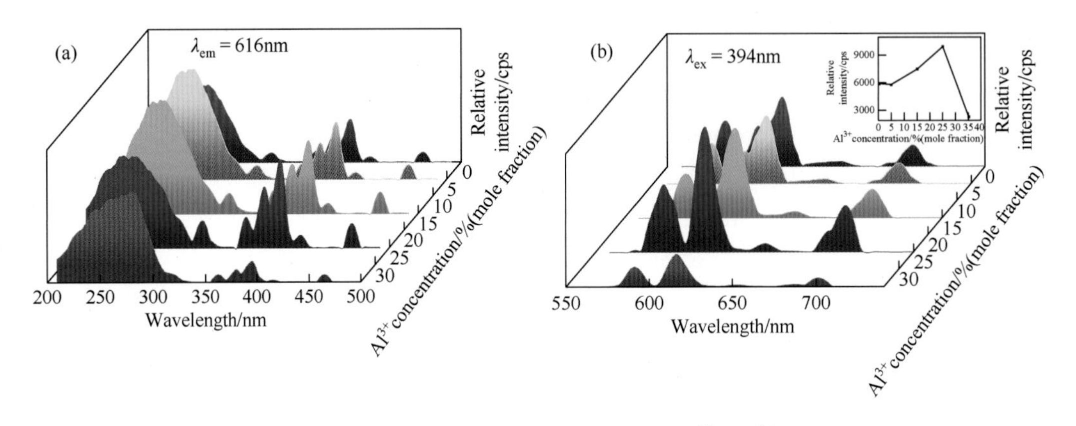


Fig.1-79 The excitation (a) and emission (b) spectra of $LaBSiO_5$:0.30$Eu^{3+}$, $yAl^{3+}$ ($y$ = 0, 0.05, 0.15, 0.25, 0.35). The inserted figure is the concentration dependence of the relative integral emission intensity of $LaBSiO_5$:0.30$Eu^{3+}$, $yAl^{3+}$

The concentration dependence of the relative integral emission intensity of $LaBSiO_5$: 0.30$Eu^{3+}$, $yAl^{3+}$ is shown in the inserted figure of Fig.1-79. The sample $LaBSiO_5$:0.3$Eu^{3+}$, 0.25$Al^{3+}$ is of the strongest red emission among these five phosphors, and its emission intensity is about 1.7 times higher than that of $LaBSiO_5$:0.30$Eu^{3+}$. The CIE chromaticity coordinates for the phosphor $LaBSiO_5$:0.3$Eu^{3+}$,0.25$Al^{3+}$ are $x$ = 0.638, $y$ = 0.362, in terms of the corresponding emission spectrum.

To evaluate the luminescent properties of the red phosphor, the emission intensity of $LaBSiO_5$:0.3$Eu^{3+}$,0.25$Al^{3+}$ is compared with that of $Y_2O_2S$:0.05$Eu^{3+}$ under 395 nm light excitation. Fig.1-80 exhibits the excitation and emission spectra of $LaBSiO_5$:0.3$Eu^{3+}$,0.25$Al^{3+}$ and $Y_2O_2S$:0.05$Eu^{3+}$. As discussed above, the main emission line of $LaBSiO_5$:0.3$Eu^{3+}$ is the $^5D_0$-$^7F_2$ transition of $Eu^{3+}$ at 616 nm. Other transitions from the $^5D_J$ levels to $^7F_J$ ground states, such as $^5D_0$-$^7F_J$ lines in the 570-720 nm range and $^5D_1$-$^7F_J$ transitions in the 520-570 nm range are weaker. $Y_2O_2S$:0.05$Eu^{3+}$ has a hexagonal structure and the point symmetry of the yttrium site is $C_{3V}$ (3 m). $Eu^{3+}$ is expected to occupy the $Y^{3+}$ site in $Y_2O_2S$:0.05$Eu^{3+}$. The main emission peaks at 626 nm of $Y_2O_2S$:0.05$Eu^{3+}$ are ascribed to $Eu^{3+}$ transition from the $^5D_0$ to $^7F_2$ transition. Other transitions from the $^5D_J$ ($J$ = 0, 1, 2, 3) levels to $^7F_J$ ($J$ = 0, 1, 2, 3, 4, 5, 6) ground states are lower. From Fig.1-80, the emission intensity of $LaBSiO_5$:0.3$Eu^{3+}$,0.25$Al^{3+}$ under 395 nm light excitation is about 3.8 times stronger than that of $Y_2O_2S$:0.05$Eu^{3+}$.

Our purpose is to obtain a highly efficient red-emitting phosphor for the near-UV InGaN-based white LEDs. Thus, single red LEDs were fabricated by combining the prepared red phosphors with InGaN chips. Fig.1-81 is the EL spectra of the InGaN LED chip and the red

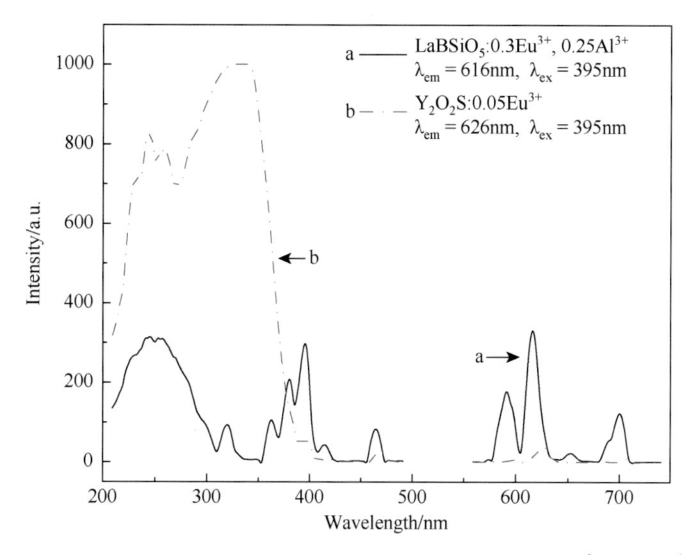


Fig.1-80 The excitation and emission spectra of $LaBSiO_5$:0.3$Eu^{3+}$, 0.25$Al^{3+}$ (a) and $Y_2O_2S$:0.05$Eu^{3+}$ (b)

LEDs-based on red phosphors $LaBSiO_5$:0.3$Eu^{3+}$ and $LaBSiO_5$:0.3$Eu^{3+}$, 0.25$Al^{3+}$. The emission band at 395 nm in the spectrum a is attributed to the emission of the InGaN chip. Compared with the spectrum a, the emission peaks at 616 nm and 702 nm in spectra (b and c) are due to the emissions of red phosphor excited by the emission of the InGaN LED chip. The bright red light from these two red LEDs can be observed by naked eyes (Fig.1-81). The emission of the InGaN chip (395 nm) can still be observed in Fig.1-81 (b and c). This is advantageous to obtain a w-LED by combining this phosphor with appropriate blue and green phosphors. These results suggest that $LaBSiO_5$:0.3$Eu^{3+}$,0.25$Al^{3+}$ is a promising red light component for the near-UV InGaN-based white LEDs.

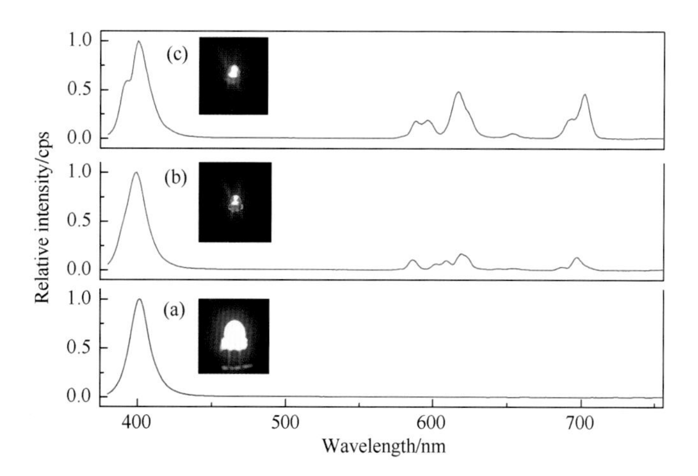


Fig.1-81 EL spectra of the InGaN LED chip (a), the red LED-based on $LaBSiO_5$:0.3$Eu^{3+}$ (b), the red LED-based on $LaBSiO_5$:0.3$Eu^{3+}$,0.25$Al^{3+}$ (c) under 20 mA current excitation. The inserted figures are the images of these LEDs

### 1.4.2 $Y_2O_2S:Eu^{3+}$ nanocrystals prepared by the molten salt synthesis

$Y_2O_2S:Eu^{3+}$ is a classical red-emitting phosphor, which is mainly prepared by the conventional solid-state reaction. The size of particles obtained by the solid-state method is about 3 μm with irregular morphology. To prepare nanocrystals of $Y_2O_2S:Eu^{3+}$ with uniform size, there are several successful attempts to prepare $Y_2O_2S:Eu^{3+}$ with the wet chemical synthesis.

Wet chemical methods, such as the sol-gel route, have been widely developed to prepare the luminescent materials. However, the sol-gel process often requires expensive organic precursors and solvents. Then a low-temperature technique called molten salt synthesis (abbreviated as MSS), is beginning to attract a great deal of interest. MSS is one simple and versatile approach to obtain single-phase powders at lower temperatures with shorter reaction times and little residual impurities compared with the conventional solid-state reactions. The ionic fluxes of molten salts possess high reactivity toward different inorganic species and relatively low melting points, which is convenient for the preparation of inorganic materials. As the reaction medium, the inorganic molten salts exhibit favorable physicochemical properties, such as a greater oxidizing potential, high mass transfer, high thermal conductivity, as well as relatively lower viscosities and densities compared to the conventional solvents. Furthermore, MSS is one effective way to prepare nano-scale shape-controlled materials.

In this part, a series of red-emitting phosphors $Y_2O_2S:Eu^{3+}$ were prepared by MSS with NaCl as the molten salt. Different surfactants, such as polyoxyethylene (9) nonyl phenyl ether (NP-9), polyoxyethylene (5) nonyl phenyl-ether (NP-5), sodium Lauryl Ether Sulfate (SLES), octyl phenol together ethylene (10) ether oxygen (OP-10), and sodium dodecylbenzene sulfonate (LAS) were used to control the particles size of $Y_2O_2S:Eu^{3+}$.

Fig.1-82 shows the XRD patterns of $Y_2O_2S:Eu^{3+}$ with different surfactants NP-10 (Y-NP10), NP-9 (Y-NP9), NP-5 (Y-NP5), SES (Y-SES), OP-10 (Y-OP10), and LAS (Y-LAS).

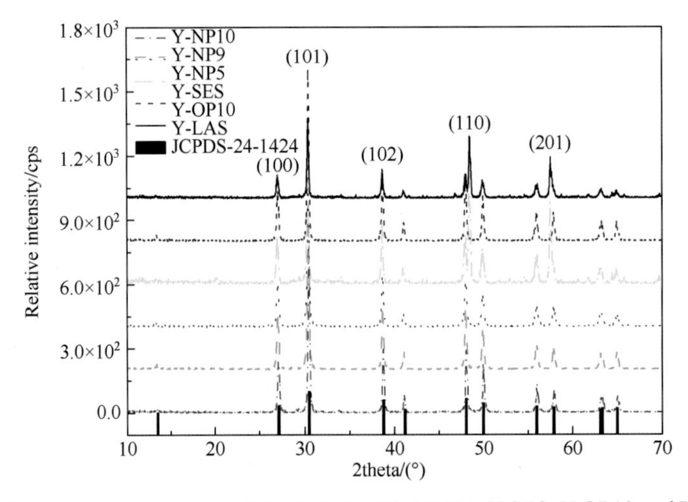


Fig.1-82 The XRD patterns of Y-NP10, Y-NP9, Y-NP5, Y-SES, Y-OP10 and Y-LAS

These XRD patterns of the samples are consistent with that given in the JCPDS card 24-1424 ($Y_2O_2S$). These phosphors are of a single-phase, and no $Y_2O_3$ phase can be observed. The result shows that single-phase $Y_2O_2S$: $Eu^{3+}$ can be easily prepared by MSS at a lower temperature.

A series of chemical reactions are proposed as following:

$$Y(NO_3)_3 \cdot 6H_2O + Eu\,(NO_3)_3 \cdot 6H_2O + 6NaOH \longrightarrow Y(OH)_3 + Eu(OH)_3 + 6NaNO_3 + 12H_2O \quad (1\text{-}1)$$

$$6NaOH + (2x + 1)S \longrightarrow 2Na_2S_x + Na_2SO_3 + 3H_2O \quad (1\text{-}2)$$

$$Na_2S_x + Y(OH)_3 + Eu(OH)_3 + Na_2SO_3 + 6NaNO_3 + NaCl \longrightarrow Y_2O_2S{:}Eu^{3+} + \text{flux residue} \quad (1\text{-}3)$$

In the conventional solid-state reactions for $Y_2O_2S$, the temperature for the transformation of $Y_2O_3$ into $Y_2O_2S$ is about 1100 ℃ with flux. In this molten salt reactions, the reaction temperature is 850 ℃. The reason may be as follows. Firstly, the ionic fluxes of molten salts exhibit original and rich chemical characters, which is convenient for the preparation of $Y_2O_2S$:$Eu^{3+}$. Secondly, S (sulfur) with the high chemical activity can be easily obtained from the decomposition of $Na_2S_x$ (sodium polysulfide), which shows stronger reducibility compared with S from purchase. Then $Y_2O_2S$: $Eu^{3+}$ may easily be prepared at a lower temperature.

SEM images reveal the morphology and size of as-prepared samples Y-NP10, Y-NP9, Y-NP5, Y-SES, Y-OP10 and Y-LAS, as shown in Fig.1-83. $Y_2O_2S$:$Eu^{3+}$ nanocrystals (NCs) have been successfully prepared by the MSS method with NP-10. The size of the particle is 10-20 nm [Fig.1-83(a)]. The result shows the surfactant NP-10 can form micelles and prevent $Y_2O_2S$ NCs growth effectively. But their crystallinity is lower. Y-NP9 and Y-NP5 show good crystallinity, and their sizes are about 100 nm and 150 nm, respectively. We also prepared $Y_2O_2S$:$Eu^{3+}$ with other surfactants. Y-SES and Y-OP10 particles exhibit regular morphology with about 500 nm and 200 nm. The sample of Y-LAS shows high crystallinity, and its average particle size is about 1 μm.

Fig.1-83 SEM images of Y-NP10, Y-NP9, Y-NP5, Y-SES, Y-OP10 and Y-LAS

Fig.1-84 shows the excitation spectra of Y-NP9, Y-NP5, Y-SLES, Y-OP10 and Y-LAS under monitored 627 nm. As shown in Fig.1-84, these five samples show similar excitation bands with different intensities. The bands from 300 nm to 380 nm are all due to the $S^{2-}\rightarrow Eu^{3+}$ CT transition in $Y_2O_2S$. The narrow peaks (every weak) beyond 380 nm can be attributed to intra-configurational 4f-4f transitions of $Eu^{3+}$ ions.

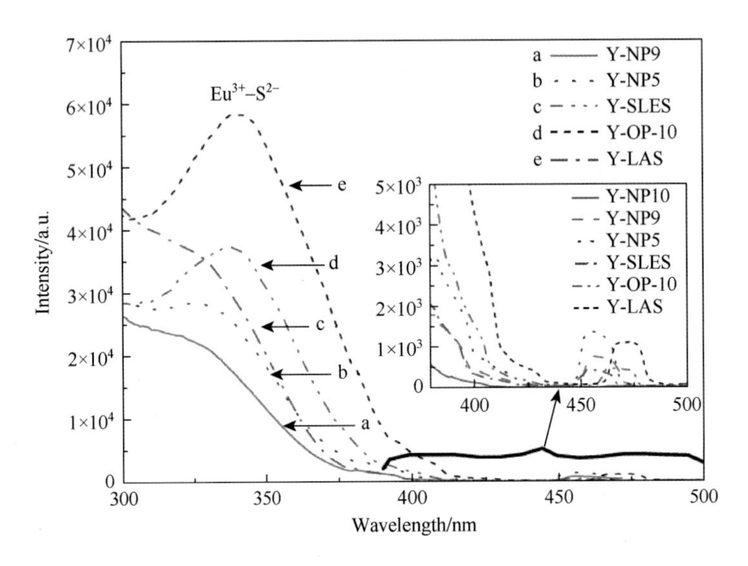


Fig.1-84　The excitation spectra ($\lambda_{em}$ = 627 nm) of Y-NP9, Y-NP5, Y-SLES, Y-OP10 and Y-LAS. The inserted figure is the enlarged spectra from 390 nm to 500 nm

Some differences also can be found in these curves. Firstly, the excitation of Y-LAS shows the strongest intensity among these six samples under monitored 627 nm. Consulting the SEM micrographs, it can be seen that Y-LAS shows the best crystallinity and the largest particle size, this is convenient for the UV absorption. Secondly, the center of the CT band ($S^{2-}\rightarrow Eu^{3+}$) of Y-LAS shows a slight shorter-wavelength shifting in nano-scale phosphor, due to the quantum size effect of the phosphor nanoparticles. The reason is that the nano size of the phosphor increases the kinetic energy of the electrons and results in a larger bandgap. Thus, higher energy is required to excite the phosphor.

Y-NP9, Y-NP5, Y-SES, Y-OP10 and Y-LAS have similar emission spectra, as shown in Fig.1-85. $Y_2O_2S:Eu^{3+}$ has a hexagonal structure and the point symmetry of the yttrium site is $C_{3V}$ (3 m). $Eu^{3+}$ is expected to occupy the $Y^{3+}$ site in $Y_2O_2S:Eu^{3+}$. The main emission peaks at 627 nm and 616 nm of $Y_2O_2S:Eu^{3+}$ are ascribed to $Eu^{3+}$ transition from $^5D_0$ to $^7F_2$ and its strongest peak is at 627 nm. Other transitions from the $^5D_J$ ($J$ = 0, 1, 2, 3) levels to $^7F_J$ ($J$ = 0, 1, 2, 3, 4, 5, 6) ground states are weaker. The bright red light of Y-NP9 can be observed under 365 nm excitation. According to the crystal-field theory, the $J$ = 0 state does not split ($A_1$), and the $J$ = 1 state splits into two Stark levels ($A_2$ and E) resulting in one and two emission bands, respectively. This observed behavior indicates only one emitting $Eu^{3+}$ symmetry site. From

Fig.1-85, no peak arising from a $^5D_0$–$^7F_2$ transition at 611 nm is observed suggesting the absence of oxide ($Y_2O_3$) impurity. This result is corresponding to that of the XRD patterns.

In general, two factors, either higher UV absorption or higher quantum efficiency will result in higher emission intensity. Comparing the excitation spectra in Fig.1-84 with the emission spectra in Fig.1-85, we think that the change of emission intensity may be the main result of altering UV absorption around 400 nm in the present case. The nanoparticles yield more defects and crystallographic distortions, which results in the weakest emission intensity of phosphor nanoparticles.

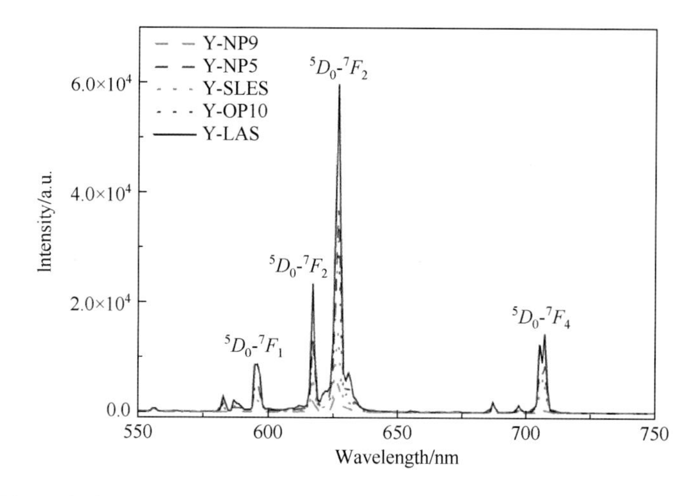


Fig.1-85 The emission spectrum ($\lambda_{ex}$ = 350 nm) of Y-NP9, Y-NP5, Y-SLES, Y-OP10 and Y-LAS

Fig.1-86 is the EL spectrum of the red LEDs of the near-UV InGaN-based on Y-NP9 under 20 mA current excitation. The emission peak at 395 nm is due to the emission of the InGaN chip.

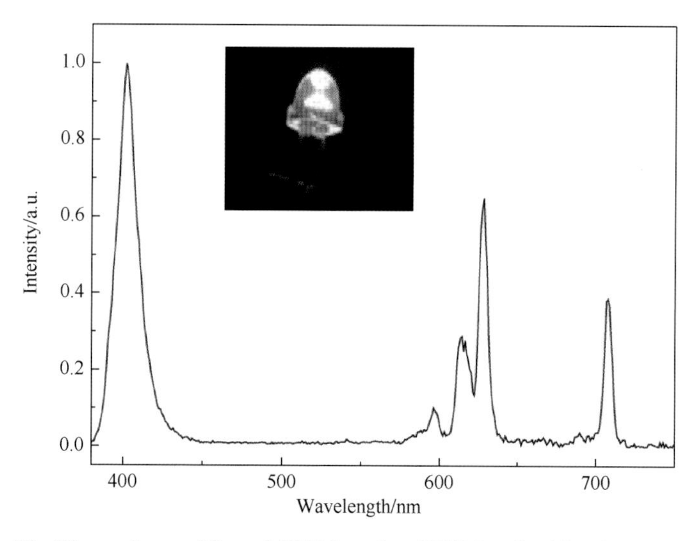


Fig.1-86 The EL spectrum of the red LED-based on Y-NP9 under 20 mA current excitation. The inserted figure is the image of the LED

The sharp peaks at 595 nm, 613 nm, 627 nm and 706 nm are due to the emissions of red phosphor. This red LED exhibits intense red emission, which can be observed by the naked eyes (Fig.1-86). The CIE chromaticity coordinates of the red LED are $x = 0.502$, $y = 0.252$. The intensive emission of the InGaN chip can still be observed in Fig.1-86. It is advantageous to obtain a w- LED by combining this phosphor with appropriate blue and green phosphors.

### 1.4.3 $Y_2O_3:Eu^{3+}$ prepared by mimicking wood tissue

Cubic $Y_2O_3:Eu^{3+}$ is one of the most important commercial red phosphors used in fluorescent lights, cathode ray tubes (CRT), plasma display panel (PDP) and field emission display (FED). Many methods for the synthesis of the $Eu^{3+}$-doped $Y_2O_3$ phosphor have been reported, including the solution combustion method, the sol-gel method, the spray pyrolysis method, the hydrothermal method and the precipitation method.

Recently, more interests are focused on the design of inorganic materials with wood templates. Wood tissue shows a hierarchical porous structure of unidirectionally oriented cells. Its porous structures range from the millimeter scale via the micrometer scale to the nanometer scale. Different kinds of wood share different kinds of morphology and arrangement of the cells, such as large vessel cells dominate in hardwoods, while softwoods are composed of tracheal cells with significantly less variation of pore channel geometry. Then the wood tissue is an interesting template for the fabrication of inorganic materials. By mimicking different kinds of wood tissue, inorganic materials with different morphology can be prepared.

The bio-templating approach offers a simple route for patterning inorganic materials of general interest with environment-friendly characteristics. Hence we prepared the red phosphor $Y_2O_3:Eu^{3+}$ by mimicking the wood tissue of Nepal Alder and investigated its PL properties in this element.

The cubic $Y_2O_3:Eu^{3+}$ has been successfully prepared by mimicking the wood tissue of Nepal Alder. In the synthesized process, $Y^{3+}$ and $Eu^{3+}$ ions are infiltrated into the wood tissue and precipitated by the $OH^-$, which was ionized by $NH_3 \cdot H_2O$. $Y_2O_3:Eu^{3+}$ can be obtained by decomposing the mixtures of $Y(OH)_3$ and $Eu(OH)_3$. The detailed process can be described as follows:

$$NH_3 \cdot H_2O \longrightarrow NH_4^+ + OH^- \tag{1-4}$$

$$Y^{3+} + OH^- \longrightarrow Y(OH)_3 \tag{1-5}$$

$$Eu^{3+} + OH^- \longrightarrow Eu(OH)_3 \tag{1-6}$$

$$Y(OH)_3 + Eu(OH)_3 \longrightarrow Y_2O_3:Eu^{3+} \tag{1-7}$$

The XRD pattern of $Y_2O_3:Eu^{3+}$ fired at 500 ℃ is shown in Fig.1-87. It is consistent with the JCPDS card No. 41-1105 [$Y_2O_3$]. This reveals that the phosphor shares the cubic structure with the single phase. The EDS spectrum (Fig.1-87) shows the sample is composed of $Y_2O_3:Eu^{3+}$ without any impurities, and $Eu^{3+}$ has been effectively built into the $Y_2O_3$ host lattice and occupies the $Y^{3+}$ site. This result is in agreement with the XRD pattern. The composition analysis shows an approximate atom ratio of $Y^{3+}$ to $Eu^{3+}$ is about 10 : 1.

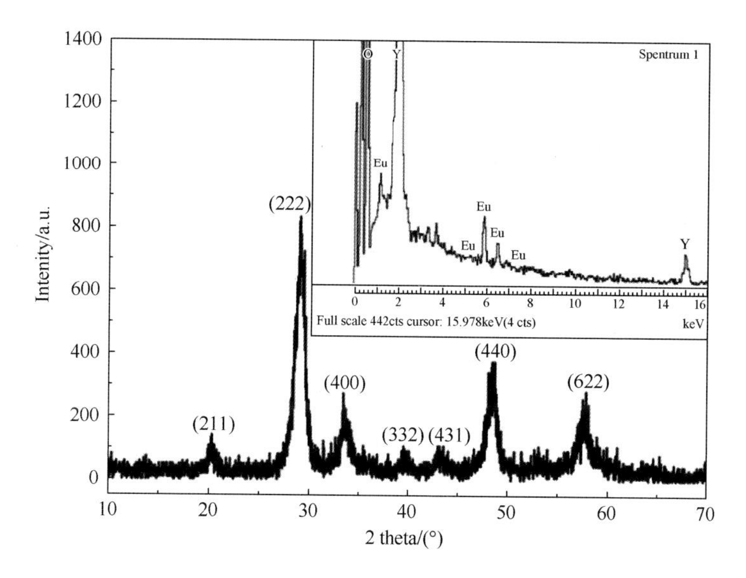


Fig.1-87 The XRD pattern of $Y_2O_3$:$Eu^{3+}$. The inseted figure is the EDS spectrum

Fig.1-88 is the SEM images of $Y_2O_3$:$Eu^{3+}$ fired at 500 ℃. $Y_2O_3$:$Eu^{3+}$ shows a flocculent shape like the wood tissue. The wood tissue is comprised of some thin fibers, whose average grain size is about 1 μm [Fig.1-88(b)].

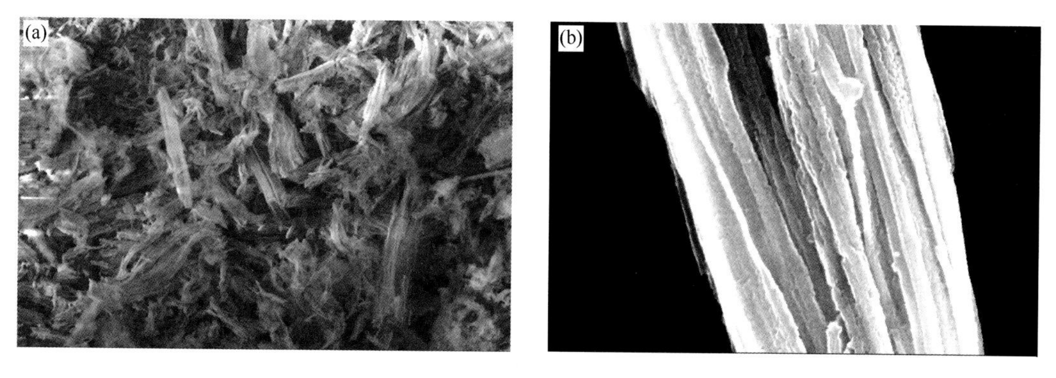


Fig.1-88 SEM images of $Y_2O_3$:$Eu^{3+}$ fired at 500 ℃

The excitation and emission spectra of $Y_2O_3$:$Eu^{3+}$ are shown in Fig.1-89, and they are similar to those in references. The broad excitation band from 230 nm to 290 nm is attributable to the charge transfer (CT) from the 2p orbital of $O^{2-}$ to the 4f orbital of $Eu^{3+}$. The sharp lines at the range from 350 nm to 550 nm are intra-configurational 4f-4f transitions of $Eu^{3+}$ ions in the host lattices. Curve b is the emission spectrum of $Y_2O_3$:$Eu^{3+}$ under 254 nm excitation. The bright red light can be observed under 254 nm excitation (Fig.1-88). The strongest peak is at 611 nm, which is due to the $^5D_0 \rightarrow {}^7F_2$ transition in $C_2$ symmetry for $Eu^{3+}$. Other f-f transitions of $Eu^{3+}$, such as $^5D_0 \rightarrow {}^7F_J$ transitions at 570-720 nm and $^5D_1 \rightarrow {}^7F_J$ transitions at 520-570 nm are very weak.

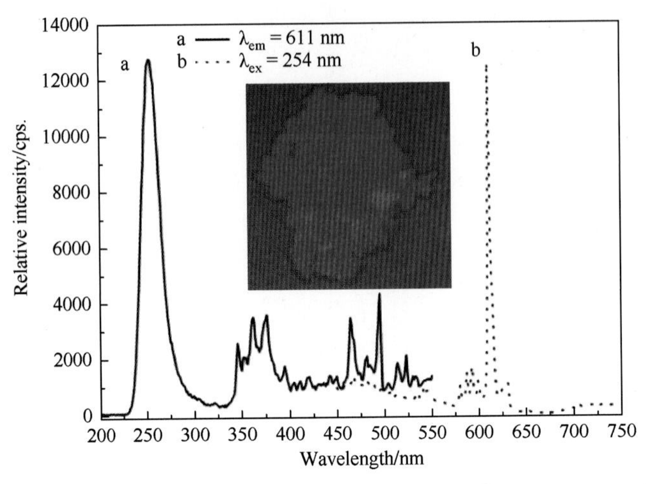


Fig.1-89 The excitation (a) and emission spectra (b) of $Y_2O_3:Eu^{3+}$ ($\lambda_{ex} = 254$ nm, $\lambda_{em} = 611$ nm). The inserted figure is the picture of $Y_2O_3$: $Eu^{3+}$ under 254 nm excitation

The decay curve for the $^5D_0 \rightarrow {}^7F_2$ transition of the $Eu^{3+}$ of $Y_2O_3:Eu^{3+}$ is shown in Fig.1-90. It is well fitted into the single-exponential function, and the lifetime $\tau$ value is calculated to be 0.908 ms.

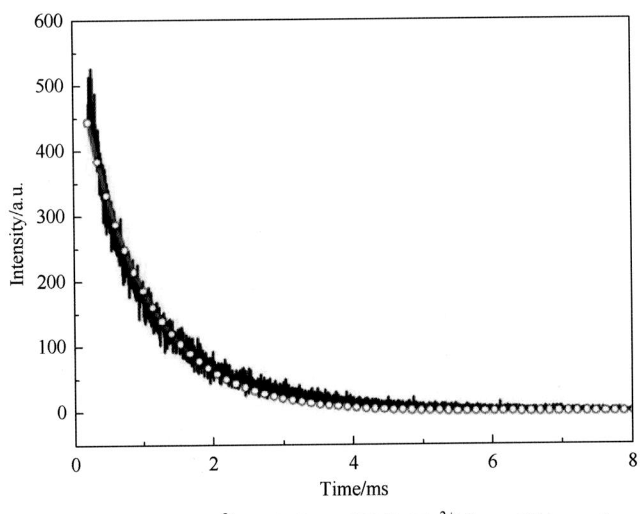


Fig.1-90 Decay curve of the $Eu^{3+}$ emission of $Y_2O_3:Eu^{3+}$ ($\lambda_{ex} = 254$ nm, $\lambda_{em} = 611$ nm)

## 1.5 $Eu^{3+}$ complexes for LEDs

### 1.5.1 Eu(MBPTFA)$_3$Phen

There are several strong advantages for rare-earth organic complexes, such as high color purity with narrow half-peak widths, emissive wavelengths based on f-f transitions of lanthanide ion seldom change after ligand modification. However, it is difficult to find high-performance Eu(III) complexes that can be efficiently excited by the 400 nm near-UV light since the excitation efficiency of rare-earth complexes in this wavelength range decreases rapidly. The challenge lies in the synthesis of a europium complex that meets all the three requirements: ①there is a suitably expanded $\pi$-conjugated system in the molecule so that the excitation wavelength shifts red till about 395 nm; ②good coordination stability to the europium (III) ion; ③high stability to heat and UV irradiation. To the best of our knowledge, few formal articles on rare-earth organic complexes applied in illuminating LEDs were reported.

To overcome these problems, the $\beta$-diketone ligand, 1-(4'-methoxy-biphenyl-4yl-)-4, 4,4,-trifluoro-butane-1,3-dione (marked as MBPTFA) was used as the first ligand and 1,10-phenanthroline (marked as Phen) as the second ligand. Thus, a ternary europium (III) complex, $Eu(MBPTFA)_3Phen$ was prepared. The thermal stability and PL properties of the complex were investigated. At last, the single red LED was fabricated by incorporation of the Eu(III) complex with a 395 nm emitting InGaN chip.

The ligand MBPTFA was used as the first ligand and 1, 10-phenanthroline (Phen) as the second ligand, $Eu(MBPTFA)_3Phen$ was prepared by following the method, which was prepared by the following method reported in the literature. The elemental analysis datum for $Eu(MBPTFA)_3Phen$ ($EuC_{63}H_{44}N_2O_9F_9$) are as follows. Found (calculated)/%: C 59.80 (59.67), H 3.44 (3.50), N 2.17 (2.20). The structure of the Eu complex $Eu(MBPTFA)_3Phen$ is shown in Fig.1-91.

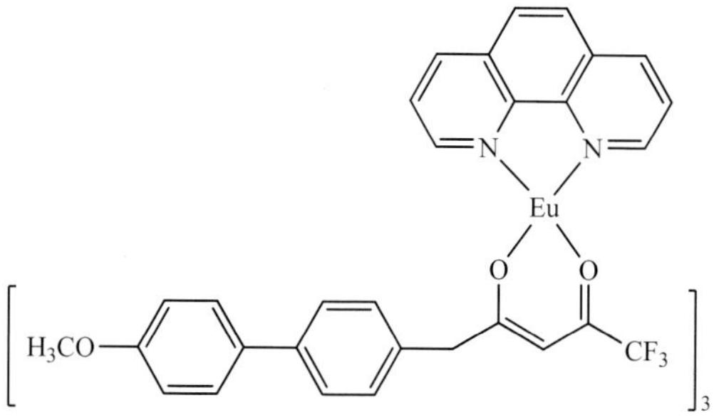

Fig.1-91 The molecular structure of the europium (III) ternary complex

The thermogravimetric analysis (TGA) curve (Fig.1-92) shows that the decomposition temperature of the $Eu^{3+}$ complex is 344 ℃. Such high thermal stability is favorable to form a stable film of the $Eu^{3+}$ complex as the emitter in the fabrication of LEDs.

Phosphorescence spectra for the ethanol solutions ($1\times10^{-5}$ mol·L$^{-1}$) of the corresponding Gd(III) binary complexes with the ligand MBPTFA were measured at 77 K. The excitation wavelength was set at the maximum UV-absorption wavelength of the Gd(III) complexes. The lowest triplet-state level energies of the ligand were determined by the shortest wavelength transition in the phosphorescence spectrum to be 23931 cm$^{-1}$ (417 nm).

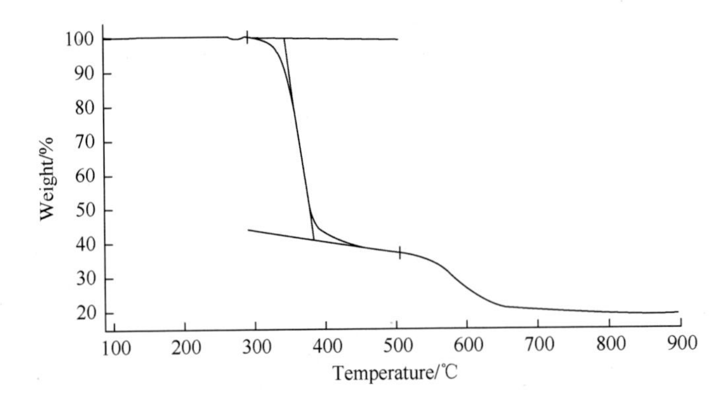


Fig.1-92　The TG curve of the complex Eu (MBPTFA)$_3$Phen in $N_2$ atmosphere

The excitation and emission spectra of the $Eu^{3+}$ complex powder are shown in Fig.1-93. The complex molecule exhibits a sharp excitation band from 385 nm to 400 nm [Fig.1-93(a)], and this result indicates that the energy absorbed by the ligands has been efficiently transferred to the central $Eu^{3+}$ ions in the complex. Several red-emitting peaks can be observed from Fig.1-93 under the excitation of 395 nm. This is ascribed to the intensely sensitized characteristic luminescence of $Eu^{3+}$ ions. The strongest emission is locating at 612 nm, which is due to the $^5D_0$ - $^7F_2$ transitions of $Eu^{3+}$ in the complex.

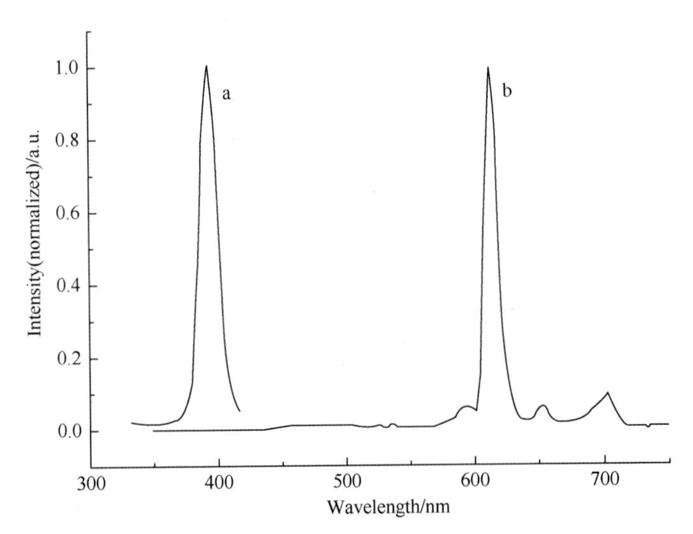


Fig.1-93　The excitation and emission spectra of the $Eu^{3+}$ complex powders:
(a) Em of the $Eu^{3+}$ complex ($\lambda_{em}$ = 612 nm), (b) Ex of the $Eu^{3+}$ complex ($\lambda_{ex}$ = 395 nm)

The probable PL mechanism for the $Eu^{3+}$ complexes is discussed as follows. ①Energy of the exciting UV-violet light is absorbed by the ligands; ②the excited ligands decay rapidly to their lowest triplet state; ③the energy transfers from the triplet state of the ligands to the quasi-resonant energy state of the $Eu^{3+}$ ion via the intersystem; ④the excited europium ion decays to the ground state via emissions in the visible region. The triplet state energy of ligands

MBPTFA (23931 $cm^{-1}$) is higher than the $^5D_0$ lowest excited state of $Eu^{3+}$ (17300 $cm^{-1}$) matching the excited states of $Eu^{3+}$. Therefore, the energy can be efficiently transferred from the ligands MBPTFA to the central $Eu^{3+}$ ions. The ligands strongly sensitize the luminescence of $Eu^{3+}$ in the corresponding europium (III) complexes and the maximum emission peak is located at 612 nm due to the $Eu^{3+}$ electric dipole transitions ($^5D_0$ - $^7F_2$).

The luminescence quantum yield for the $Eu^{3+}$ complex was determined to be $\varphi = 0.17$ with a reference solution of quinine sulfate in 0.5 mol·$L^{-1}$ sulfuric acid ($\varphi = 0.546$) and was corrected by the refractive index of the solvent at 298 K. The luminescence quantum yield is high among rare-earth organic complexes, suggesting the synthesized complex is an excellent red-emitting phosphor.

The InGaN chip acts as an energy donor while the Eu complex is a receptor. Based on the excitation spectrum [Fig.1-93(a)], the complex molecule exhibits a sharp excitation band from 385 nm to 400 nm. This band is well overlapped with the emission spectrum of the InGaN LED chip so that the excitation wavelength is near the UV-violet and the energy emitted from the InGaN chip can be efficiently absorbed by the ligands which then transferred to the central $Eu^{3+}$ ions in the complex.

The $Eu^{3+}$ organic complex powder was mixed with the epoxide resin in a proper mortar. Then, the mixture was fitted into the little bowel of the InGaN chip, and heated to solid. At last, the epoxide resin was applied to coat the InGaN chip. The thickness of the coating is 5-8 mm, a red-emitting LED was fabricated by the combination of one InGaN chip with the synthesized Eu $(MBPTFA)_3Phen$ as a red-emitting phosphor.

Fig.1-94 shows the emission spectra of the red LED-based on $Eu(MBPTFA)_3Phen$ [Fig.1-94(b)] and the original InGaN-LED chip without phosphor [Fig.1-94(a)] both under 20 mA current operation. A sharp and intense peak at 612 nm is due to the $Eu^{3+}$ emission from $Eu(MBPTFA)_3Phen$

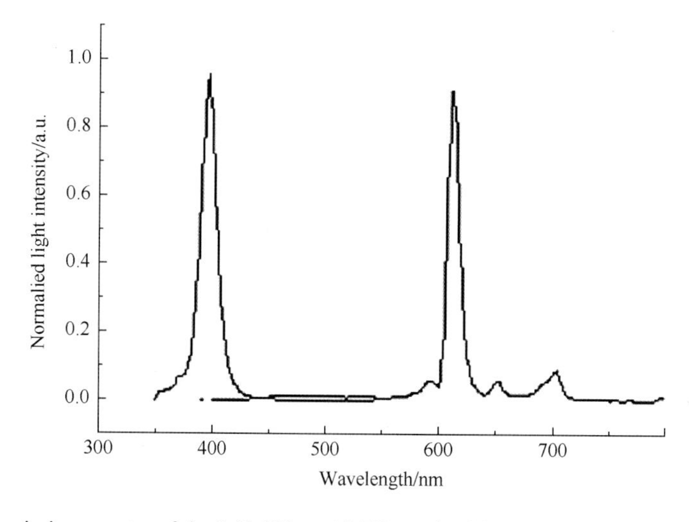


Fig.1-94 The emission spectra of the InGaN-based LEDs under 20 mA current operation: (a) the original InGaN-LED without phosphor, (b) InGaN-LED with $Eu(MBPTFA)_3Phen$ as a phosphor

complex in the red LED [Fig.1-94(b)]. The bright red light from the LED is observed by naked eyes. Its CIE chromaticity coordinates are calculated to be $x = 0.65$, $y = 0.34$, which are close to the NTSC standard values for red (0.67, 0.33). Comparison of the emission band of the InGaN chip [Fig.1-94(a)] with the excitation spectrum of the Eu complex powder [Fig.1-93(a)] confirms the matching of the complex to the InGaN chip. Only a little emission from the InGaN chip remains in the emission spectrum of the complex-LED [Fig.1-94(b)], indicating that the near-UV-violet emitted energy from the InGaN chip has been efficiently absorbed by the ligands and transfers to the central $Eu^{3+}$ ions in the complex. Intensely $Eu^{3+}$ ion characteristic red light-emitting is observed. Hence, it is considered to be a good candidate for the red component of a three-band w-LED.

### 1.5.2 [Eu$_2$(2,7-BTFDBC)$_{3-n}$(DBM)$_{2n}$(Phen)$_2$] ($n = 0, 1, 2$)

$Eu^{3+}$ $\beta$-diketone complexes are promising red-emitting phosphors because they show a wide excitation band in the UV region, strong emission intensity, and high quantum yield. For example, the complex [Eu(DBM)$_3$(phen)] is one of the famous red-emitting $Eu^{3+}$-based molecular phosphors with high quantum yield. However, a major disadvantage of this kind of complex is its low photochemical stability under ultraviolet irradiation. Meanwhile, the complex shows a weaker absorption band in the blue region, which is not suitable for the blue-emitting GaN LED chip.

There are two traditional approaches to achieve visible light excitation. One is to introduce a transition metal ion into the Eu(III) complex. Unfortunately, the luminescent efficiency of this kind of Eu(III) complexes is always very low. The other strategy is to synthesize $Eu^{3+}$ organic complexes with a suitably expanded $\pi$-conjugated system in the molecule so that the excitation wavelength shows a red-shift. Some researchers have done some good jobs to expand the $\pi$-conjugated system using the ligands Michler's ketone, phenalenone and tridentate in Eu(III) complexes. Using the carbazole group, our group has synthesized a novel dinuclear $Eu^{3+}$ organic complex with the longer $\pi$-conjugated system.

To the best of our knowledge, strong $\pi$-stacking interactions between aromatic ligands can produce a significant bathochromic-shift in fluorescent excitation and UV-vis spectra of the complexes. The intramolecular $\pi$-stacking structure has also been found in the lanthanide complexes. Thus, the $\pi$-stacking interactions could be used to achieve visible light excitation for the Eu(III) complexes. In this work, three europium complexes with aromatic ligands were synthesized with 2,7-bis(4'4'4'-trifluoro-1,3-dioxobutyl)- (9-ethyl-9H-carbazole) (marked as 2,7-BTFDBC) and/or dibenzoyl methane (marked as DBM) as the first ligand, and 1, 10-phenanthroline (Phen) as the second ligand. The broad excitation bands in the blue region have been observed for the as-synthesized complexes. The thermal stability and PL properties of the synthetic complexes were also investigated. At last, the single red LED was fabricated by the combination of the synthetic Eu(III) complex with a 460 nm emitting GaN chip.

The ligand 2,7-BTFDBC was synthesized according to our previous work. Fig.1-95 depicts the structures and synthesis route of the Eu(III) complexes 1-3. To synthesize the Eu(III) complexes, 2,7-BTFDBC and DBM were used as the first ligand and Phen as the second ligand. The raw materials including 2,7-BTFDBC, DBM, Phen, and $EuCl_3$ (the molar ratio of 3 ∶ 0 ∶ 2 ∶ 2, 2 ∶ 2 ∶ 2 ∶ 2, 1 ∶ 4 ∶ 2 ∶ 2, respectively) were dissolved in the *N*, *N*-dimethylformamide (DMF) solution. Then the pH value of the solution was carefully adjusted to 7.0-8.0 by dropping the concentrated triethylamine. Followed by stirring for 6 h at 60 ℃, the solution was cooled to room temperature. Then, the products were collected from the solution on a vacuum filter by washing with water and ethanol alternately. At last, the yellow powder was obtained after the products at 60 ℃ in a vacuum drying oven for 24 h.

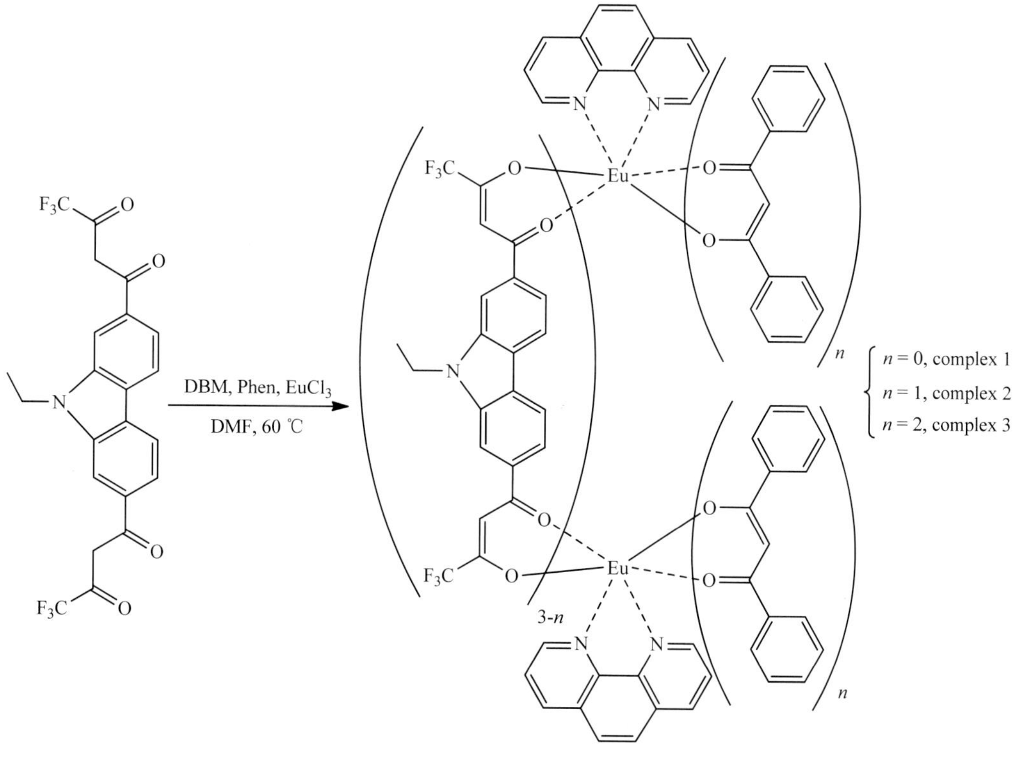

DBM, Phen, $EuCl_3$

DMF, 60 ℃

$n = 0$, complex 1

$n = 1$, complex 2

$n = 2$, complex 3

Fig.1-95 Structures and synthesis procedures of the Eu (III) complexes 1-3

The ground state geometries of the $Eu^{3+}$ complex 3 were calculated by both the density functional theory (DFT) and the semiempirical method. DFT calculations were performed using the GAMESS suite of the program without symmetry restriction. The large core effective core potential (LC-ECP), as well as the corresponding basis set suggested by Dolg et al. was used for europium. For valence orbitals, the (7s6p5d)/[5s4p3d] basis set was used. All-electron basis set 6-311+G(d, p) has been used for F, O, C, N and H.

The semi-empirical calculations on the ground state geometries of the Eu(III) complex 3 were carried out using the Sparkle/PM7 model implemented in the MOPAC2012 program. The

MOPAC keywords used in the Sparkle/PM7 calculations were PRECISE, GEO-OK, GNORM = 0.25, SCFCRT = 1.D-10 (to increase the SCF convergence criterion) and XYZ (the geometry optimizations were performed in Cartesian coordinates).

Using the intermediate neglect of differential overlap/spectroscopic-configuration interaction (INDO/S-CI) method implemented in the ZINDO program, we calculated the electronically excited states and oscillator strengths of the Eu(III) complex 3 by the representation of Eu(III) with a point charge of +3e. The results were transformed via the Swizard program (Version 4.6) into each UV spectrum using Gaussian functions with half-widths of 3000 $cm^{-1}$.

Fig.1-96 shows UV-vis absorption spectra of the free ligands (2, 7-BTFDBC, DBM, Phen) and their corresponding Eu(III) complexes in $C_2H_5OH$ solution ($1.0\times10^{-5}$ $mol\cdot L^{-1}$). The similar absorption peaks of the free ligands, located at 228 nm, 264 nm and 350 nm are also observed for the corresponding complexes after the coordination of the Eu(III). This indicates that the Eu(III)-coordination does not cause significantly the $\pi$–$\pi^*$ transitions of the free ligands. On the other hand, complex 3 exhibits the highest absorption intensity among these samples, which is attributed to the increase of the ligand DBM. This suggests that DBM in complex 3 is a good alternative for the ancillary ligand.

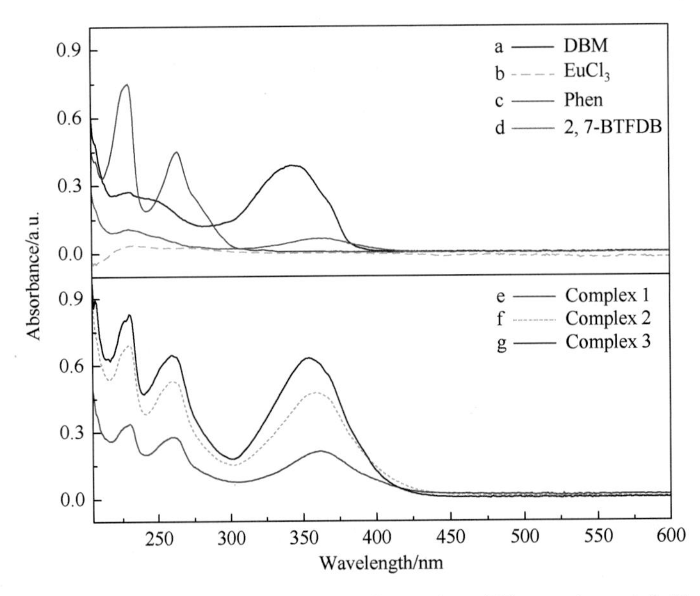


Fig.1-96 UV-visible absorption spectra of europium (III) complexes 1-3, ligands (2, 7-BTFDBC, DBM, Phen) and $EuCl_3$ in $C_2H_5OH$ solution of $1.0\times10^{-5}$ $mol\cdot L^{-1}$

To investigate the PL properties of the complexes 1-3, their powders excitation spectra by monitoring 612 nm emissions are shown in Fig.1-97. All excitation spectra exhibit one broadband between 200 nm and 500 nm, which could be attributed to the large $\pi$–conjugated system. Interestingly, the excitation intensity in the excited band has been remarkably enhanced after the ligand DBM is introduced, and reaches the highest intensity in complex 3, indicating that the ligand DBM can transfer efficiently energy to $Eu^{3+}$.

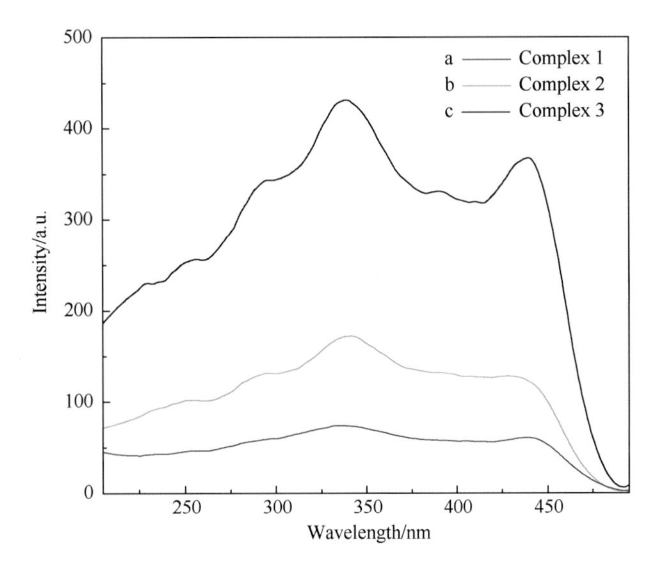


Fig.1-97 Excitation spectra of europium(III) complexes (1, 2, 3) powders by monitoring 612 nm emission

The emission spectra of the synthetic complexes 1-3 are also measured by fixing 440 nm as the excitation wavelength (Fig.1-98). The intensity of emission peaks in the vicinity of 612 nm is remarkably enhanced after the ligand DBM is introduced, and reaches the highest value in complex 3. This result is in agreement with the corresponding excitation spectra. This enhanced intensity of the transition band ($^5D_0 \rightarrow {}^7F_2$) indicates that the double $\beta$-diketone ligands complex 3 can stimulate $Eu^{3+}$ occurring the emission of bright red light. The CIE coordinates according

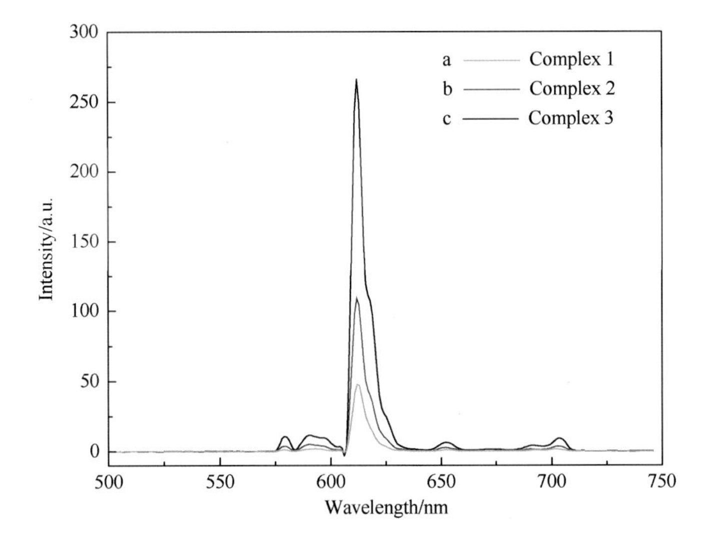


Fig.1-98 Emission spectra of europium (III) complexes powders excited by the 440 nm light

to the emission spectra of complex 3 is calculated to be $x = 0.666$, $y = 0.334$, which is very close to the NTSC standard values for red ($x = 0.67$, $y = 0.33$), suggesting that the complex 3 can be performed in the application of some special fields excited by the long-wavelength light, such as the red phosphor in the blue light excited LEDs.

Besides, Fig.1-98 shows that all complexes have similar emission spectra, and five characteristic emission peaks with different intensity can be observed in 578 nm, 591 nm, 612 nm, 652 nm, 704 nm, corresponding to the $Eu^{3+}$ energy level transition of $^5D_0 \rightarrow ^7F_0$, $^5D_0 \rightarrow ^7F_1$, $^5D_0 \rightarrow ^7F_2$, $^5D_0 \rightarrow ^7F_3$ and $^5D_0 \rightarrow ^7F_4$, respectively. The photophysical process in the Eu(III) complexes can be called a ligand-sensitized luminescence process (antenna effect), which is schematically represented in Fig.1-99.

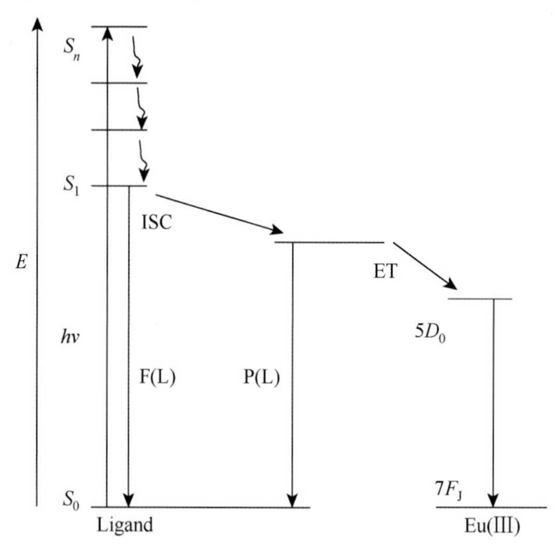


Fig.1-99 Schematic representation of the photophysical process of the Eu(III) complexes

The molecular geometry and electronic absorption spectrum of complex 3 have been calculated using theoretical calculation approaches. The DFT and semiempirical calculation (Sparkle/PM7) give the coincident predictions for the structure of the complex 3 (Fig.1-100).

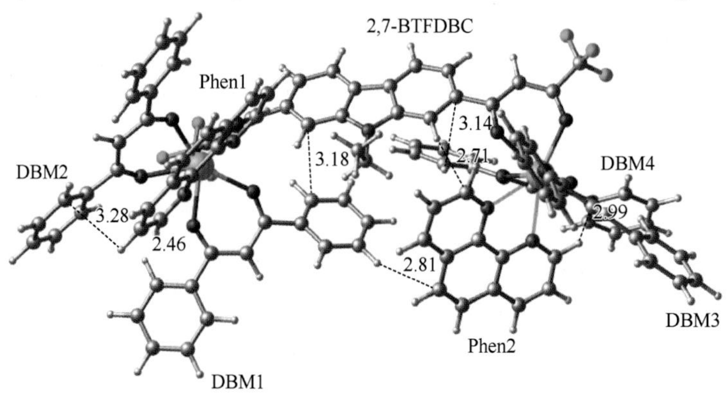


Fig.1-100 The stacking interactions in global minimum energy structure observed from the theoretical calculations on the complex 3

Each Eu(III) ion is coordinated with eight atoms, including six oxygen atoms (two are from the diketone ligand 2, 7-BTFDBC, four are from two DBM ligands), and two nitrogen atoms from the ligand Phen. The coordination structure of the central Eu(III) can be described as a distorted square antiprism. It is in line with the coordination chemistry of Eu(III) with eight coordinated structures. The electronic absorption spectrum, simulated by ZINDO/S based on the theoretical structure, is overlaid with the experimental result (black line) in Fig.1-101. The oscillator strengths and $\lambda_{max}$ are close to the experimental data, indicating that the predicted molecular geometries of the complex 3 are reasonable.

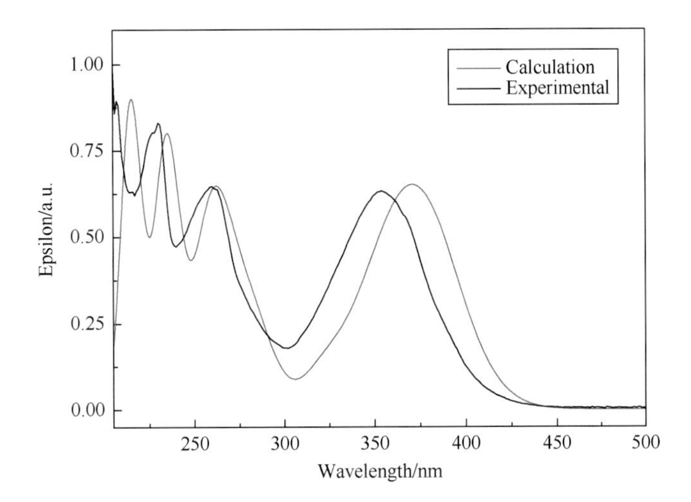


Fig.1-101 The comparison of the absorption spectra of the complex 3 from the theoretical calculation and experimental results

Meanwhile, the approximate distances among the ligands (Phen, DBM and 2, 7-BTFDBC) in complex 3 are the region between 2.57 Å and 3.89 Å (Table 1-8), which indicates that there are some kinds of π-stacking structures. Furthermore, two types of π-stacking among these ligands can be observed in complex 3: T-shaped and parallel displaced. Interestingly, the ligand Phen in complex 3 formed an obvious π-stacking structure with DBM and 2,7-BTFDBC, one parallel displaced and two T-shaped. The π-stacking structures of Phen can effectively stabilize the coordination structure, which can improve the thermal stability for the complex 3. More importantly, the π-stacking effect can produce a significant spectral bathochromic-shift in fluorescent excitation spectra, due to the lowing effect for the energy of excited states. This bathochromic-shift effect has been observed in Table 1-8. On the other hand, the remarkable enhancement intensity in excitation and emission spectra can be understood by the modification of the electron density of the ligand 2,7-BTFDBC. Compared with complex 1, the π-stacking effect in complex 3 leads to increase the electronic density in the carbazole ring and increases the electron transition probability. Therefore, the excitation and emission intensities are enhanced coincidently.

**Table 1-8. The approximate distance and types of $\pi$-stacking among the ligands of complex 3**

| | Distance /Å | Types of $\pi$-stacking |
|---|---|---|
| Phen1-DBM1 | 2.46 | T-shaped |
| Phen1-DBM2 | 3.28 | parallel displaced |
| DBM1-2, 7-BTFDBC | 3.18 | parallel displaced |
| Phen2-DBM1 | 2.81 | T-shaped |
| DBM4-2, 7-BTFDBC | 3.14 | parallel displaced |
| Phen2-DBM4 | 2.99 | T-shaped |
| Phen2-2, 7-BTFDBC | 2.71 | T-shaped |

High thermal stability is an important requirement for most applications, especially in w-LEDs. We have recorded the thermogravimetric analysis (TGA) and differential thermogravimetric (DTG) curves for complex 3 with the highest emission intensity and thermal stability. The result in Fig.1-102 indicates that the decomposition temperature of the complex 3 is 363.5 ℃, which is high enough for the luminescence application.

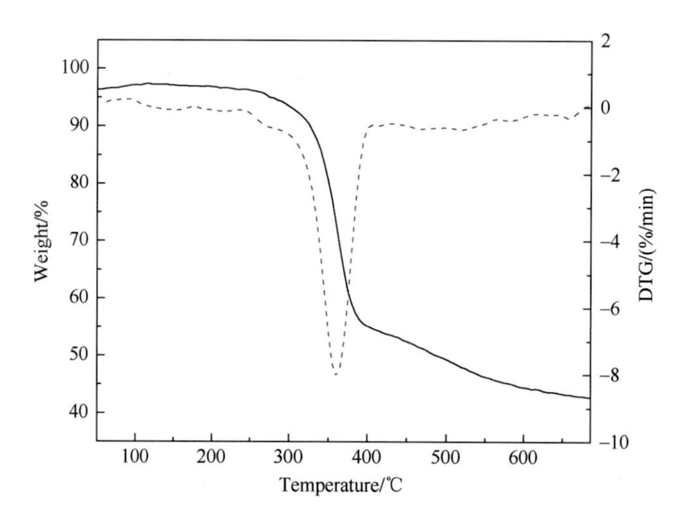


Fig.1-102 TG and DTG curves of the complex 3 in $N_2$ atmosphere

For the sake of further investigation of the potential application of the ternary complexes 1-3, all the complexes were used as a red phosphor for the fabrication of red LEDs. The emission spectra and photographs of the original 460 nm LED chip without phosphors and with the complexes under 20 mA forward bias are shown in Fig.1-103. The peak at 460 nm can be attributable to the emission of the GaN chip, while the sharp and intense red-emission peak at 612 nm is due to the emission of the complexes. This result indicates that the blue light emitting from the GaN chip has been efficiently absorbed and transferred to the central $Eu^{3+}$ ions by the ligands in the complexes. Therefore, $Eu^{3+}$ ion characteristic red light-emitting is observed. The most important observation is that the intensity of the red-light increases remarkably with the increase of the ligand DBM in the synthetic Eu(III) complexes. Thus, the complex 3 ([$Eu_2$(2,7-BTFDBC)(DBM)$_4$(Phen)$_2$]) can be used as a good component in the fabrication of w-LEDs excited by a blue LED chip.

In conclusion, several europium(III) complexes [$Eu_2$(2,7-BTFDBC)$_{3-n}$(DBM)$_{2n}$Phen$_2$] ($n = 0$, 1, 2) were synthesized. The complex [$Eu_2$(2,7-BTFDBC)(DBM)$_4$(Phen)$_2$] exhibits the highest emission intensity with the visible-light excitation. The theoretical calculation results confirm that the pendant phenyl domains in this complex engage in multiple $\pi$-stacking interactions, including T-shaped and parallel displaced stacking. Thus, the intramolecular $\pi$-stacking interactions can cause high emission intensity with visible-light excitation in europium(III) complexes, while they will help to provide a new way to design a highly efficient red phosphor with the visible light excitation.

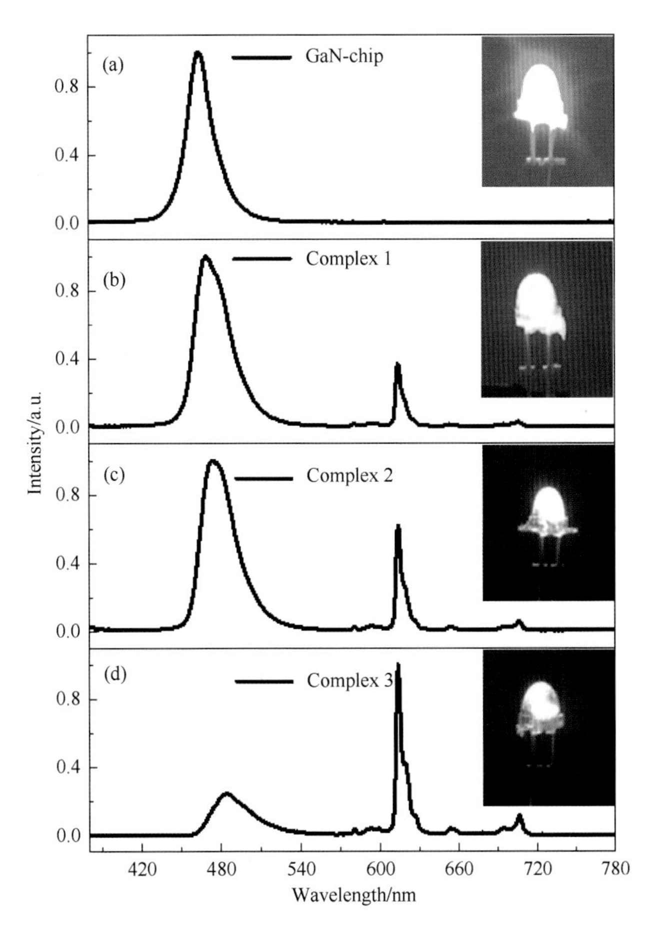


Fig.1-103 Emission spectra and photographs (right) of the initial GaN-LED without phosphor (a) and LED with phosphors 1-3, respectively (b-d). The inserted figures are the lighting photographs of four LEDs

# References


Ahmed M O, Liao J L, Chen X, et al. 2002. Anhydrous tris (dibenzoylmethanido) (o-phenanthroline) europium (III). [Eu $(DBM)_3$ (Phen) ]. Acta Crystallographica Section E, 59: m29.

Aiga F, Iwanaga H, Amano A. 2005. Density functional theory investigation of Eu (III) complexes with $\beta$-diketonates and phosphine oxides: model complexes of fluorescence compounds for ultraviolet LED devices. The Journal of Physical Chemistry A, 109 (49): 11312.

Aspinall H C. 2002. Chiral lanthanide complexes: coordination chemistry and applications. Chemical Reviews, 102 (6): 1807.

Bang J, Abboudi M, Abrams B, et al. 2004. Combustion synthesis of Eu-, Tb-and Tm-doped $Ln_2O_2S$ (Ln = Y, La, Gd) phosphors. Journal of Luminescence, 106 (3): 177.

Barkleit A, Tsushima S, Savchuk O, et al. 2011. $Eu^{3+}$-mediated polymerization of benzenetetracarboxylic acid studied by spectroscopy, temperature-dependent calorimetry, and density functional theory. Inorganic Chemistry, 50 (12): 5451.

Biju P R, Jose G, Thomas G, et al. 2004. Energy transfer in $Sm^{3+}$: $Eu^{3+}$ system in zinc sodium phosphate glasses. Optical Materials, 24 (4): 671.

Blasse G, Grabmaier B C. 1994. Luminescent materials. Berlin: Springer-Verlag.

Campos A F, Dongegá C D M, Malta O L. 1997. Theoretical estimates of vibronic transition rates for $Pr^{3+}$ in different host lattices. Journal of Luminescence. 72-74 (6): 166.

Cao X, Meng L, Li Z, et al. 2014. Large red-shifted fluorescent emission via intermolecular $\pi$-$\pi$ stacking in 4-ethynyl-1, 8-naphthalimide-based supramolecular assemblies. Langmuir, 30: 11753.

Cascales C, Blas A M, Rico M, et al. 2005. The optical spectroscopy of lanthanides $RE^{3+}$ in ABi $(XO_4)_2$ (A = Li, Na; X = Mo, W) and LiYb $(MoO_4)_2$ multifunctional single crystals: relationship with the structural local disorder. Optical Materials. 27 (11): 1672.

Cascales C, Serrano MD, Esteban-Betegón F, et al. 2006. Structural, spectroscopic, and tunable laser properties of $Yb^{3+}$-doped NaGd $(WO_4)_2$. Physical Review B, 74 (17): 3840.

Chi L, Chen H, Zhuang H, et al. 1997. Crystal structure of $LaBSiO_5$. Journal of Alloys and Compounds, 252 (1-2): L12.

Chiu C H, Liu C H, Huang S B, et al. 2007. White-light-emitting diodes using red-emitting LiEu $(WO_4)_{2-x}$ $(MoO_4)_x$ phosphors. Journal of the Electrochemical Society, 154 (7): J181.

Dang P P, Li G G, Yun X H, et al. 2021. Thermally stable and highly efficient red-emitting $Eu^{3+}$-doped $Cs_3GdGe_3O_9$ phosphors for WLEDs: non-concentration quenching and negative thermal expansion. Light: Science & Applications, 10: 29.

D'Aléo A, Picot A, Beeby A, et al. 2008. Efficient sensitization of europium, ytterbium, and neodymium functionalized tris-dipicolinate lanthanide complexes through tunable charge-transfer excited states. Inorganic Chemistry, 47 (22): 10258.

Deng Y, Feng X, Yang D, et al. 2012. Pi-pi stacking of the aromatic groups in lignosulfonates. BioResources, 7: 1145.

Deshpande A S, Burgert I, Paris O, 2006. Hierarchically structured ceramics by high-precision nanoparticle casting of wood. Small, 2 (8-9): 994.

Dhanaraj J, Jagannathan R, Trivedi D C, 2003. $Y_2O_2S$: Eu nanocrystals-synthesis and luminescent properties. Journal of Materials Chemistry, 13 (7): 1778.

Dolg M, Stoll H, Savin A, et al. 1989. Energy-adjusted pseudopotentials for the rare earth elements. Theoretica Chimica Acta, 75 (3): 173.

Elliott P I, Haslam C E, Spey S E, et al. 2006. Formation and reactivity of Ir (III) hydroxycarbonyl complexes. Inorganic Chemistry, 45 (16): 6269.

Greil P, Lifka T, Kaindl A. 1998. Biomorphic cellular silicon carbide ceramics from wood: I. processing and microstructure. Journal of the European Ceramic Society, 18 (14): 1961.

Greil P. 2001. Biomorphous ceramics from lignocellulosics. Journal of the European Ceramic Society, 21 (2): 105.

Guo C, Li B, He Y, et al. 1991. High pressure effect on the luminescence spectra of $Eu^{3+}$ in stoichiometric host microcrystals. Journal of Luminescence 48-49: 489.

He P, Wang H, Liu S, et al. 2009. An efficient europium (III) organic complex as red phosphor applied in LED. Journal of the Electrochemical Society, 156 (2): E46.

He P, Wang H, Liu S, et al. 2009. Luminescent dinuclear Eu (III) organic complex as a red-emitting phosphor for fabrication of LEDs. Electrochemical and Solid-State Letters, 12 (5): B61.

He P, Wang H, Liu S, et al. 2009. Visible-light excitable europium (III) complexes with 2, 7-positional substituted carbazole group-containing ligands. Inorganic Chemistry, 48 (23): 11382.

He P, Wang H, Yan H, et al. 2010. A strong red-emitting carbazole based europium (III) complex excited by blue light. Dalton Transactions, 39 (38): 8919.

Hu Y, Zhuang W, Ye H, et al. 2005. A novel red phosphor for white light-emitting diodes. Journal of Alloys and Compounds, 390 (1): 226.

Huang J, Loriers J, Porcher P, et al. 1984. Crystal field effect and paramagnetic susceptibility of $Na_5Eu$ $(MoO_4)_4$ and $Na_5Eu$ $(WO_4)_4$. Journal of Chemical Physics, 80 (12): 6204.

Jadhav A P, Kim C W, Cha H G, et al. 2009. Effect of different surfactants on the size control and optical properties of $Y_2O_3$: $Eu^{3+}$ nanoparticles prepared by coprecipitation method. The Journal of Physical Chemistry C, 113 (31): 13600.

Janiak C. 2000. A critical account on $\pi$-$\pi$ stacking in metal complexes with aromatic nitrogen-containing ligands. Dalton Transactions, 3885.

Jia W, Perez-Andújar A, Rivera I. 2003. Energy transfer between $Bi^{3+}$ and $Pr^{3+}$ in doped $CaTiO_3$. Journal of the Electrochemical Society, 150 (7): H161.

Jing X, Ireland T, Gibbons C, et al. 1999. Control of $Y_2O_3$: Eu spherical particle phosphor size, assembly properties, and performance for FED and HDTV. Journal of the Electrochemical Society, 146 (12): 4654.

Jung K Y, Han K H. 2005. Densification and photoluminescence improvement of $Y_2O_3$ phosphor particles prepared by spray pyrolysis. Electrochemical and Solid State Letters, 8 (2): H17.

Jüstel T. 1998. New Developments in the field of luminescent materials for lighting and displays. Angewandte Chemie International Edition, 37 (22): 3084.

Kawahara Y, Petrykin V, Ichihara T, et al. 2006. Synthesis of high-brightness sub-micrometer $Y_2O_2S$ red phosphor powders by complex homogeneous precipitation method. Chemistry of Materials. 18 (26): 6303

Kim H, Kwon I E, Park C H, et al. 2000.Phosphors for plasma display panels. Journal of Alloys and Compounds, 311 (1): 33.

Kodaira C A, Brito H F, Felinto M C F C. 2003. Luminescence investigation of $Eu^{3+}$ ion in the $RE_2$ $(WO_4)_3$ matrix (RE = La and Gd) produced using the Pechini method. Journal of Solid State Chemistry, 171 (1-2): 401.

Koren A B, Curtis M D, Francis A H, et al. 2003. Intermolecular interactions in $\pi$-stacked conjugated molecules. synthesis, structure, and spectral characterization of alkyl bithiazole oligomers. Journal of the American Chemical Society, 125 (17) 5040.

Kostova M H, Zollfrank C, Batentschuk M, et al. 2009. Bioinspired design of $SrAl_2O_4$: $Eu^{2+}$ phosphor. Advanced Functional Materials, 19 (4): 599.

Kumar A, Kumar J. 2011. Perspective on europium activated fine-grained metal molybdate phosphors for solid-state illumination. Journal of Materials Chemistry, 21 (11): 3788.

Laberty-Robert C, Ansart F, Deloget C, et al. 2001. Powder synthesis of nanocrystalline $ZrO_2$-8%$Y_2O_3$ via a polymerization route. Materials Research Bulletin, 36 (12): 2083.

Lee G H, Kim T H, Yoon C, et al. 2008. Effect of Local Environment and $Sm^{3+}$-codoping on the luminescence properties in the $Eu^{3+}$-doped potassium tungstate phosphor for white LEDS. Journal of Luminescence, 128 (12): 1922.

Lee S, Seo S. 2002. Optimization of yttrium aluminum garnet: $Ce^{3+}$ phosphors for white light-emitting diodes by combinatorial chemistry method. Journal of the Electrochemical Society, 149 (11): J85.

Lenaerts P, Ryckebosch E, Driesen K, et al. 2005. Study of the luminescence of tris (2-thenoyltrifluoroacetonato) lanthanide (III) complexes covalently linked to 1, 10-phenanthroline-functionalized hybrid sol-gel glasses. Journal of Luminescence, 114 (1): 77.

Leonyuk N I, Belokoneva E L, Bocelli G, et al. 1999. High-temperature crystallization and X-ray characterization of $Y_2SiO_5$, $Y_2Si_2O_7$

and $LaBSiO_5$. Journal of Crystal Growth, 205 (3): 361.

Leonyuk N I, Belokoneva E L, Bocelli G, et al. 2015. Crystal growth and structural refinements of the $Y_2SiO_5$, $Y_2Si_2O_7$ and $LaBSiO_5$ single crystals. Crystal Research and Technology, 34 (9): 1175.

Li X, Yang C, Liu Q S, et al. 2020. Enhancement of luminescence properties of $SrAl_2Si_2O_8$: $Eu^{3+}$ red phosphor. Ceramics International, 46 (11): 17376.

Lin Y, Zhang Z, Zhang F, et al. 2000. Preparation of the ultrafine $SrAl_2O_4$: Eu, Dy needle-like phosphor and its optical properties. Materials Chemistry and Physics, 65 (1): 103.

Liu F, Tang C, Chen Q Q, et al. 2009. Supramolecular $\pi$-$\pi$ stacking pyrene-functioned fluorenes: toward efficient solution-processable small molecule blue and white organic light-emitting diodes. The Journal of Physical Chemistry C, 113 (11): 4641.

Liu J, Cano-Torres J M, Cascales C, et al. 2005. Growth and continuous-wave laser operation of disordered crystals of $Yb^{3+}$: $NaLa(WO_4)_2$ and $Yb^{3+}$: NaLa $(MoO_4)_2$. Physica Status Solidi A, 202 (4): R29.

Liu L, Li W, Hong Z, et al. 1997. Europium complexes as emitters in organic electroluminescent devices. Synthetic metals, 91 (1-3). 267.

Liu W, Farrington G C, Chaput F, et al. 1996. Synthesis and electrochemical studies of spinel phase $LiMn_2O_4$ cathode materials prepared by the Pechini process. Journal of the Electrochemical Society, 143 (3): 879.

Lu W, Chan M C, Cheung K K, et al. 2001. $\pi$-$\pi$ Interactions in organometallic systems. crystal structures and spectroscopic properties of luminescent mono-, bi-, and trinuclear trans-cyclometalated platinum (II) complexes derived from 2, 6-diphenylpyridine. Organometallics, 20 (12): 2477.

Mahalley B N, Pode R B, Gupta P K. 1999. Synthesis of $GdVO_4$: Bi, Eu red phosphor by combustion process. Physica Status Solidi (a). 177 (1): 293.

Méndez-Blas A, Rico M, Volkov V, et al. 2007.Crystal field analysis and emission cross sections of $Ho^{3+}$ in the locally disordered single-crystal laser hosts $M^+$ Bi $(XO_4)_2$ ($M^+$ = Li, Na; X = W, Mo). Physical Review B, 75 (17): 174208

Meyer J, Tappe F. 2015. Photoluminescent materials for solid-state lighting: state of the art and future challenges. Advanced Optical Materials. 3 (4): 424.

Nakamura S, Fasol G. 1997. The blue laser diode: GaN-based light emitters and laser. Berlin: Springer.

Nakamura S, Senoh M, Mukai T. 1993. High-power InGaN/GaN double-heterostructure violet light-emitting diodes. Applied Physics Letters, 62: 2390.

Neeraj S, Kijima N, Cheetham A K. 2004. Novel red phosphors for solid-state lighting: the system $NaM(WO_4)_{2-x}$ $(MoO_4)_x$: $Eu^{3+}$ (M = Gd, Y, Bi).Chemical Physics Letters. 387 (1-3): 2.

Niittykoski J, Aitasalo T, Hölsä J, et al. 2004.Effect of boron substitution on the preparation and luminescence of $Eu^{2+}$ doped strontium aluminates. Journal of Alloys and Compounds, 374 (1-2): 108.

Ozawa L. 1981.Excitation into charge transfer band of $Y_2O_2S$: Eu. Journal of the Electrochemical Society, 128 (11): 2484.

Pan J, Yau L, Chen L, et al. 1988. Studies on spectra properties of $Na_5Eu$ $(WO_4)_4$ luminescent crystal. Journal of Luminescence, 40-41 (2): 856.

Pappalardo R G, Walsh J, Hunt R B. 1983. Cerium-activated halophosphate phosphors I. strontium fluoroapatites. Journal of the Electrochemical Society, 130 (10): 2087.

Park J H, Lee D H, Shin H S, et al. 1996. Transition of the particle-growth mechanism with temperature variation in the molten-salt method. Journal of the American Ceramic Society, 79 (4): 1130.

Parker C A, Rees W T. 1960. Correction of fluorescence spectra and measurement of fluorescence quantum efficiency. Analyst, 85 (1013): 587.

Peng C Y, Zhang H J, Yu J B, et al. 2005. Synthesis, characterization, and luminescence properties of the ternary europium complex covalently bonded to mesoporous SBA-15. The Journal of Physical Chemistry B, 109 (32): 15278.

Peng T, Yang H, Pu X, et al. 2004. Combustion synthesis and photoluminescence of $SrAl_2O_4$:Eu, Dy phosphor nanoparticles. Materials Letters. 58 (3-4): 352.

Pimputkar S, Speck J S, DenBaars S P, et al. 2009. Prospects for LED lighting. Nature Photonics, 3: 180.

Pode R B, Dhoble S J. 1997. Photoluminescence in $CaWO_4$: $Bi^{3+}$, $Eu^{3+}$ Material. Physica Status Solidi, 203 (2): 571.

Pust P, Schmidt P J, Schnick W. 2015. A revolution in lighting. Nature Materials, 14: 454.

Rajomahan R K, Annapurna K, Buddhudu S. 1996. Fluorescence spectra of $Eu^{3+}$: $Ln_2O_2S$ (Ln = Y, La, Gd) powder phosphors. Materials Research Bulletin. 31 (11): 1355.

Rambo C R, Andrade T, Fey T, et al. 2008. Microcellular $Al_2O_3$ ceramics from wood for filter applications. Journal of the American Ceramic Society, 91 (3): 852.

Rao R R. 1996. Preparation and characterization of fine-grain yttrium-based phosphors by sol-gel process. Journal of the Electrochemical Society, 143 (1): 189.

Rico M, Liu J, Griebner U, et al. 2004. Tunable laser operation of ytterbium in disordered single crystals of Yb: NaGd $(WO_4)_2$. Optics Express, 12 (22): 5362.

Ridley J E, Zerner M C. 1976. Triplet states via intermediate neglect of differential overlap: benzene, pyridine and the diazines. Theoretica chimica acta, 42 (3): 223.

Ronda C R. 1995. Phosphors for lamps and displays: an applicational view. Journal of Alloys and Compounds, 225 (1-2): 534.

Schmidt M W, Baldridge K W, Boatz J A, et al. 1993. General atomic and molecular electronic structure system. Journal of Computational chemistry, 14 (11): 1347.

Shahedipour F S, Ulmer M P, Wessels B W, et al. 2002. Efficient GaN photocathodes for low-level ultraviolet signal detection. IEEE Journal of Quantum Electronics, 38 (4): 333.

Sheu J K, Chang S J, Kuo C H, et al. 2003. White-light emission from near UV InGaN-GaN LED chip precoated with blue/green/red phosphors. IEEE Photonics Technology Letters, 15 (1): 18.

Shi F, Meng J, Ren Y. 1996. Structure and luminescent properties of three new silver lanthanide molybdates. Journal of Solid State Chemistry, 121 (1): 236 .

Shin Y, Liu J, Chang J H, et al. 2001. Hierarchically ordered ceramics through surfactant-templated sol-gel mineralization of biological cellular structures. Advanced Materials, 13 (10): 728.

Shionoga S, Yen W M. 1999. Phosphor handbook. Boston: CRC Press.

Singh A, Singh S, Mishra H, et al. 2010. Structural, thermal, and fluorescence properties of Eu $(DBM)_3$Phenx complex doped in PMMA. The Journal of Physical Chemistry B, 114 (41): 13042.

Stanzl-Tschegg S E. 2009. Fracture properties of wood and wood composites. Advanced Engineering Materials, 11 (7): 600.

Stedman N J, Cheetham A K, Battle P D. 1994. Crystal structures of two sodium yttrium molybdates: NaY $(MoO_4)_2$ and $Na_5Y$ $(MoO_4)_4$. Journal of Materials Chemistry, 4 (5): 707.

Stedman N J, Cheetham A K, Battle P D. 1994. $YMoO_4$ revisited: the crystal structure of $Y_5Mo_4O_{18}$. Journal of Materials, 4 (9): 1457.

Steigerwald D A, Bhat J C, Collins D, et al. 2002. Illumination with solid state lighting technology. IEEE Journal of Selected Topics in Quantum Electronics, 8 (2): 310.

Subbotin K A, Zharikov E V, Smirnov V A. 2002. Yb-and Er-Doped single crystals of double tungstates NaGd $(WO_4)_2$, NaLa $(WO_4)_2$, and NaBi $(WO_4)_2$ as active media for lasers operating in the 1.0 and 1.5 μm ranges. Optic and Spectroscopy, 92 (4): 601.

Sun D H, Zhang J L, Liu Y L, et al. 2011. Synthesis and characterization of [Eu $(DBM)_3$phen]$Cl_3$@$SiO_2$-$NH_2$ composite nanoparticles. Journal of Nanoscience and Nanotechnology, 11 (11): 9656.

Toradi C C, Page C, Brixner L H, et al. 1987. Structure and luminescence of some $CsLnW_2O_8$ compounds. Journal of Solid State Chemistry, 69 (1): 171.

Trunov V K, Berezina T A, Evdokimov A A, et al. 1978. Li (Na) -rare earth tungstates and molybdates with scheelite structure. Russian Journal of Inorganic Chemistry, 23: 1465.

Tseng Y H, Chiou B S, Peng C C, et al. 1988. Spectral properties of $Eu^{3+}$-activated yttrium oxysulfide red phosphor. Thin Solid Films, 330 (2): 173.

Van Deun R, Nockemann P, Fias P, et al. 2005. Visible light sensitisation of europium (III) luminescence in a 9-hydroxyphenal-1-one complex. Chemical Communications, 590.

Van Vliet J P M, Blasse G, Brixner L H. 1988. Luminescence properties of the system $Gd_{1-x}Eu_x(IO_3)_3$. Journal of the Electrochemical Society, 135 (6): 1574-1578.

Vila L D D, Stucchi E B, Davolos M R. 1997. Preparation and characterization of uniform, spherical particles of $Y_2O_2S$ and $Y_2O_2S$: Eu. Journal of Materials Chemistry, 7 (10): 2113.

Volkov S V. 1990. Chemical reactions in molten salts and their classification. Chemical Society Reviews, 19 (1): 21.

Voron'ko Y K, Veshnyakova M A, Lomonova E E, et al. 2003. Growth and luminescent properties of NaGd $(WO_4)_2$: $Yb^{3+}$ crystals. Inorganic Materials, 39 (13): 1308.

Voron'ko Y K, Zharikov E V, Lis D A, Et al. 2004. Spectroscopic investigations of NaGd $(WO_4)_2$ and NaLa $(MoO_4)_2$ single crystals, doped by $Yb^{3+}$ Ions. Proceedings SPIE, 5478: 60.

Wakefield G, Holland E, Dobson P J, et al. 2001. Luminescence properties of nanocrystalline $Y_2O_3$: Eu. Advanced Materials, 13: 1557.

Wang M, Fan X, Xiong G. 1995. Luminescence of $Bi^{3+}$ ions and energy transfer from $Bi^{3+}$ ions to $Eu^{3+}$ ions in silica glasses prepared by the sol-gel process. Journal of Physics and Chemistry of Solids, 56 (6): 859.

Wang X, Li Y D. 2002. Synthesis and characterization of lanthanide hydroxide single-crystal nanowires. Angewandte Chemie International Edition, 41 (24): 4790.

Wang Z, Xiang N, Wang Q, et al. 2010. A red europium (III) ternary complex for InGaN-based light-emitting diode. Journal of Luminescence, 130 (1): 35.

Werts M V, Duin M, Hofstraat J, et al. 1999. Bathochromicity of Michler's ketone upon coordination with lanthanide (III) $\beta$-diketonates enables efficient sensitisation of $Eu^{3+}$ for luminescence under visible light excitation. Chemical Communications, 0: 799.

Wu M H, Chen B L, He C, et al. 2022. A high quantum yield red phosphor $NaGdSiO_4$:$Eu^{3+}$ with intense emissions from the $^5D_0 \rightarrow {}^7F_{1,2}$ transition. Ceramics International, 48 (16): 23213.

Wu G S, Lin Y, Yuan X Y, et al. 2004. A novel synthesis route to $Y_2O_3$: Eu nanotubes. Nanotechnology, 15: 568.

Xiang N J, Lee T H, Gong M L, et al. 2006. Synthesis of 2-phenylquinoline-based ambipolar molecules containing multiple 1, 3, 4-oxadiazole spacer groups. Synthetic Metals, 156 (2-4): 270.

Xiang N J, Leung L M, So S K, et al. 2006. Red InGaN-based light-emitting diodes with a novel europium (III) tetrabasic complex as mono-phosphor. Materials Letters, 60 (23): 2909.

Xu C Y, Zhang Q, Zhang H, et al. 2005. Synthesis and characterization of single-crystalline alkali titanate nanowires. Journal of the American Chemical Society, 127 (33): 11584.

Xu L, Wei B, Zhang Z, et al. 2006. Synthesis and luminescence of europium doped yttria nanophosphors via a sucrose-templated combustion method. Nanotechnology, 17 (17): 4327.

Xue Y N, Xiao F, Zhang Q Y. 2011. Enhanced red light emission from $LaBSiO_5$: $Eu^{3+}$, $R^{3+}$ (R = Bi or Sm) phosphors. Spectrochimica Acta Part A: Molecular and Biomolecular Spectroscopy, 78 (2): 607-611.

Yamanoto H, Seki S, Ishiba T. 1991. The Eu site symmetry in AEu $(MoO_4)_2$ (A = Cs or Rb) generating saturated red luminescence. Journal of Solid State Chemistry, 94 (2): 396.

Yan B, Lei F. 2010. Molten salt synthesis, characterization and luminescence of $ZnWO_4$: $Eu^{3+}$ nanophosphors. Journal of Alloys and Compounds, 507 (2): 460.

Yang C, Fu L M, Wang Y, et al. 2004. A highly luminescent europium complex showing visible-light-sensitized red emission: direct observation of the singlet pathway. Angewandte Chemie International Edition, 43 (38): 5010.

Yang J, Lin J. 2010. Sol-gel synthesis of nanocrystalline $Yb^{3+}/Ho^{3+}$-doped $Lu_2O_3$ as an efficient green phosphor. Journal of the Electrochemical Society, 157 (12): K273.

Yang Y S, Gong M L, Li Y Y, et al. 1994. Effects of the structure of ligands and their $Ln^{3+}$ complexes on the luminescence of the central $Ln^{3+}$ ions. Journal of Alloys and Compounds, 207-208: 112.

Yen W M, Shionoya S, Yamamoto H. 2006. Phosphor Handbook (second edition). Boston: CRC Press.

Yu H, Li T, Chen B, et al. 2013. Preparation of aligned Eu (DBM)$_3$phen/PS fibers by electrospinning and their luminescence properties. Journal of colloid and interface science, 400: 175.

Yum J, Seo S, Lee S, et al. 2003. $Y_3Al_5O_{12}$: $Ce_{0.05}$ Phosphor coatings on gallium nitride for white light-emitting diodes. Journal of the Electrochemical Society, 150 (2): H47.

Zerner M C, Loew G H, Kirchner R F, et al. 1980. An intermediate neglect of differential overlap technique for spectroscopy of transition-metal complexes. ferrocene. Journal of the American Chemical Society, 102 (2): 589.

Zhang J, Ning J, Liu X, et al. 2003. Synthesis of ultrafine YAG: Tb phosphor by nitrate-citrate sol-gel combustion process. Materials Research Bulletin, 38 (7): 1249.

Zhang W W, Xu M, Zhang W P, et al. 2003. Site-selective spectra and time-resolved spectra of nanocrystalline $Y_2O_3$: Eu. Chemical Physics Letters, 376 (3-4): 318.

Zhao S, Chen L, Zhao H, et al. 2004. Laser-diode pumped passively Q-switched $Nd^{3+}$: NaY $(WO_4)_2$ laser with $Cr^{4+}$: YAG saturable absorber. Optical Materials, 27 (3): 481.

# Chapter II $Mn^{4+}$-activated red phosphors for white light-emitting diodes

## 2.1 $Mn^{4+}$-activated fluorides red phosphors for blue LED chips

### 2.1.1 Introduction

Over the past decade, w-LEDs have been a topic of interest to scientists and engineers due to their high energy efficiency, long lifetime, and environmentally friendly properties. Generally, the most popular w-LED is fabricated from a blue GaN chip and yellow-emitting YAG phosphor, which can absorb the blue light excited from the GaN chip to emit the white light with high luminous efficiency. However, the white light possessed by this kind of w-LED lacks a red emission component, which results in it hard for indoor lighting application because of the highly correlated color temperature (CCT>6, 000 K) and the low color rendering index ($R_a$<80). In general, a suitable red-emitting phosphor for a blue LED chip should meet the following requirements: ①the phosphor is of high stability; ②the production-cost is low; ③the phosphor exhibits strong absorption around 460 nm and strong red-emitting with the appropriate chromaticity coordinates near the NTSC standard values. At present, rare-earth ions $Eu^{2+}$ or $Ce^{3+}$ activated (oxy) nitride compounds red phosphors have been developed and applied in the w-LEDs system. However, these (oxy) nitride phosphors still have several drawbacks. Firstly, a large part of the red emission is beyond 650 nm, which decreases the luminous efficiency (LE) of radiation. Secondly, the preparation needs extremely harsh conditions such as high temperature and reducing atmosphere, which increase their production cost. Finally, such phosphors are easily oxidized at high temperatures. Hence it is urgent to exploit new red phosphors with high color-purity and low production-cost for warm w-LEDs.

$Mn^{4+}$-doped materials, especially, aforementioned red phosphors doped $Mn^{4+}$ have been widely investigated because of their distinct $^2E \rightarrow ^4A_2$ transitions in symmetrically octahedral crystal field, resulting in the phosphor broad adsorption bands from 380 nm to 490 nm and sharp emission peaks at 610-760 nm. A typical example is $Mn^{4+}$-activated $Mg_2TiO_4$ red phosphor. However, its emission is located between 650 nm and 720 nm, which are too far red-shifted for efficient warm w-LEDs with high LE. To overcome this drawback, a series of narrow red-emitting fluorides activated by $Mn^{4+}$ ions have been produced. For example, $K_2SiF_6$:$Mn^{4+}$, which can be as a component on w-LED, shows a strong broad excitation band in the blue region and a narrow-band emission in the red region. $Na_3AlF_6$:$Mn^{4+}$ almost has no red emission degradation at 160 ℃, suggesting that this red phosphor shows a promising application in high power w-LEDs. It is known that $Mn^{4+}$ is sensitive to the surrounding environment and hard to be controlled. Hence, many synthesis routes have been tried to prepare $Mn^{4+}$ doped fluoride complexes. The group of Adachi reported a series of red fluoride phosphors $K_2XF_6$:$Mn^{4+}$ (X = Sn, Ge, Si) prepared through a wet chemical etching route in the aqueous $KMnO_4$ and HF mixed solution. A co-precipitation method to prepare $Na_2SiF_6$:$Mn^{4+}$, $K_2SiF_6$:$Mn^{4+}$, $K_2GeF_6$:$Mn^{4+}$ and $Rb_2SiF_6$:$Mn^{4+}$ was established by Liu and his co-authors with the use of $H_2O_2$ as a reductant to reduce $Mn^{7+}$ to $Mn^{4+}$. Pan and her colleagues demonstrated a hydrothermal method for the synthesis of $Mn^{4+}$ doped fluorides, such as $BaSiF_6$:$Mn^{4+}$, $BaTiF_6$:$Mn^{4+}$ and $K_2TiF_6$:$Mn^{4+}$ phosphors. We have also prepared $BaGeF_6$:$Mn^{4+}$ and $K_2XF_6$:$Mn^{4+}$ (X = Si, Ti, Ge) red phosphors by the hydrothermal method and co-precipitation method, respectively. $Mn^{4+}$ doped fluorides can also be prepared by an ion-exchange method, e.g. the group of Chen successfully prepared $K_2TiF_6$:$Mn^{4+}$ and $K_2SiF_6$:$Mn^{4+}$ red phosphors by the ion exchange method with $K_2MnF_6$ powders in HF solution at room temperature.

In this section, a lot of fluorides doped $Mn^{4+}$ were prepared by different methods. The influence of the reaction parameters, such as the concentration of starting materials, the reaction temperature and time, on the PL properties of these red phosphors have been investigated systematically. The obtained products emit intensive red emission under the blue light excitation with high thermal-quenching resistance, implying these fluoride phosphors are potential red components for warm w-LEDs.

### 2.1.2 Preparation methods

#### 2.1.2.1 Hydrothermal method

All source materials in this work, including acetone, absolute alcohol, $KHF_2$, $KMnO_4$, NaF/KF/RbF/CsF, HF (40%), $SiO_2$/$GeO_2$/$TiO_2$, $H_2O_2$ (30%) and $BaCO_3$ are analytical grade and used without any further purification. In a typical synthesis for $BaGeF_6$:$Mn^{4+}$ (marked as BGFM), 0.494 g of $BaCO_3$, 0.262 g of $GeO_2$ and 0.079 g of $KMnO_4$ were added into a Teflon cup containing 50.0 mL of HF (40%) solution with magnetic stirring for 30 min. The hydrothermal synthesis was conducted at 180 ℃ for 8 h. After the autoclave was cooled down naturally to room temperature, the resulting solid product was collected carefully from the cup, washed extensively with distilled water and methanol several times and dried at 60 ℃ for 12 h.

#### 2.1.2.2 Co-precipitation method

The co-precipitation method is an ordinary method to synthesize the fluoride phosphors.

Alkali metal fluorides, such as NaF, KF and CsF were used as precipitators. In a typical synthesis for $Na_2XF_6:Mn^{4+}$ (X = Si, Ge, Ti) red phosphors, 2.5 mmol $XO_2$ (X = Si, Ge, Ti) was adequately dissolved in 5 mL 40% HF solution. Then 0.5 mmol $NaMnO_4·H_2O$ was added into the above transparent solution. When the solution was heated to 70 ℃, 5 mmol NaF was slowly added into the solution, which was kept at 70 ℃ for 30 min under continuous stirring. Finally, the precipitates were washed with methanol several times and dried at 80 ℃ for 2 h.

2.1.2.3 Ion exchange method

For the ion exchange method, $K_2MnF_6$ used as the $Mn^{4+}$ source was synthesized according to the method described in the literature. Specifically, 20 g $KHF_2$ and 2 g $KMnO_4$ were dissolved in HF (40%) solution. The yellow powder $K_2MnF_6$ was precipitated by slowly dropping $H_2O_2$ (30%). After fast filtering and washing with acetone, the yellow powder was oven-dried at 80 ℃ for 1 h. In a typical synthesis for $Cs_2TiF_6:Mn^{4+}$, we started with a 50 mL plastic beaker containing 10 mL magnetically stirring 40% HF solution in which 0.3994 g $TiO_2$ was added until completely dissolved. Then 0.1236 g $K_2MnF_6$ and 1.6709 g CsF were put into the colorless transparent solution in order. The formation of $Cs_2TiF_6:Mn^{4+}$ phosphor was carried out at room temperature. After 30 min magnetically stirring and a quick cooling process in the salt-ice bath, the precipitates were collected, washed with methanol several times and dried at 80 ℃ for 1 h.

### 2.1.3 Structures, morphologies, PL properties and applications

2.1.3.1 $BaGeF_6:Mn^{4+}$

The XRD patterns of the hydrothermally synthesized $BaGeF_6$ with and without $Mn^{4+}$ doping are shown in Fig.2-1, along with its corresponding standard diffraction card. Obviously, in Fig.2-1(a), all the diffraction peaks that appear for the as-prepared product agree well with the standard JCPDS card of $BaGeF_6$ (No. 74-0924, space group $R$-3$m$, $a = b = c = 4.83$ nm), which indicates each product is a single pure phase of $BaGeF_6$. Fig.2-1(b) displays the XRD pattern of BGFM hydrothermally treated with 10.0 mmol·$L^{-1}$ $KMnO_4$ and 40% HF solution at 180 ℃ for 8 h. As similar to curve a, it is also of a single pure phase, which means the doping of $Mn^{4+}$ does not change the crystal structure of this $BaGeF_6$ host. That is because $Mn^{4+}$ not only has the same valence state as $Ge^{4+}$ but also the identical ionic radius with $Ge^{4+}$ (0.53 Å, CN = 6 for $Mn^{4+}$ and $Ge^{4+}$). This results in a charge balance between $Ge^{4+}$ and $Mn^{4+}$ in BGFM and a similar profile between Fig.2-1(a) and Fig.2-1(b).

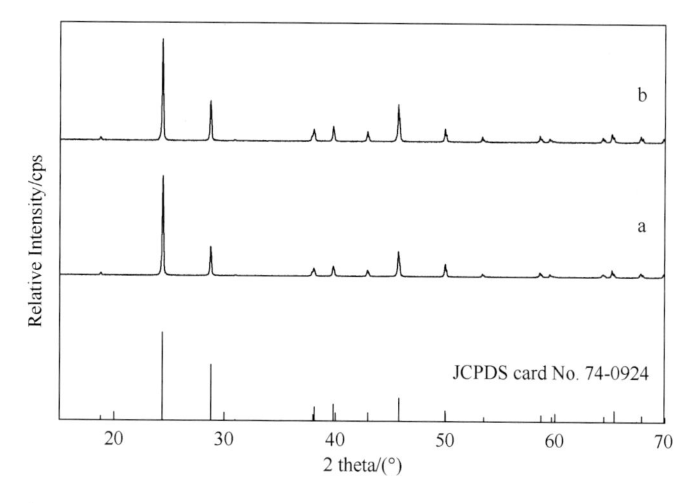


Fig.2-1 Representative XRD patterns of the (a) $BaGeF_6$ host and (b) BGFM product prepared at 180 ℃ for 8 h with 10.0 mmol·$L^{-1}$ $KMnO_4$ and 40% HF solution

To determine whether $Mn^{4+}$ ions were doped into the $BaGeF_6$ matrix, the luminescent properties of the above obtained BGFM product were examined and shown in Fig.2-2(a) It demonstrated two broadband excitation peaks, according to the Tanabe-Sugano diagram, which

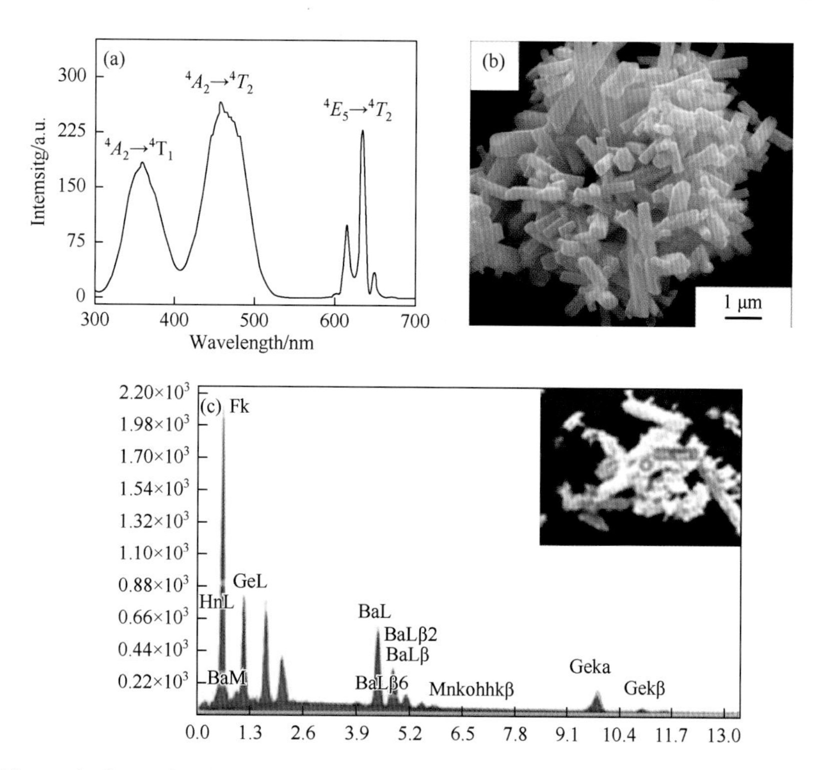


Fig.2-2 (a) The excitation and emission spectra with a typical photo image excited under a UV lamp, (b) SEM picture, (c) EDS data of the as-synthesized BGFM red phosphor obtained from 10.0 mmol·$L^{-1}$ $KMnO_4$ and 40% HF solution at 180 ℃ for 8 h

are attributable to the spin-allowed and parity-forbidden transitions from the ground state $^4A_{2g}$ to the excited states $^4T_{2g}$ and $^4T_{1g}$ of $Mn^{4+}$ in the UV and blue region, respectively. The corresponding emission spectrum exhibits three strong peaks locating in the red region, which is a typical property of $Mn^{4+}$ ion-activated phosphors as reported previously. This strongly indicates that the $Mn^{4+}$ ions have been doped into the $BaGeF_6$ matrix.

Fig.2-2(b) and Fig.2-2(c) exhibit the representative SEM and EDS results of the particulate BGFM red phosphor prepared from the hydrothermal treatment of $BaCO_3$, $GeO_2$ and $KMnO_4$ in the 40% HF solution. The SEM picture illustrates that many rod-shaped crystals with smooth surfaces are present in the as-synthesized BGFM product. Closely observing the tips of these rods, each rod has a diameter of about 0.2 μm and a length of around 2 μm. The morphologies of all obtained BGFM products prepared from the different concentrations of HF are similar to Fig.2-2(b) (Fig.2-3).

Fig.2-3　SEM images of $BaGeF_6:Mn^{4+}$ products obtained with 10.0 $mmol\cdot L^{-1}$ $KMnO_4$, (a) 10%, (b) 20% and (c) 30% HF solution at 180 ℃ for 8 h

The EDS spectra of Fig.2-2(c) recognized the peaks of F, Ge, Ba and Mn elements, which also can be identified from the XPS profile (Fig.2-4), suggesting that the Mn element has

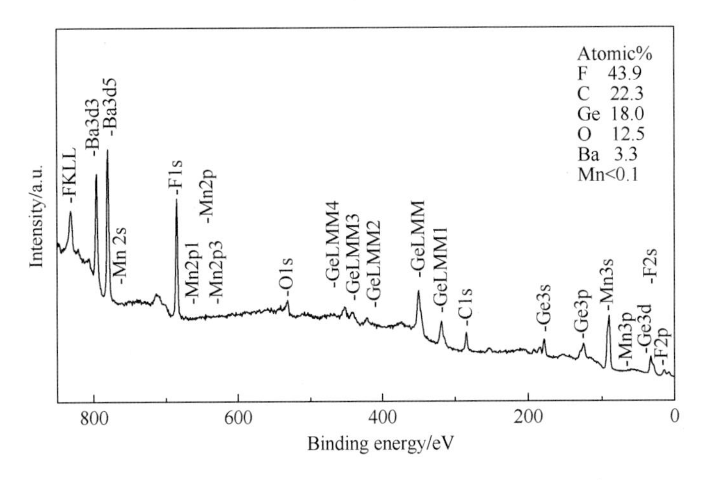


Fig.2-4　XPS spectrum of red phosphor $BaGeF_6:Mn^{4+}$

been indeed doped into the matrix lattice to occupy the Ge lattice site. This result further verifies that the $Mn^{4+}$ ions are active centers in $BaGeF_6$ to emit the red light under the blue light excitation. The molar concentration of $Mn^{4+}$ in this product, which is also indicated by the XPS result, is very low. Moreover, the absence of oxygen peak in the XPS result implies that there is no presence of $MnO_2$ or other manganese oxides in the hydrothermal reaction, which is identical to the previous work.

Because of the same ionic radius and charge balance, $Mn^{4+}$ ions can substitute the octahedral $Ge^{4+}$ ionic sites in $BaGeF_6$ to form the BGFM red phosphor, resulting in its strong absorption in the blue region. This is also observed in the diffuse reflection spectra (DRS) (Fig.2-5), which reveals a strong adsorption peak at 457 nm. This phenomenon is a strong indication of its potential application in blue GaN chips.

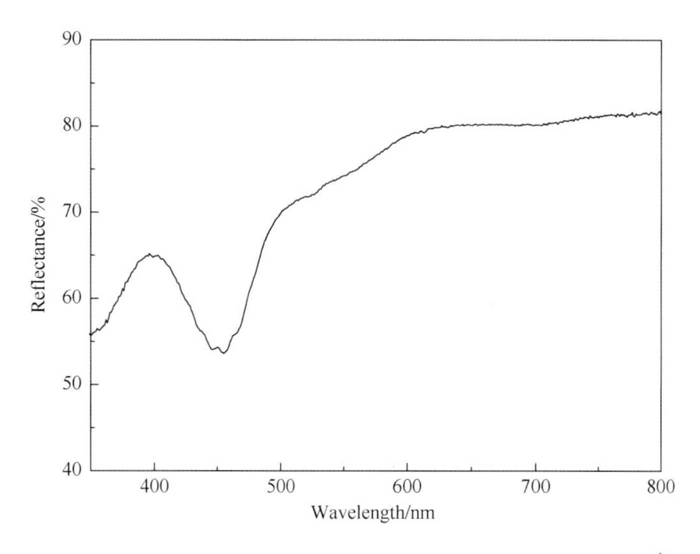


Fig.2-5 Diffuse reflection spectrum of red phosphor $BaGeF_6:Mn^{4+}$

For a crystal prepared from the hydrothermal route, it is necessary to investigate the influence of some general reaction parameters, such as the concentration of starting materials, the reaction temperatures and times on its PL properties. Therefore, different concentrations of HF (10%-40%) were employed to investigate the PL properties of BGFM red phosphors. Fig.2-6 exhibits the excitation and emission spectra prepared with different HF concentrations. It is clear that the shapes of these excitation spectra are similar, each of which is composed of two broadband peaks. As illustrated above, the broad excitation bands are attributed to the spin-allowed and parity-forbidden transitions $^4A_2$-$^4T_{1g,2g}$ of $Mn^{4+}$ respectively. The stronger excitation locates at 460 nm, which matches perfectly with the blue emission of the GaN chip (460 nm).

Fig.2-6(b) shows a series of corresponding emission spectra of the BGFM phosphors under the 460 nm light excitation. Three emission peaks locating at 615 nm, 634 nm and 649 nm strongly indicate the emitted light is red, which are identified as the anti-Stokes and Stokes transitions between

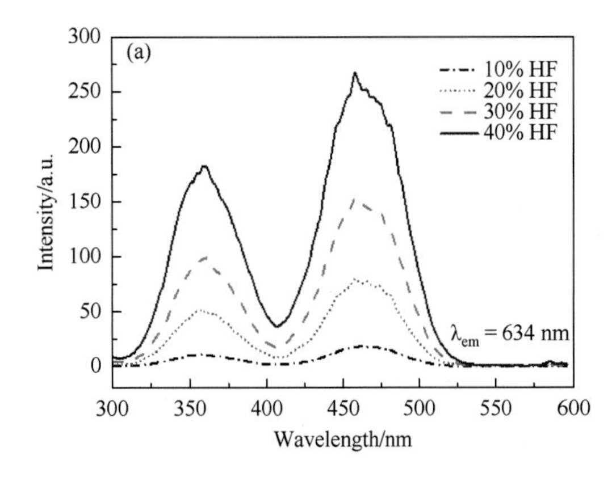


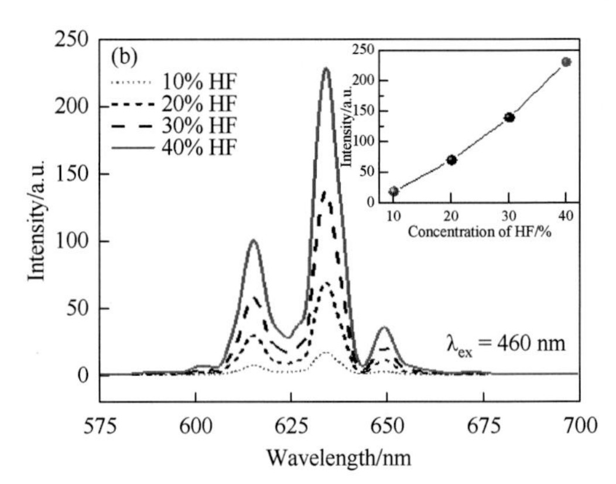


Fig.2-6 (a) Excitation and (b) emission spectra of red phosphor BGFM obtained at 180 ℃ for 8 h with 10.0 mmol·L$^{-1}$ $KMnO_4$ and (a) 10%, (b) 20%, (c) 30% and (d) 40% HF. The inserted figure is the relationship between HF concentration and the relative emission intensity of BGFM

$^2E$ and $^4A_2$ levels of $Mn^{4+}$ respectively. Among all the emission peaks, the strongest peak locates at 634 nm. Moreover, Fig.2-6(b) exhibits the effect of HF concentration on the emission intensity of BGFM. The emission intensity increases with the HF concentrations until the content reaches 40%, at which the obtained BGFM phosphor emits stronger red light than that of other samples. That may because the higher HF concentration is beneficial for the formation of a stable $MnF_6^{2-}$ group. With the highest HF concentration, the obtained BGFM phosphor emits the brightest red light. Therefore, the following experiments are conducted in this 40% HF environment. The relationship between the HF concentration and the relative emission intensity of BGFM phosphor is illustrated in the insert of Fig.2-6(b).

As we know, the doping amount of a phosphor plays a very important role in its PL properties. But in this work, it is hard to accurately determine the doping amount of $Mn^{4+}$ in BGFM products. Thus from a practical point of view, it is very important to investigate the effect of $KMnO_4$ concentration on the PL properties of the as-synthesized BGFM red phosphors. It can be observed that the emission spectra in Fig.2-7 also share a similar shape. Without the addition of $KMnO_4$, no emission peak can be observed. Adding a small amount of $KMnO_4$ into the reaction system, the emission intensity sharply increases. Furthermore, the strongest peak locating at 634 nm is very close to the ideal red (620 nm) in terms of the LE of w-LEDs. With a continual increase of $KMnO_4$ concentration, the emission contrarily drops, which may be due to the concentration quenching of $Mn^{4+}$ in $BaGeF_6$ crystal lattice. The relationship between the $KMnO_4$ concentration and the relative emission intensity of BGFM phosphor is inserted in Fig.2-7.

With the addition of $KMnO_4$, the emission peaks gradually increase until the content of $KMnO_4$ reaches 10.0 mmol·L$^{-1}$, at which the BGFM emits the strongest red light and the CIE chromaticity coordinates are $x = 0.695$, $y = 0.305$. This value is close to the "ideal red" of the NTSC ($x = 0.67$, $y = 0.33$, Fig.2-8).

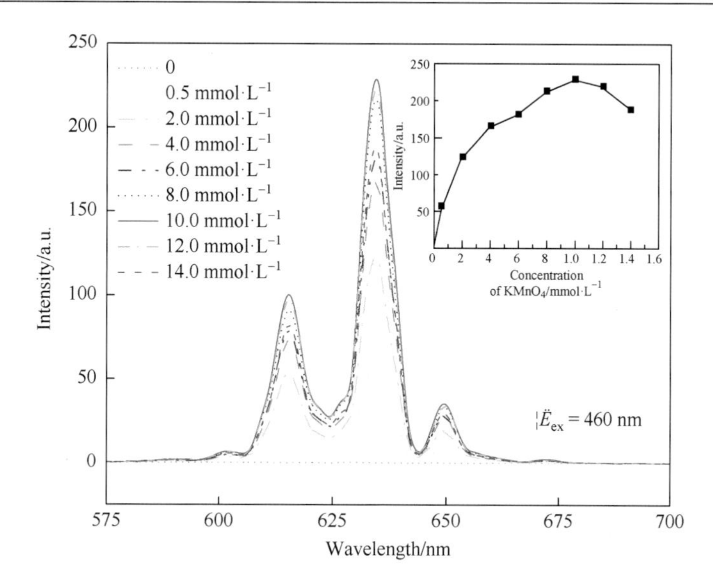


Fig.2-7 Emission spectra of BGFM red phosphors obtained from 40% HF with different concentration of $KMnO_4$ at 180 ℃ for 8 h. The inserted figure is the relationship between the concentration of $KMnO_4$ and the relative emission intensity of BGFM

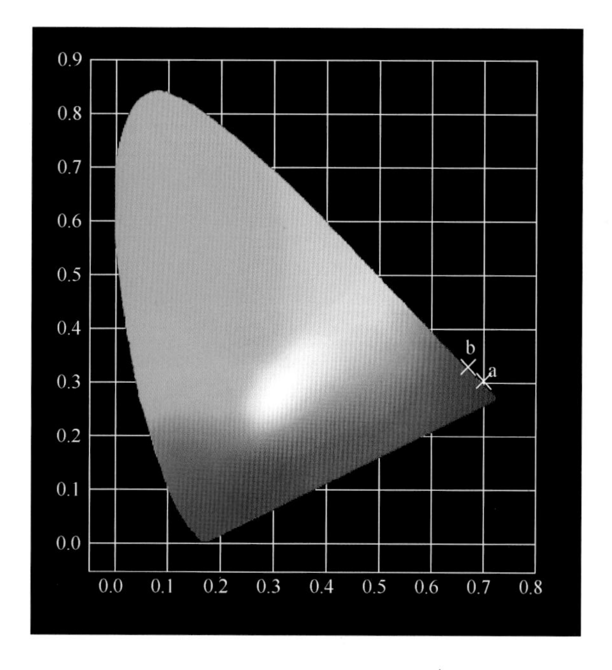


Fig.2-8 CIE chromaticity diagram for (a) $BaGeF_6:Mn^{4+}$ and (b) NTSC "ideal red"

Fig.2-9 shows the effect of the synthesis temperature on the PL property of the obtained BGFM phosphor. Similar to as illustrated above, all BGFM phosphors emit red light as well, and the emission intensity can be enhanced by increasing the reaction temperature from 100 ℃ to 180 ℃, which is probably due to the improved crystallization at a higher temperature.

However, with the further increase of the reaction temperature, the emission intensities decrease as shown in Fig.2-9. Some structures tend to deteriorate at a higher temperature.

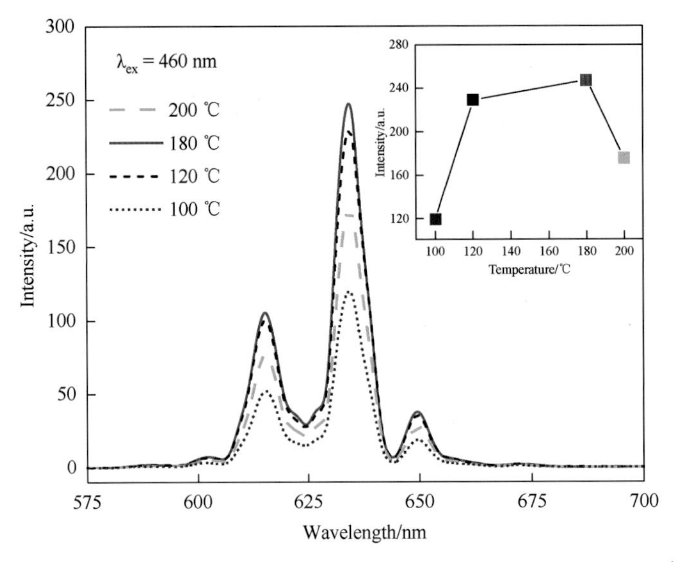


Fig.2-9　Emission spectra of BGFM red phosphors obtained from 40% HF and 10.0 mmol·$L^{-1}$ $KMnO_4$ at 100 ℃, 120 ℃, 180 ℃ and 200 ℃ for 8 h. The inserted figure is the relationship between the reaction temperature and the relative emission intensity of BGFM

The reaction time influenced little on the PL properties of the BGFM phosphors (Fig.2-10) and the best reaction period is 8 h. Therefore, according to the results obtained above, the optimum reaction condition is 10.0 mmol·$L^{-1}$ $KMnO_4$, 40% HF, reacted at 180 ℃ for 8 h.

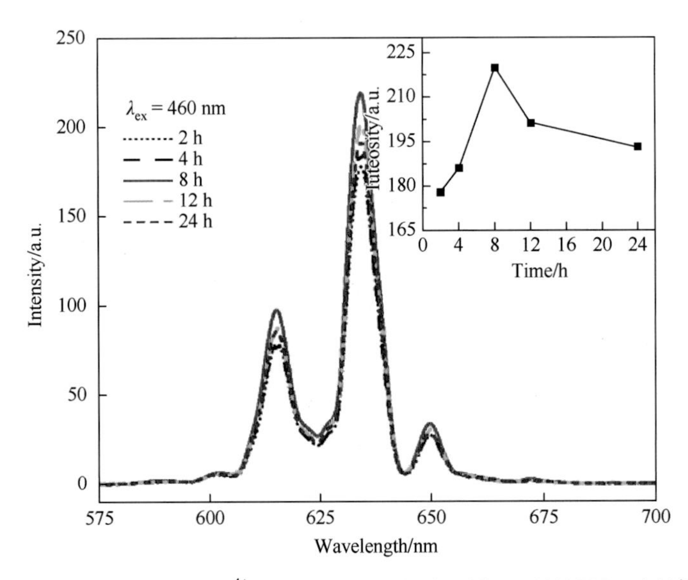


Fig.2-10　Emission spectra of $BaGeF_6$:$Mn^{4+}$ red phosphors obtained from 40% HF and 10.0 mmol·$L^{-1}$ $KMnO_4$ at 180 ℃ for 2 h, 4 h, 8 h, 12 h and 24 h. The inserted figure is the effect of the reaction time on the relative emission intensity of BGFM

Fig.2-11 shows the EL spectra of the GaN chip, the LED-based on BGFM, YAG, and the mixture of YAG and BGFM under 20 mA current excitation. The peak at 460 nm can be attributable to the emission of the GaN chip [Fig.2-11(a)]. While the peaks at 634 nm are due to the emission of BGFM phosphor [Fig.2-11(b)], which can make the bright red light be observed by naked eyes [Fig.2-11(b)]. The difference between the two w-LEDs is shown in Fig.2-11 (c, d). The red emission in Fig.2-11(d) is very clear, which means the addition of BGFM is favorable for improving the $R_a$ and CCT levels of the YAG type w-LED.

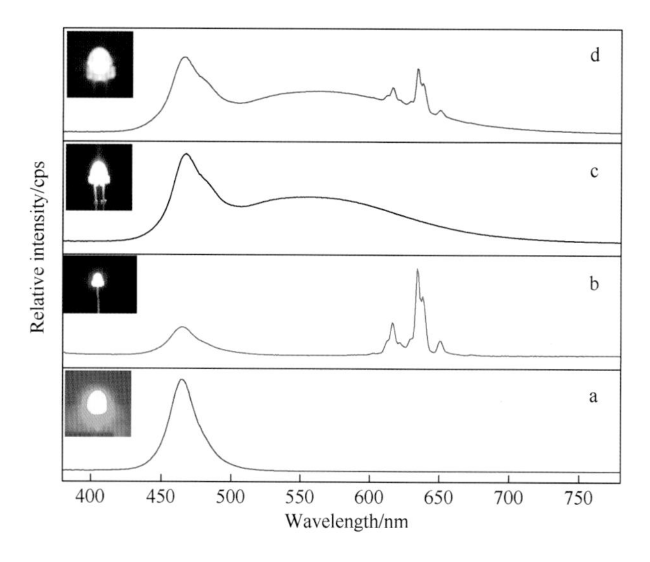


Fig.2-11 EL spectra of (a) the GaN chip, the LED-based on (b) BGFM, (c) YAG, and (d) the YAG-BGFM mixture under 20 mA current excitation. The inserted figures are the corresponding light images

The performance is summarized in Table 2-1. Briefly, the white light emitted from the GaN-YAG-BGFM type w-LED is warmer than that from the GaN-YAG system (4210 K of CCT, 84 of $R_a$, 52.21 lm/W of LE vs. 6283 K of CCT, 76 of $R_a$, 45.21 lm/W of LE). It is worth to point out that the LE has also been enhanced with the use of BGFM.

**Table 2-1 Performance of the GaN-based w-LEDs coated with (1) YAG, (2) the mixture of YAG and BGFM at 20 mA forward current and 5 V reverse voltage**

| No. of LEDs samples | CCT/K | $R_a$ | LE/(lm/W) | CIE $(x, y)$ |
|---|---|---|---|---|
| 1 | 6283 | 76.0 | 45.21 | (0.3129, 0.3660) |
| 2 | 4210 | 84 | 52.21 | (0.3693, 0.3608) |

In summary, a class of $BaGeF_6:Mn^{4+}$ red phosphors with micro-rod morphology have been prepared from $BaCO_3$, $GeO_2$ and $KMnO_4$ in the HF solution using a one-pot hydrothermal route. The influence of synthesis conditions, including the concentrations of $KMnO_4$ and HF, the synthesis temperatures and times, on the PL properties of the $BaGeF_6:Mn^{4+}$ product, have been investigated in detail. The optimal $BaGeF_6:Mn^{4+}$ red phosphor can absorb the broadband blue light and emit the red light efficiently, which resulting in that the YAG-BGFM based w-LEDs display warmer white light than that of the YAG type. Therefore, the improved optical performance of w-LEDs may find potential application in indoor lighting.

#### 2.1.3.2 $BaSiF_6:Mn^{4+}$

In this part, we present a comprehensive investigation of the synthesis parameters, including the concentration of $KMnO_4$ and HF, the synthesis temperatures and times on the PL of $BaSiF_6:Mn^{4+}$ (marked as BSFM) red phosphors. The composition and morphology are studied. The PL properties of the as-prepared products are highly dependent on the concentrations of HF and $KMnO_4$, as well as the reaction times. The obtained pink powder phosphors intensely emit the red light under the blue light illumination. The w-LED prepared from the mixture of YAG and $BaSiF_6:Mn^{4+}$ shows higher LE and lower CCT than that only with YAG.

Fig.2-12 presents the XRD patterns of the as-synthesized $BaSiF_6$ and $BaSiF_6:Mn^{4+}$ at 180 ℃ for 8 h with different $KMnO_4$ concentrations. Curve a is the XRD pattern of $BaSiF_6$ without $Mn^{4+}$, which is in agreement with the JCPDS card (No. 15-0736, $a = b = 7.185$ Å, $c = 7.01$ Å, space group $R$-3$m$). This result shows the obtained sample is of a single phase. XRD patterns of the $BaSiF_6:Mn^{4+}$ samples are also similar to that of the $BaSiF_6$, and no impurity peaks can be observed. This indicates that these samples are also of a single phase, and the introduction of $Mn^{4+}$ ion does not change the crystal structure of this $BaSiF_6$ host.

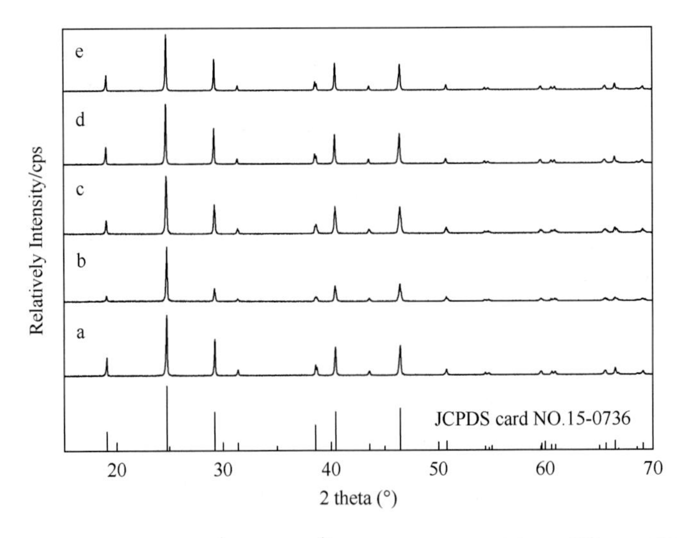


Fig.2-12 XRD patterns of the resulting $BaSiF_6:Mn^{4+}$ products prepared from different $KMnO_4$ concentration: 0 (a), 2 (b), 4 (c), 8 (d) and 12 $mmol \cdot L^{-1}$ (e) at 180 ℃ for 6 h

To furtherly analyze the crystal structure of these samples, the Rietveld refinements of $BaSiF_6$ and $BaSiF_6:Mn^{4+}$ were performed on the TOPAS Academic software. The XRD patterns of the experimental, calculated, and different results are shown in Fig.2-13. Meanwhile, the final refined structural parameters are summarized in Table 2-2 and Table 2-3.

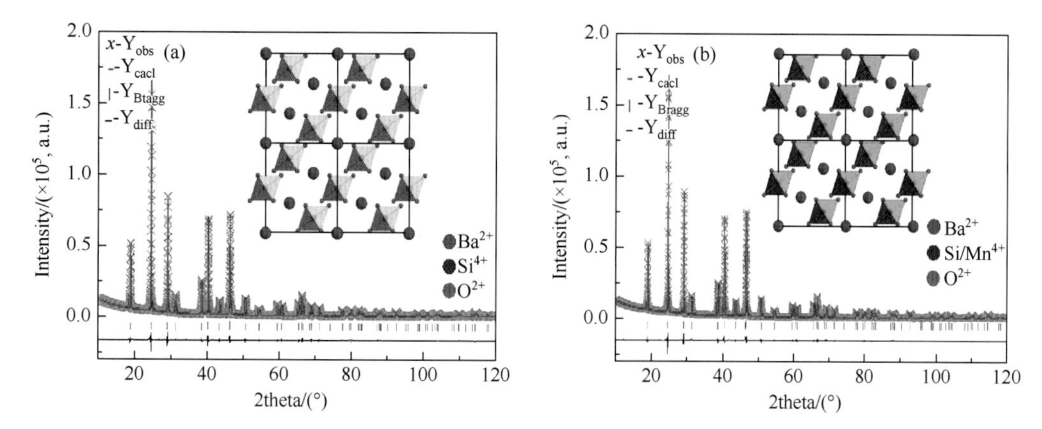


Fig.2-13 Experimental (crosses) and calculated (red solid line) XRD patterns and their difference (black solid line) for $BaSiF_6$ (a) and $BaSiF_6:Mn^{4+}$ (b) at room temperature. The blue ticks indicate the Bragg reflection positions. The inseted figures show the projections viewed from [010]

According to the refinement results, $Mn^{4+}$ can occupy the crystal site of $Si^{4+}$ to coordinate with six $F^-$ anions forming stable $MnF_6^{2-}$ octahedra, which resulted from the same valence state and similar ionic radius between $Mn^{4+}$ (0.53 Å, CN = 6) and $Si^{4+}$ (0.40 Å, CN = 6). The unit cell volume of $BaSiF_6:Mn^{4+}$ (313.82 $Å^3$) is a little larger than that of $BaSiF_6$ (313.77 $Å^3$), which is due to the larger radius of the doping $Mn^{4+}$ ions.

**Table 2-2 Crystal data of $BaSiF_6$ from the Rietveld refinement**

| Atom | Site | $x$ | $y$ | $z$ | Occupancy | $U_{iso}$ ($Å^2$) |
|---|---|---|---|---|---|---|
| Ba | 3a | 0 | 0 | 0 | 1 | 0.73 (3) |
| Si | 3b | 0 | 0 | 1/2 | 1 | 0.84 (5) |
| F | 18 h | 0.2182 (2) | 0.1091 (1) | 0.3587 (2) | 1 | 0.99 (4) |

Space group: $R\bar{3}m$ , $a$ = 7.1913 (0) Å, $c$ = 7.0058 (1) Å, $\alpha = \beta$ = 90°, $\gamma$ = 120°, $z$ = 3, cell volume: 313.77 (0) $Å^3$. $R_{wp}$ = 4.71%, $R_p$ = 3.27%, $R_{Bragg}$ = 1.12%

**Table 2-3 Crystal data of $BaSiF_6:Mn^{4+}$ from the Rietveld refinement**

| Atom | Site | $x$ | $y$ | $z$ | Occupancy | $U_{iso}$ ($Å^2$) |
|---|---|---|---|---|---|---|
| Ba | 3a | 0 | 0 | 0 | 1 | 0.63 (2) |
| Si | 3b | 0 | 0 | 1/2 | 0.92 (0) | 1.24 (6) |
| Mn | 3b | 0 | 0 | 1/2 | 0.08 (0) | 1.24 (6) |
| F | 18 h | 0.2190 (0) | 0.1095 (0) | 0.3584 (2) | 1 | 0.90 (4) |

Space group: $R\bar{3}m$, $a$ = 7.1912 (0) Å, $c$ = 7.0072 (0) Å, $\alpha = \beta$ = 90°, $\gamma$ = 120°, $z$ = 3, cell volume: 313.82 (0) $Å^3$. $R_{wp}$ = 4.07%, $R_p$ = 2.91%, $R_{Bragg}$ = 1.22%

The morphology of the as-prepared products was conducted on SEM and representative results are shown in Fig.2-14(a). It can be easily found that the product consists of a large number of quasi-uniform micrometer rods, and their surfaces are smooth. We inspected the tips of these rods closely. Each rod has a diameter of about 1 μm and a length of around 5 μm. Fig.2-14(b) is the corresponding EDS profile, in which the existence of the Mn element can be recognized. This furtherly verifies that $Mn^{4+}$ ions have occupied the lattice sites of $Si^{4+}$ ions to activate $BaSiF_6$ emitting red light under the blue light illumination. Furthermore, the O element cannot be found from the EDS spectrum, which implies that $MnO_2$ is not produced in the hydrothermal reaction.

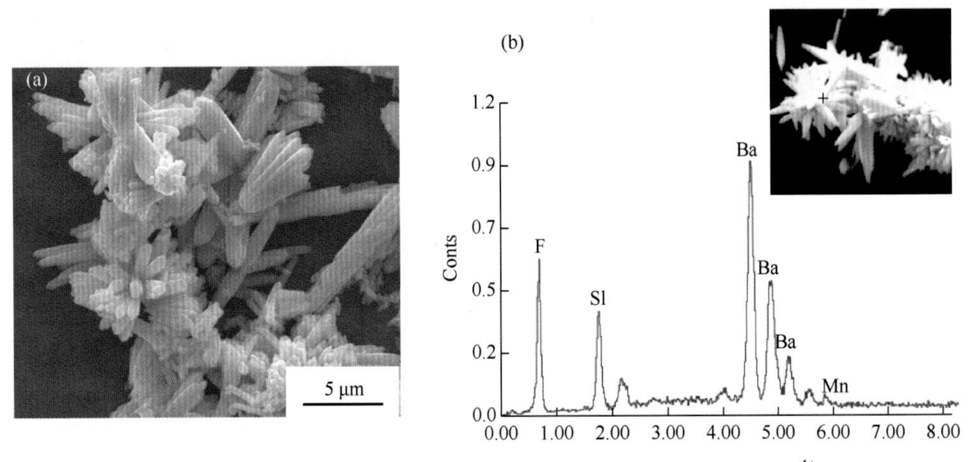


Fig.2-14 SEM image (a) and EDS profile (b) of the as-prepared $BaSiF_6$:$Mn^{4+}$ product

To comprehensively investigate the influence of synthesis parameters, such as the concentrations of $KMnO_4$ and HF, reaction temperatures and times on the PL properties of $BaSiF_6$:$Mn^{4+}$ products, a series of experiments have been done. Fig.2-15 exhibits the excitation and

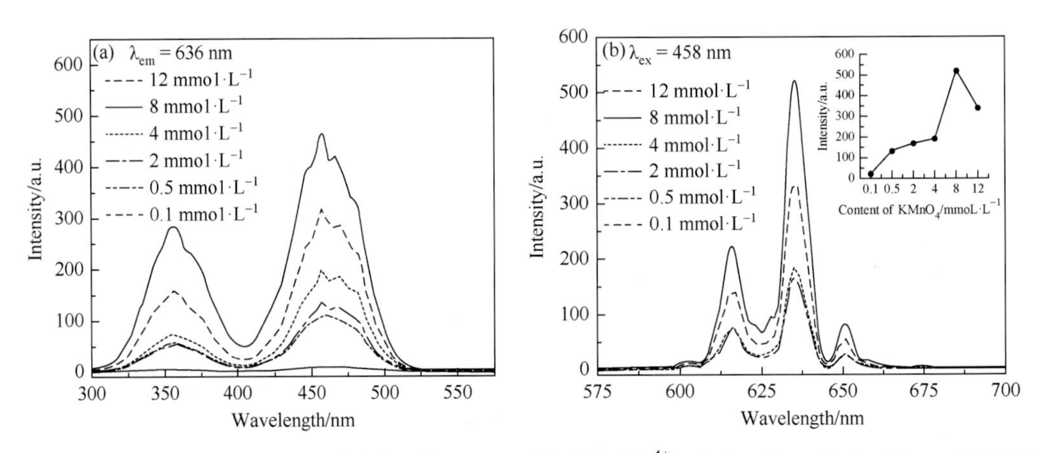


Fig.2-15 The excitation (a) and emission (b) spectra of $BaSiF_6$:$Mn^{4+}$ red phosphors obtained from 40% HF with different concentrations of $KMnO_4$ at 180 ℃ for 6 h. The inserted figure is the relationship between the $KMnO_4$ concentration and relative emission intensity

emission spectra of $BaSiF_6:Mn^{4+}$ red phosphor fabricated from different $KMnO_4$ concentrations (0.1 $mmol·L^{-1}$, 0.5 $mmol·L^{-1}$, 2 $mmol·L^{-1}$, 4 $mmol·L^{-1}$, 6 $mmol·L^{-1}$, 8 $mmol·L^{-1}$, 10 $mmol·L^{-1}$, 12 $mmol·L^{-1}$). In Fig.2-15(a), it can be easily found that the shapes of these excitation spectra are similar, which are consisted of two broadband excitation peaks locating in the UV (356 nm) and blue (458 nm) region respectively. According to the Tanabe-Sugano diagram, they are attributed to the spin-allowed transitions from the ground state $^4A_2$ to the excited states $^4T_2$ and $^4T_1$ of $Mn^{4+}$ respectively. It should be noted that the stronger excitation band matches very well with the blue emission of one GaN chip (460 nm).

Fig.2-15(b) shows their corresponding emission spectra under the 458 nm light excitation. There are three emission peaks in the range from 600 nm to 660 nm, which strongly indicates that the emitted light is red. The first peak (612 nm) is due to anti-Stokes vibronic sidebands of the excited state $^2E$ of $Mn^{4+}$ and the latter (636 nm and 651 nm) peaks are identified as its $^2E_g \rightarrow ^4A_2$ transition, in which the middle emission peak locating at 636 nm is the strongest. Moreover, Fig 2-15(b) displays the influence of the $KMnO_4$ concentration on the emission property of $BaSiF_6:Mn^{4+}$ phosphors. It can be observed that the emission intensity increases with the increase of the $KMnO_4$ concentration until it reaches 8 $mmol·L^{-1}$, at which the $BaSiF_6:Mn^{4+}$ phosphor emits the strongest red light. Its CIE chromaticity coordinates are $x = 0.64$, $y = 0.30$, which are close to the "ideal red" of the NTSC values ($x = 0.67$, $y = 0.33$). Continually increase $KMnO_4$ concentration to 12 $mmol·L^{-1}$, the emission contrarily drops. This is an obvious concentration quenching phenomenon in $BaSiF_6$ crystal lattice. The relationship between the $KMnO_4$ concentration and the relative emission intensity is inserted in Fig.2-15(b).

Different HF concentrations from 10% to 40% were employed to furtherly investigate the luminescent property of $BaSiF_6:Mn^{4+}$. Fig.2-16 shows a series of emission spectra of these $BaSiF_6:Mn^{4+}$ phosphors under the 458 nm light excitation. When a low concentration of 10% HF is employed, only a negligible emission peak at 636 nm can be observed. However, with the addition of HF, the emission intensity at 636 nm sharply increased. When the highest concentration HF (40%) is used, the emission intensity reaches the top, which is about 104 times, 30 times and 2.5 times higher than that of the other three samples. That is because the higher HF concentration is beneficial for the formation of a stable $MnF_6^{2-}$ group. Otherwise, it may be hydrolyzed into $MnO_2$, which results in the product color dark red and emission intensity decreasing. This phenomenon suggests that the emission property of this $BaSiF_6:Mn^{4+}$ red phosphor is highly dependent on the employed HF concentration, which plays an important role in obtaining the high brightness $BaSiF_6:Mn^{4+}$ red phosphor. The relationship between the HF concentration and the relative emission intensity is inserted in Fig.2-16.

In the above section, we conclude the optimum concentrations of $KMnO_4$ and HF for the fabrication of $BaSiF_6:Mn^{4+}$ red phosphor. However, for the hydrothermal synthesis, the reaction temperature and time also play an important role in obtaining the high brightness red phosphor. Fig.2-17 exhibits the effect of the reaction temperature on the emission intensity of

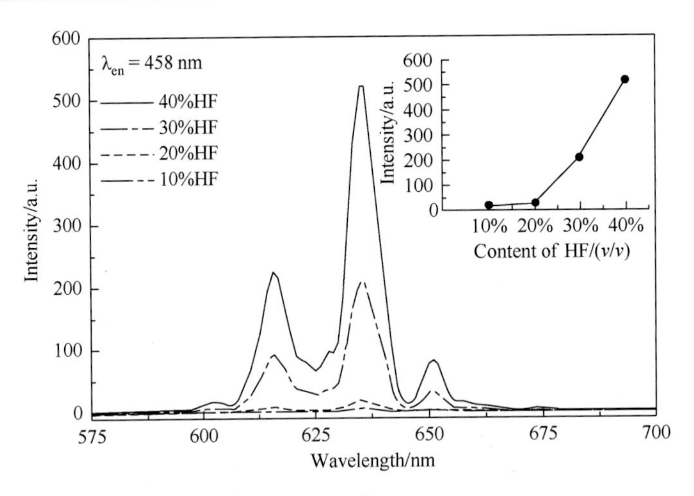


Fig.2-16 Emission spectra of red phosphors $BaSiF_6:Mn^{4+}$ obtained at 180 ℃ for 6 h with 8 mmol·$L^{-1}$ $KMnO_4$ and 10% (a), 20% (b), 30% (c) and 40% (d) HF. The inserted figure is the relationship between the HF concentration and the relative emission intensity

$BaSiF_6:Mn^{4+}$ with the optimum concentration of starting materials. When the reaction temperature is higher than 100 ℃, the emission intensities of these red phosphors are very close. It means the synthesis temperature mildly influenced the formation of $BaSiF_6:Mn^{4+}$ crystals when the synthesis temperature is higher than the boiling point of the reaction mixture. However, when a lower temperature (such as 90 ℃) is adopted, the emission intensity diminishes sharply. This may because the low reaction temperature (<100 ℃) is not favorable for the substitution of $Mn^{4+}$ for $Si^{4+}$ ions.

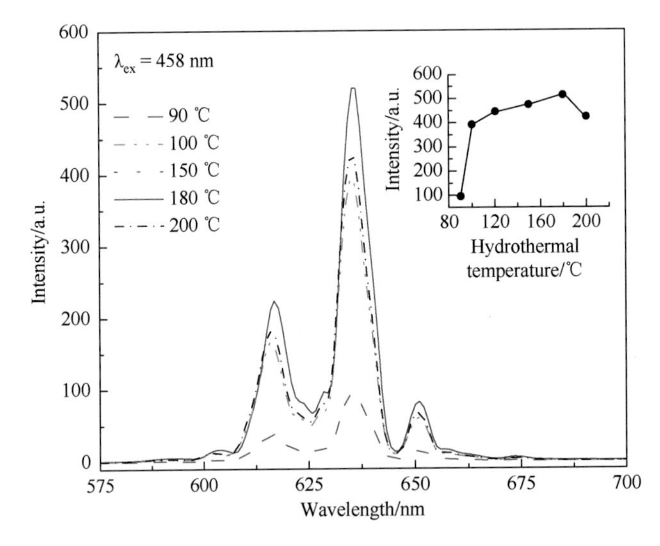


Fig.2-17 Emission spectra of $BaSiF_6:Mn^{4+}$ red phosphors obtained from 40% HF and 8 mmol·$L^{-1}$ $KMnO_4$ at 90 ℃, 100 ℃, 150 ℃, 180 ℃ and 200 ℃ for 6 h. The inserted figure is the relationship between the reaction temperature and the relative emission intensity

Similarly, Fig.2-18 shows the influence of reaction time (2 h, 6 h, 12 h, 24 h) on the emission property of $BaSiF_6:Mn^{4+}$ phosphor. With the increase of the reaction time from 2 h to 6 h, the emission intensity rises sharply, which means that a longer reaction time leads to a stronger intensity. When the reaction time is prolonged continually, the emission intensity drops dramatically. Therefore, according to the above investigation, the optimum reaction condition is 8 $mmol·L^{-1}$ $KMnO_4$, 40% HF, and reacts at 180 ℃ for 6 h.

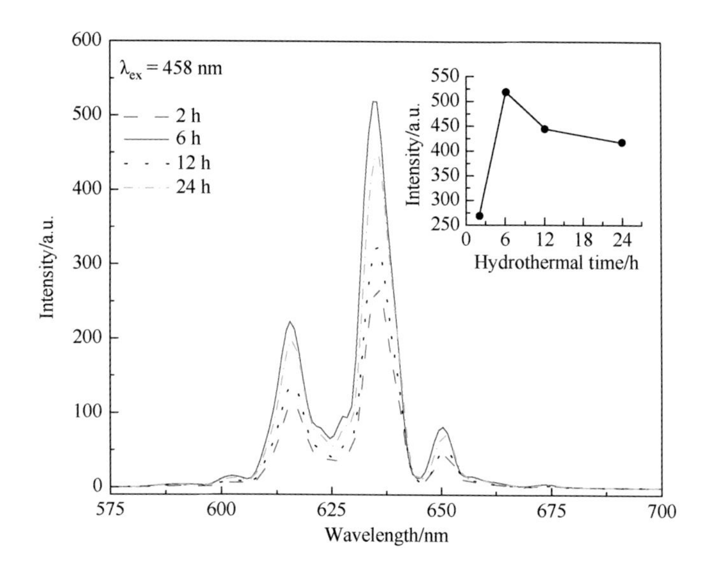


Fig.2-18 Emission spectra of $BaSiF_6:Mn^{4+}$ red phosphors obtained from 40% HF and 8 $mmol·L^{-1}$ $KMnO_4$ at 180 ℃ for 2 h (a), 6 h (b), 12 h (c) and 24 h (d). The inserted figure is the relationship between the reaction temperature and the relative emission intensity

Furthermore, the influence of temperature on the PL properties of this optimum product is shown in Fig.2-19. Obviously, with the increasing temperature, it can be found that the emission peak position does not shift. Up to 160 ℃, over 100% of the integrated emission intensity can be preserved compared with that at 20 ℃. This result indicates that the as-prepared $BaSiF_6:Mn^{4+}$ exhibits excellent thermal stability.

Fig.2-20 displays the EL of the GaN chip, the LED-based on $BaSiF_6:Mn^{4+}$, YAG and the mixture of YAG and $BaSiF_6:Mn^{4+}$ under 20 mA current excitation. The emission band at 460 nm can be attributable to the emission of the GaN chip. Compared with curve a, the emission peak at 636 nm in curve b is due to the emission of red phosphor excited by the emission of the GaN LED chip. The bright red light [Fig. 2-20(b)] and the emission of the GaN chip can be observed by naked eyes, this is beneficial for obtaining a w-LED by merging it with yellow phosphor YAG. Curves c and d are the EL spectra of the w-LEDs fabricated from YAG as well as the mixture of YAG and $BaSiF_6:Mn^{4+}$. The obvious red emission on Fig.2-20(d) means that the addition of $BaSiF_6:Mn^{4+}$ red phosphor is favorable for obtaining a warm w-LED.

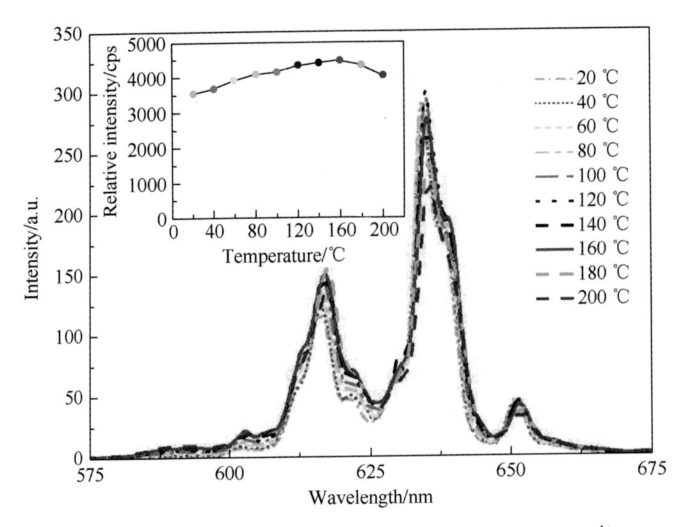


Fig.2-19 Temperature-dependent thermal luminescence spectra of the $BaSiF_6:Mn^{4+}$ red phosphor. The inserted figure is the relationship between the temperature and the relative emission intensity

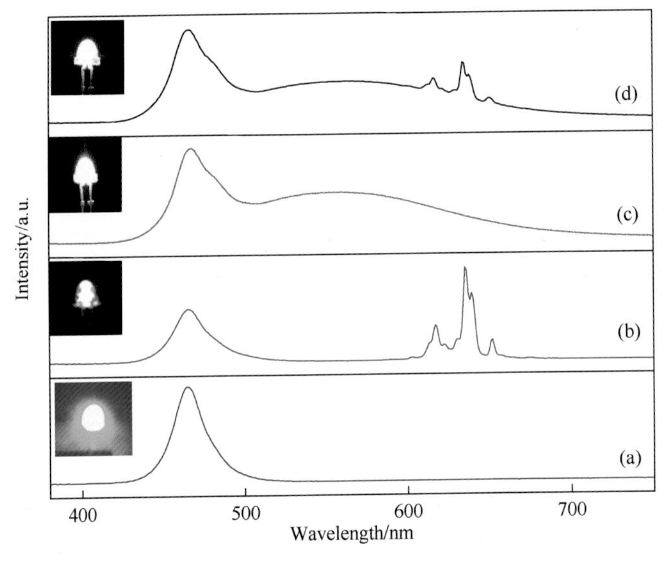


Fig.2-20 Electro-luminescent spectra of the GaN LED chip (a), the LED based on $BaSiF_6:Mn^{4+}$ (b), YAG (c) and $BaSiF_6:Mn^{4+}$-YAG mixture (d) under 20 mA current excitation. The inserted figures are the images of the LEDs

The performance of the w-LEDs is summarized in Table 2-4. It should be noted that $R_a$ and CCT levels of the prepared w-LEDs were improved after the addition of $BaSiF_6:Mn^{4+}$ red phosphor.

In summary, a series of $BaSiF_6:Mn^{4+}$ red phosphors with micro-rod morphology are synthesized by a hydrothermal route. The influence of synthesis parameters, including $KMnO_4$ and HF concentrations, the synthesis temperatures as well as the reaction times on its PL properties has been systematically investigated. The optimum reaction condition is 8 $mmol{\cdot}L^{-1}$ $KMnO_4$,

**Table 2-4 Performance of the GaN-based w-LEDs coated with (1) YAG, (2) YAG and $BaSiF_6:Mn^{4+}$ at 20 mA forward current as well as 5 V reverse voltage**

| No. of LEDs Samples | CCT/K | $R_a$ | LE/(lm/W) | CIE ($x$, $y$) |
|---|---|---|---|---|
| 1 | 6283 | 76.0 | 45.21 | (0.313, 0.366) |
| 2 | 5903 | 82 | 51.73 | (0.323, 0.359) |

40% HF, and reacts at 180 ℃ for 6 h. The resulting $BaSiF_6:Mn^{4+}$ phosphor shows intense red emission with broadband excitation in the blue region. The w-LEDs based on this $BaSiF_6:Mn^{4+}$ red phosphor presents warmer white light than that of the only one YAG component. Therefore, it is considered a good candidate for improving the optical performance of the indoor lighting w-LEDs.

#### 2.1.3.3 $BaTiF_6:Mn^{4+}$

In this part, we adopted an indirect method to systematically investigate the influence of some general and basic synthesis parameters, such as the temperatures, the time, the $KMnO_4$ and HF concentrations on PL properties of $BaTiF_6:Mn^{4+}$ (donated as BTFM). The optimum reaction condition for the preparation process has been determined. Moreover, the application of this red phosphor on LED devices has been studied.

Fig.2-21 shows the XRD patterns of the hydrothermally synthesized $BaTiF_6$ with and without the existence of $KMnO_4$ in the starting materials, along with its standard diffraction card. All the diffraction peaks in Fig.2-21(a) are indexed to a pure rhombohedral $BaTiF_6$ phase (JCPDS No. 76-0269, space group $R$-3$m$, $a = b = 7.368$ nm, $c = 7.252$ nm). With the existence of

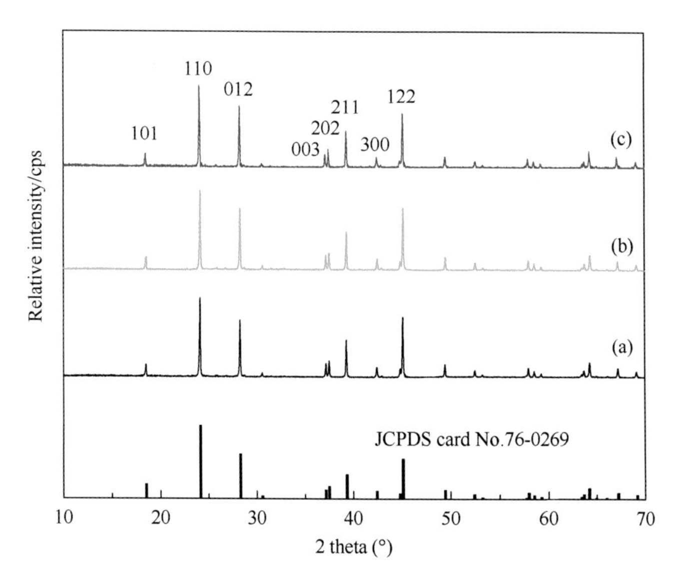


Fig.2-21 XRD patterns of the red phosphors BTFM obtained from 40% HF without $KMnO_4$ (a), 8 $mmol \cdot L^{-1}$ $KMnO_4$ in 10% HF (b) and 40% (c) by the hydrothermal process at 180 ℃ for 8 h

8.0 mmol·$L^{-1}$ $KMnO_4$, the observed XRD patterns are identical with curve a and no impurity peaks can be detected, which implies that the obtained BTFM products are also of a single pure phase. These may be due to the similar ionic radius and same valance state between $Ti^{4+}$ and $Mn^{4+}$ (0.605 Å, CN = 6 vs. 0.53 Å, CN = 6), which means that $Mn^{4+}$ ions have successfully occupied the octahedral sites of $Ti^{4+}$ without changing the lattice structure of $BaTiF_6$. Furthermore, Fig.2-21(b) illustrates that the BTFM product can be prepared with the lower content of HF.

The SEM images of the pink BTFM phosphors are shown in Fig.2-22. The red phosphor synthesized from 10% HF solution shows a rod-like morphology, particle size of roughly 10 μm length and 2 μm width [Fig.2-22(a)]. While the sample prepared with the 40% HF solution also

Fig.2-22　SEM images of as-synthesized BTFM phosphors obtained at 180 ℃ for 8 h in the 10% (a) and 40% HF (b), EDS spectrum of BTFM in the 40% HF (c), XPS spectrum of BTFM in the 40% HF (d). The inserted figure is the EDS spot

has a petal-like morphology, particle size less than 10 μm in length and about 5 μm in width [Fig.2-22(b)]. All of the particles gather to one end and form like flowers in full bloom. These may attribute to the high concentration of fluoride ion which limits the growth orientation of BTFM particles and results in its larger size in the 40% HF solution.

The EDS result of red phosphor BTFM is shown in Fig.2-22(c). The result indicates that the crystal of BTFM phosphor is composed of barium (Ba), titanium (Ti), fluorine (F) and manganese (Mn). Element Si in the spectrum is from the silicon wafer used for the observation of SEM and EDS. Fig.2-22(d) shows the XPS survey spectrum for the pink powder, the elements of barium (Ba), titanium (Ti), fluorine (F), together with manganese (Mn) species, have been detected on the sample. The element composition is in agreement with the results of EDS. Other oxygen and carbon elements detected in XPS may be mainly attributed to the adsorption of $CO_2$, $CH_4$, etc.

Fig.2-23 shows the excitation and emission spectra of BTFM in different HF solutions. These samples share similar excitation spectra with broad absorption bands in the UV (350 nm) and blue (466 nm) regions, which are attributed to the ${}^4A_2 \rightarrow {}^4T_1$, ${}^4A_2 \rightarrow {}^4T_2$ transitions of $Mn^{4+}$, respectively. Fig.2-23(b) is the emission spectra of these phosphors. There are three sharp peaks at 613 nm, 634 nm and 648 nm, respectively, which are ascribed to the electric-dipole-forbidden transition ${}^2E_g \rightarrow {}^4A_2$ of $Mn^{4+}$. These sharp peaks are caused by vibrational modes of the $MnF_6^{2-}$ octahedral. Thus the weak electron-phonon coupling for the ${}^2E_g \rightarrow {}^4A_2$ transition can be well handled. With the increasing HF concentration, the emission intensity dramatically increases, which means that the emission property of BTFM highly dependent on the HF concentration. When the HF concentration is up to 40%, the strongest emission is obtained. This may be due to that the higher HF concentration is beneficial for the formation of $MnF_6^{2-}$ octahedron. Therefore, 40% is an optimum HF concentration in our following hydrothermal procedure to obtain the highest brightness red-emitting phosphor.

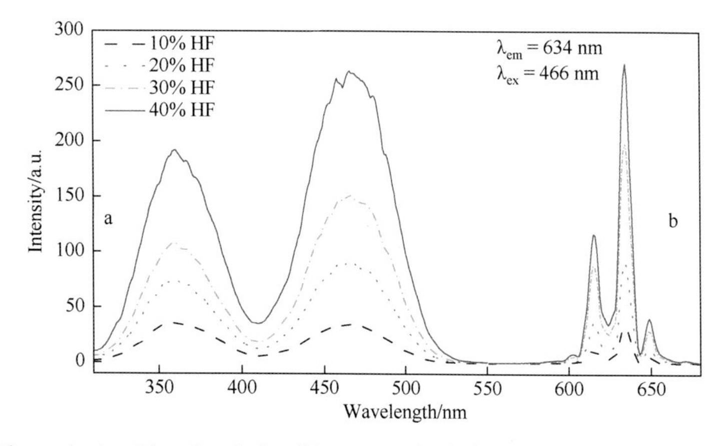


Fig.2-23 The excitation (a) and emission (b) spectra of red phosphor BTFM obtained from different concentrations of HF solution with 8 mmol·L$^{-1}$ $KMnO_4$ at 180 ℃ for 8 h

Another important factor to affect the luminescent properties of BTFM is the concentration of $KMnO_4$ solution. Fig.2-24 displays the luminescent spectra of the BTFM products obtained from different $KMnO_4$ concentrations. The excitation and emission intensities increase with the addition of $KMnO_4$ until its concentration reaches 8.0 mmol·L$^{-1}$. The luminescent intensity decreases when $KMnO_4$ is more than 8.0 mmol·L$^{-1}$, which is attributable to the $Mn^{4+}$ concentration quenching effect in transition-metal doped phosphors. Ion-ion interaction causes cross-relaxation energy transfer and non-radiative relaxation, which makes luminescence intensity decreasing especially when activator ions are close in the lattice.

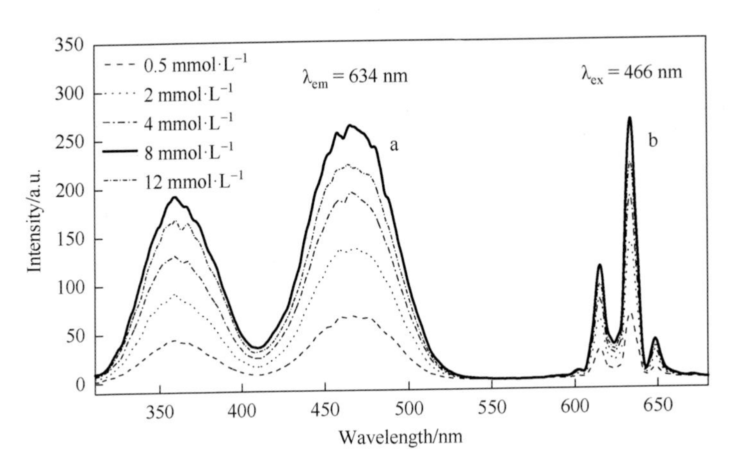


Fig.2-24 Excitation (a) and emission (b) spectra of BTFM red phosphors obtained from 40% HF solution with different concentration of $KMnO_4$ at 180 ℃ for 8 h

Besides, the effect of the hydrothermal time and temperature is also investigated in our research. Fig.2-25(a) shows the emission spectra of red phosphor BTFM obtained from different hydrothermal times. The luminescent intensity of BTFM increases when the reaction time extends to 8 h. When the reaction time exceeds 12 h, the emission spectra of BTFM decrease immediately. Meanwhile, Fig.2-25(b) shows the emission spectra of red phosphor BTFM obtained from different hydrothermal temperatures in the 40% HF solution with 8 mmol·L$^{-1}$ $KMnO_4$ for 8 h. In this hydrothermal procedure, the temperature determines the redox reaction rate of $KMnO_4$. The sample obtained at 180 ℃ possesses the strongest emission intensity.

According to the above experimental results, the optimized synthesis parameters are as following: the $KMnO_4$ concentration of 8 mmol·L$^{-1}$, the HF concentration of 40%, the reaction temperature of 180 ℃, and the reaction time of 8 h. Fig.2-26(a) is the PL spectra of BTFM under the optimum reaction condition. The sample exhibits an intense excitation band in the blue region and intense red emission with the appropriate CIE coordinates ($x = 0.695$, $y = 0.305$). The bright red light can be observed from the red phosphor under the blue light [Fig.2-26(a)].

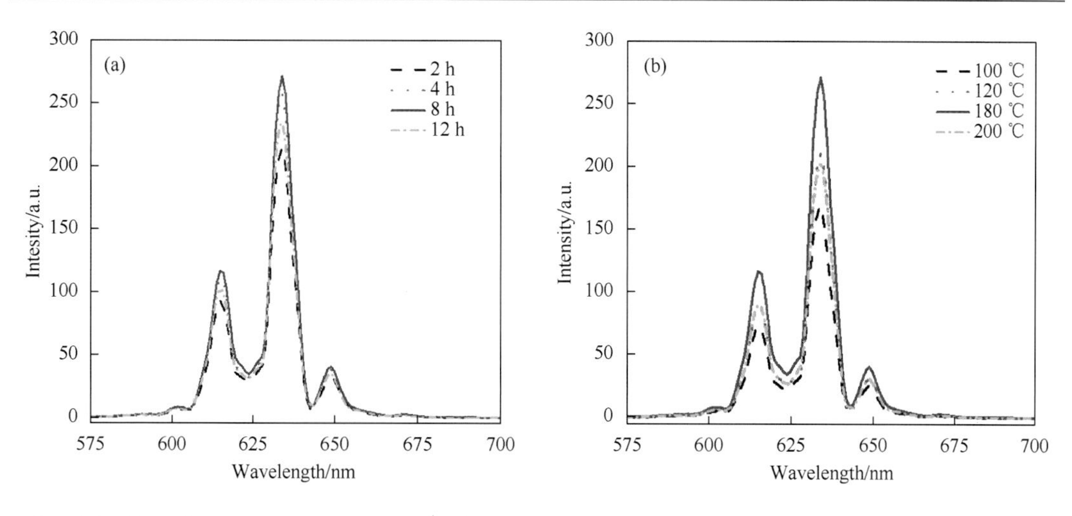


Fig.2-25 Emission spectra of $BaTiF_6$: $Mn^{4+}$ phosphors with different times (a) and different temperatures (b)

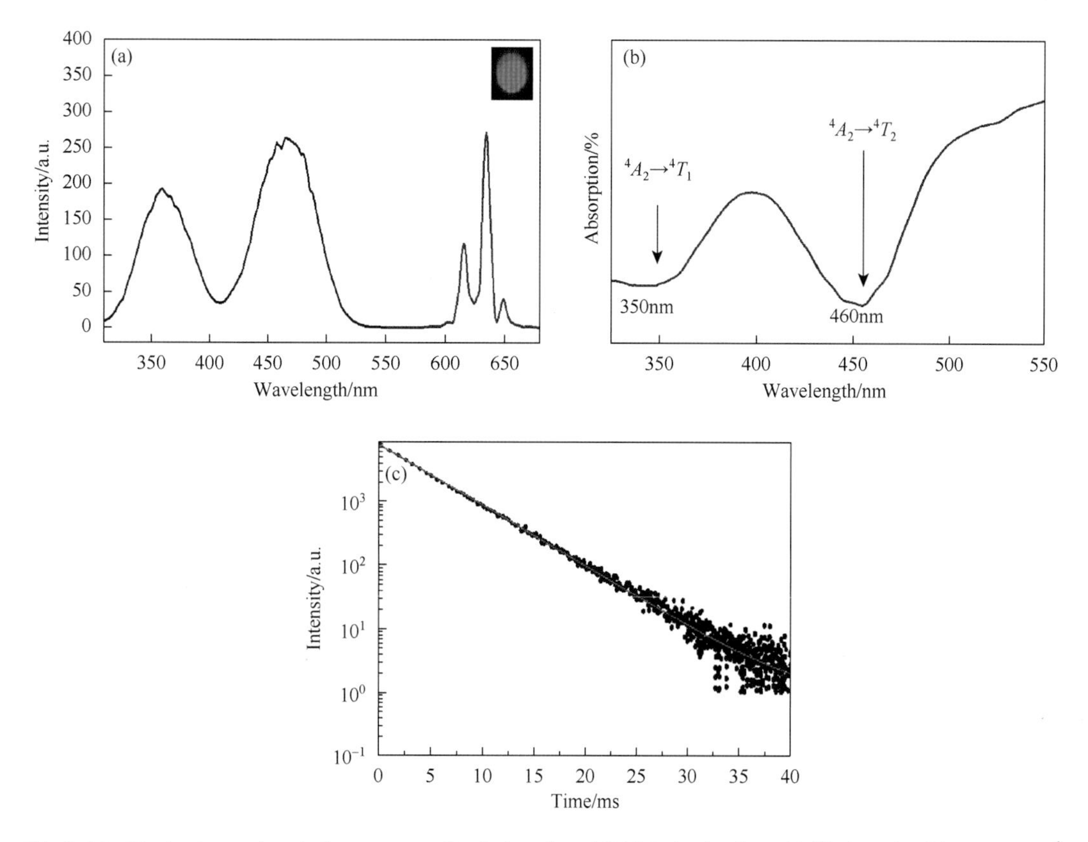


Fig.2-26 Excitation and emission spectra of red phosphors BTFM obtained at 180 ℃ for 8 h with 8 mmol·$L^{-1}$ $KMnO_4$ in the 40% HF solution (a), the diffuse reflection spectrum of red phosphor BTFM (b) and semi-logarithmic plot of the BTFM emission decay curve c

Fig.2-26(b) is the DRS spectrum of the optimized BTFM product. Two obvious absorption bands around 350 nm and 460 nm can be observed. This result is consistent with that of its excitation spectrum. The decay curve of the optimized BTFM red phosphor is shown in Fig.2-26(c). It is well fitted into the single-exponential function, and the lifetime $\tau$ value is 4.5 ms.

As shown in Fig.2-27, the TG and DSC curves of BTFM show that its thermal decomposition temperature ($T_d$) is about 440 ℃. Such high decomposition temperature suggests that this phosphor has high thermal stability and it is enough to be applied in LED devices.

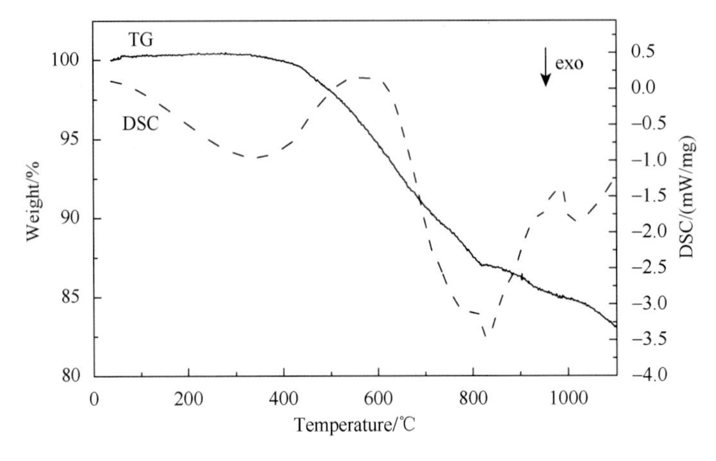


Fig.2-27  TG and DSC curves of BTFM

In the above section, BTFM red phosphor displays a strong excitation band in the blue region, sharp red emission with high color purity and high decomposition temperature, so it may be a promising phosphor for the GaN-LED backlighting. To investigate its PL properties in LED devices, the single red LED was fabricated by coating the red phosphor on a blue commercial GaN chip. The emission spectrum of the red LED with 20 mA current excitation is exhibited in Fig.2-28. The intense red-emitting peaks at 613 nm, 634 nm and 648 nm are due to the emission of BTFM excited by the emission of the blue LED chip. Compared with Fig.2-28(b), no emission of the LED chip can be found in Fig.2-28(a). This result shows this red phosphor can be efficiently excited by a blue LED chip. The CIE chromaticity coordinates according to the EL spectrum of the single LED are calculated to be $x = 0.660$, $y = 0.312$, which are close to the NTSC standard values for red ($x = 0.67$, $y = 0.33$). The red LED exhibits intense red light, which can be observed by naked eyes. Hence, it may find application in the blue LED backlighting.

In summary, a series of BTFM red-emitting phosphors have been successfully synthesized by a hydrothermal method. To obtain the optimum product, the hydrothermal synthesis parameters have been investigated in detail and the optimum synthesis condition has been confirmed. The optimal BTFM red phosphor exhibits one broad excitation band in the blue region and sharp emission in the red region. The single red LED-based on BTFM shows intense red emission with high color-purity, so it may be a promising red phosphor for the GaN-LED backlighting.

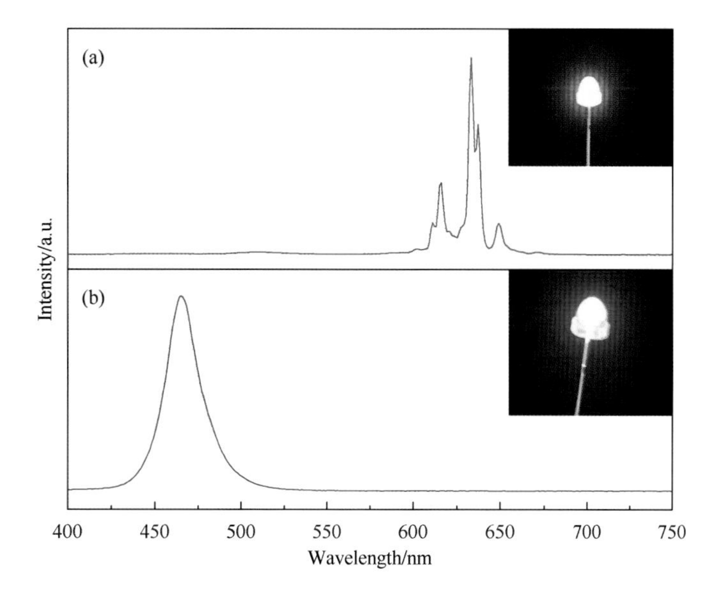


Fig.2-28 Electro-luminescent spectra of the red LED-based on BTFM (a) and GaN chip (b) under 20 mA current excitation. The inserted figures are the corresponding light images

2.1.3.4 $Na_2XF_6:Mn^{4+}$ (X = Si, Ge, Ti)

Liu demonstrated the co-precipitation approach to synthesize $Na_2SiF_6:Mn^{4+}$ from a silicon fluoride solution using $H_2O_2$ to reduce $Mn^{7+}$ to $Mn^{4+}$ at room temperature. This approach to synthesizing $Na_2SiF_6:Mn^{4+}$ is very feasible. However, the growth of $Na_2SiF_6:Mn^{4+}$ is hard to control, because the redox reaction between $H_2O_2$ and $NaMnO_4$ in the HF solution is very fast. So, red phosphors $Na_2XF_6:Mn^{4+}$ (X = Si, Ge, Ti) (marked as NXFM) were synthesized by the co-precipitation method without any reductants. All the obtained products exhibit intensive red light with high color-purity under the blue light excitation, which may find application in warm w-LEDs.

In the current work, NXFM phosphors were prepared via the co-precipitation method, and $MnF_6^{2-}$ was obtained by the decomposition of $Na_2MnO_4$ in the HF solution without $H_2O_2$. The chemical reaction is represented as follows:

$$(1-x)\, XO_2 + (4+x)\, HF + (2-x)\, NaF + xNaMnO_4 \longrightarrow Na_2X_{1-x}Mn_xF_6 + 0.75xO_2 + (2+0.5x)\, H_2O \tag{2-1}$$

The phases of as-prepared NXFM are investigated by XRD, and the representative results are shown in Fig.2-29. The XRD patterns of $Na_2SiF_6:Mn^{4+}$ and $Na_2GeF_6:Mn^{4+}$ are consistent with the corresponding JCPDS cards of $Na_2SiF_6$ (No. 33-1280) and $Na_2GeF_6$ (No.35-0814). No impurity can be found from the XRD patterns. This result indicates that the obtained phosphors $Na_2SiF_6:Mn^{4+}$ and $Na_2GeF_6:Mn^{4+}$ are of the single-phase with a similar hexagonal structure (space group of *P*321). In this hexagonal structure, $Si^{4+}/Ge^{4+}$ occupies the center of the $SiF_6^{2-}/GeF_6^{2-}$ octahedron. $Na_2TiF_6:Mn^{4+}$ also has a single-phase according to its XRD pattern compared with the JCPDS card of $Na_2TiF_6$ (No. 15-0581). In the current work, $Na_2TiF_6:Mn^{4+}$ has a hexagonal structure with a space group of *P*-3*m*1. This crystal structure is different from the trigonal structure of $Na_2TiF_6:Mn^{4+}$ reported by Adachi. In this structure of $Na_2TiF_6$, $Ti^{4+}$ is also coordinated by six $F^-$ to form a $TiF_6^{2-}$ octahedron. Since $Mn^{4+}$ has the same charge as $Si^{4+}$, $Ge^{4+}$ and $Ti^{4+}$, meanwhile its radius (53 pm, CN = 6) is close to that of $Si^{4+}$ (40 pm, CN = 6), $Ge^{4+}$ (53 pm, CN = 6) and $Ti^{4+}$ (61 pm, CN = 6), the doped $Mn^{4+}$ ions do not change the structure of $Na_2XF_6$ hosts. Then $Mn^{4+}$ will occupy the site of the $XF_6^{2-}$ (X = Si, Ge, Ti) octahedron.

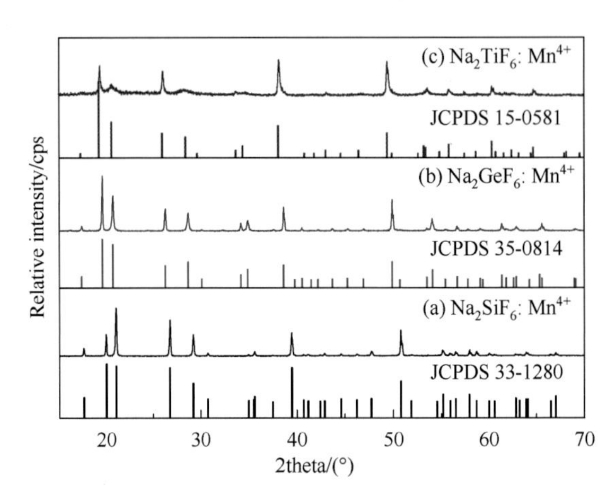


Fig.2-29 XRD patterns of (a) $Na_2SiF_6:Mn^{4+}$, (b) $Na_2GeF_6:Mn^{4+}$ and (c) $Na_2TiF_6:Mn^{4+}$

The morphology and composition of the obtained NXFM products are examined by the SEM and EDS analysis, and the representative results are shown in Fig.2-30. The obtained products exhibit an irregular morphology with smooth surfaces. Closely inspecting the particle size distribution among them, it can be found that the $Na_2SiF_6:Mn^{4+}$ product displays an apparent larger size (about 20 μm) than that of $Na_2GeF_6:Mn^{4+}$ (about 0.8 μm) and $Na_2TiF_6:Mn^{4+}$ (about 0.5 μm) products. Three curves in Fig.2-30 are their corresponding EDS results. It can be observed the peaks belonging to F, Si, K and Mn elements from Fig.2-30(a2), which indicates that the Mn element has been indeed doped into the matrix lattice to occupy the lattice site of Si. The doping amount of $Mn^{4+}$ in $Na_2SiF_6:Mn^{4+}$ product is 1.9% (mole fraction), measured from the AAS analysis. Similar results can be obtained from $Na_2TiF_6:Mn^{4+}$ and $Na_2GeF_6:Mn^{4+}$ products in Fig.2-30(b2-c2), whose doping amounts of $Mn^{4+}$ are 2.4% (mole fraction) and 1.0% (mole fraction), respectively. Moreover, the absence of oxygen peak in the EDS spectra implies that there is no $MnO_2$ produced during this precipitation process.

Fig.2-31 is the excitation and emission spectra of NXFM. These phosphors exhibit similar excitation spectra with two broad bands located at 460 nm and 355 nm, which can be assigned to the $^4A_2 \rightarrow ^4T_2$ transition and $^4A_2 \rightarrow ^4T_1$ transition of $Mn^{4+}$, respectively. The red emissions of these phosphors between 600 nm and 650 nm are due to the spin-forbidden $^2E \rightarrow ^4A_2$ transition of $Mn^{4+}$. The strongest emission peak is located at 627 nm, which has some blue-shifts compared with

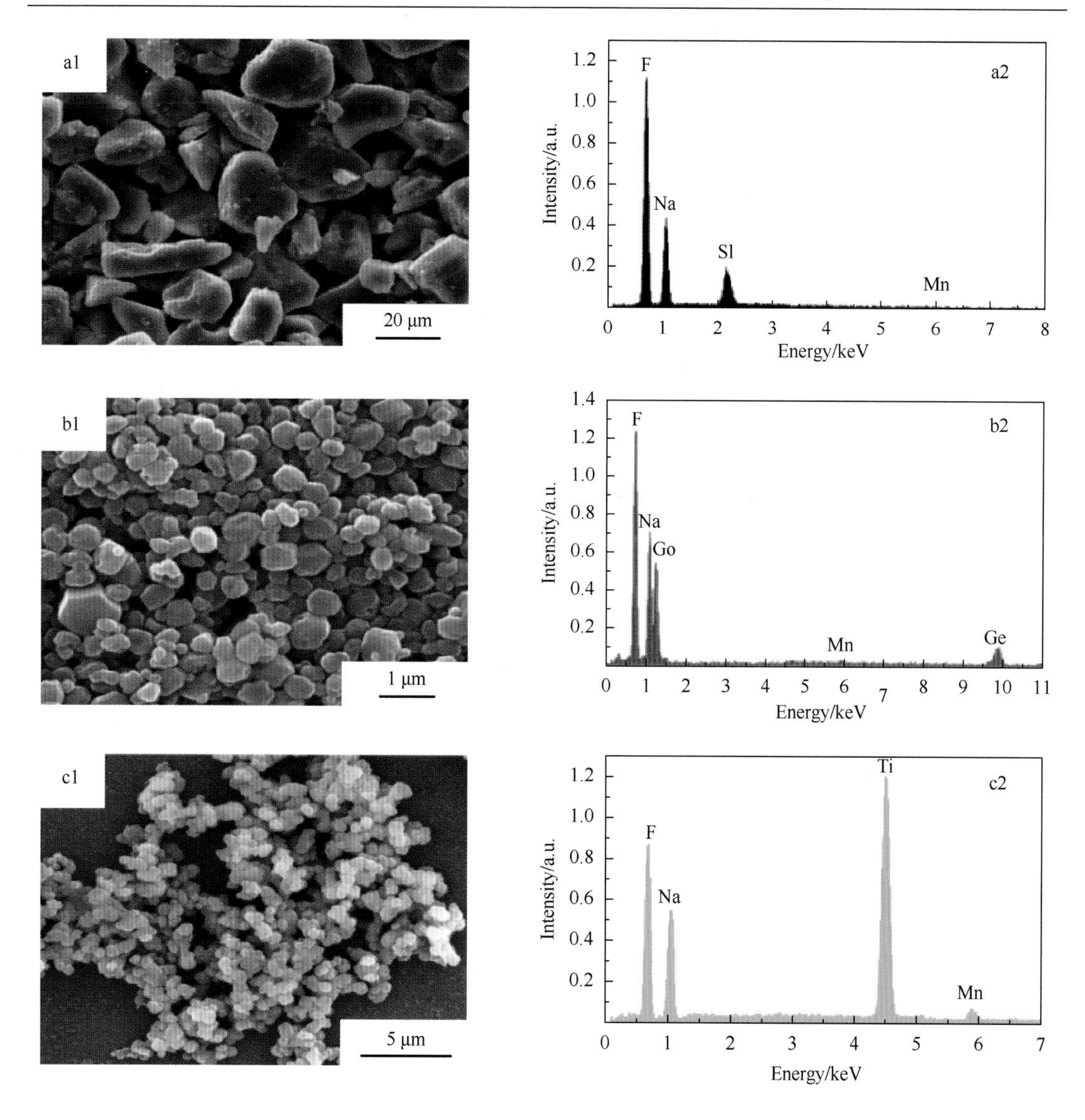


Fig.2-30 SEM images and EDS spectra of ($a_1$, $a_2$) $Na_2SiF_6:Mn^{4+}$, ($b_1$, $b_2$) $Na_2GeF_6:Mn^{4+}$ and ($c_1$, $c_2$) $Na_2TiF_6:Mn^{4+}$

that of $K_2TiF_6:Mn^{4+}$ (632 nm). The emission peaks located at 609 nm, 617 nm, 627 nm, 631 nm, 642 nm are ascribed as the transitions of the $v_6$ ($t_{2u}$), zero phonon line (ZPL), $v_6$ ($t_{2u}$), $v_4$ ($t_{1u}$) and $v_3$ ($t_{1u}$) vibronic modes, respectively. Intense ZPL emission of these phosphors can be found in these emission spectra. This result is different from those of $K_2XF_6:Mn^{4+}$ and $BaXF_6:Mn^{4+}$. Compared with $Na_2GeF_6:Mn^{4+}$ and $Na_2TiF_6:Mn^{4+}$, $Na_2SiF_6:Mn^{4+}$ exhibits the strongest red emission under the 465 nm excitation. The CIE coordinates according to the emission spectra of $Na_2SiF_6:Mn^{4+}$, $Na_2GeF_6:Mn^{4+}$ and $Na_2TiF_6:Mn^{4+}$ are calculated to be ($x = 0.684$, $y = 0.315$), ($x = 0.685$, $y = 0.315$) and ($x = 0.683$, $y = 0.316$), respectively. These values are very close to the NTSC standard values for red ($x = 0.67$, $y = 0.33$).

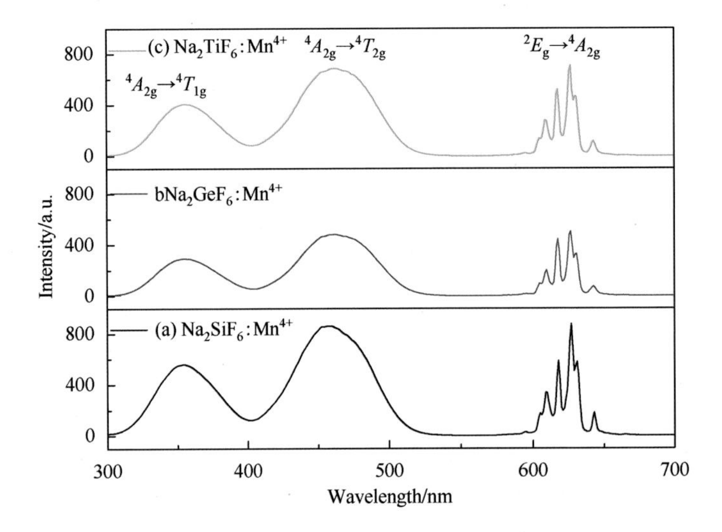


Fig.2-31 The excitation ($\lambda_{em}$ = 627 nm) and emission ($\lambda_{ex}$ = 465 nm) spectra of (a) $Na_2SiF_6$:$Mn^{4+}$, (b) $Na_2GeF_6$:$Mn^{4+}$ and (c) $Na_2TiF_6$:$Mn^{4+}$

Fig.2-32 is the DRS of $Na_2SiF_6$ and $Na_2SiF_6$:$Mn^{4+}$. Comparing with $Na_2SiF_6$, $Na_2SiF_6$:$Mn^{4+}$ exhibits two obvious absorption bands at 350 nm and 465 nm in curve b, which are due to the absorption bands of $Mn^{4+}$ in $Na_2SiF_6$:$Mn^{4+}$. This result is in agreement with the excitation spectrum of $Na_2SiF_6$:$Mn^{4+}$.

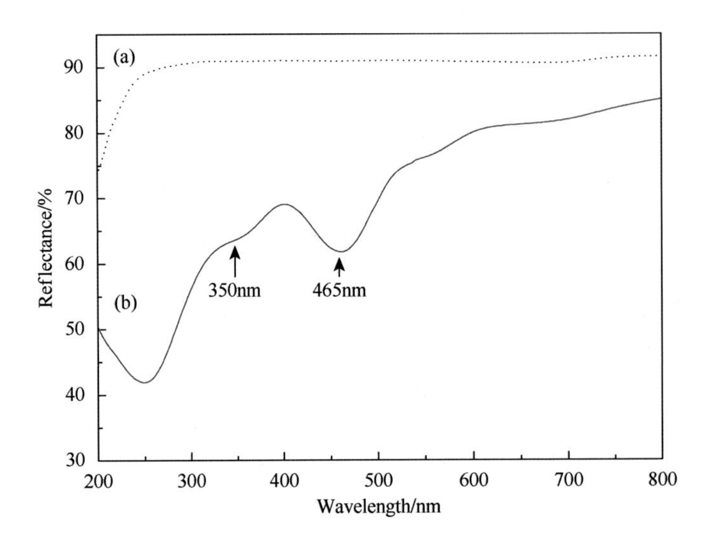


Fig.2-32 Diffuse reflectance spectra of (a) $Na_2SiF_6$ and (b) $Na_2SiF_6$:$Mn^{4+}$

Fig.2-33 shows the decay curves for the $^2E_g$→$^4A_2$ transition of $Mn^{4+}$ in $Na_2SiF_6$:$Mn^{4+}$, $Na_2GeF_6$:$Mn^{4+}$ and $Na_2TiF_6$:$Mn^{4+}$. These curves are fitted into a single-exponential function, and their lifetime $\tau$ values are 3.74 ms, 3.68 ms and 2.68 ms, respectively.

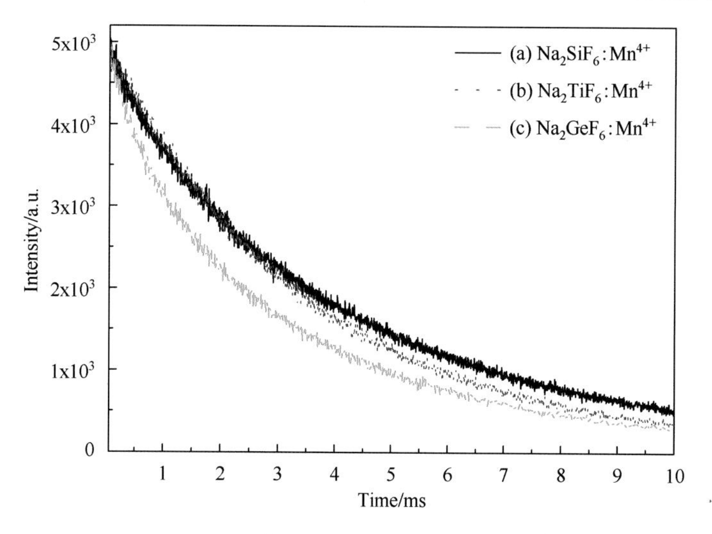


Fig.2-33 Decay curves of the $Mn^{4+}$ emission in (a) $Na_2SiF_6$:$Mn^{4+}$, (b) $Na_2TiF_6$:$Mn^{4+}$ and (c) $Na_2GeF_6$:$Mn^{4+}$($\lambda_{ex}$ = 465 nm; $\lambda_{em}$ = 627 nm)

Since LED devices work usually at a temperature close to 150 ℃, thermal stability is an important parameter for phosphors. As shown in Fig.2-34, TG and DSC curves of $Na_2SiF_6$:$Mn^{4+}$ show that its thermal decomposition temperature ($T_d$) is about 400 ℃. Such high decomposition temperature suggests that this phosphor has high thermal stability and it is enough to be applied in w-LED devices.

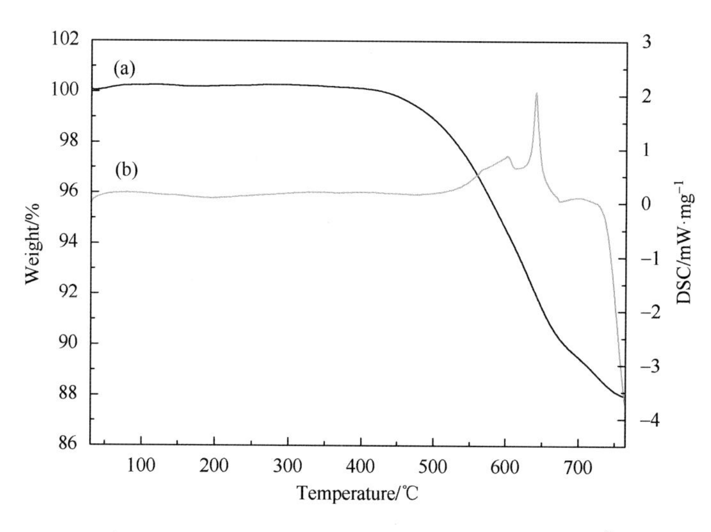


Fig.2-34 (a) TG and (b) DSC curves of $Na_2SiF_6$:$Mn^{4+}$

To investigate the potential applications of these red phosphors, the single red LEDs and w-LEDs were fabricated by combining blue GaN chips, the red phosphors NXFM with or without commercial YAG. Fig.2-35 is the EL spectra of the single red LEDs-based on NXFM phosphors under 20 mA current excitation. The emission of an LED chip is at 465 nm, and its half-peak width is about 20 nm. Compared with the spectrum a, the emission of the LED chip cannot be observed in the spectrum b, indicating that $Na_2SiF_6:Mn^{4+}$ can efficiently absorb the emission of the LED chip. The red emission between 600 nm and 650 nm is due to the emission of $Na_2SiF_6:Mn^{4+}$. The bright red light can be observed from this red LED-based on $Na_2SiF_6:Mn^{4+}$. The red LEDs with $Na_2GeF_6:Mn^{4+}$ and $Na_2TiF_6:Mn^{4+}$ share similar EL spectra. Different from the spectrum b, the emission of the LED chip still exits in the EL spectra of red LEDs-based on $Na_2GeF_6:Mn^{4+}$ and $Na_2TiF_6:Mn^{4+}$. This result shows that the absorption efficiency of $Na_2GeF_6:Mn^{4+}$ and $Na_2TiF_6:Mn^{4+}$ is lower than that of $Na_2SiF_6:Mn^{4+}$, which is in agreement with their excitation spectra.

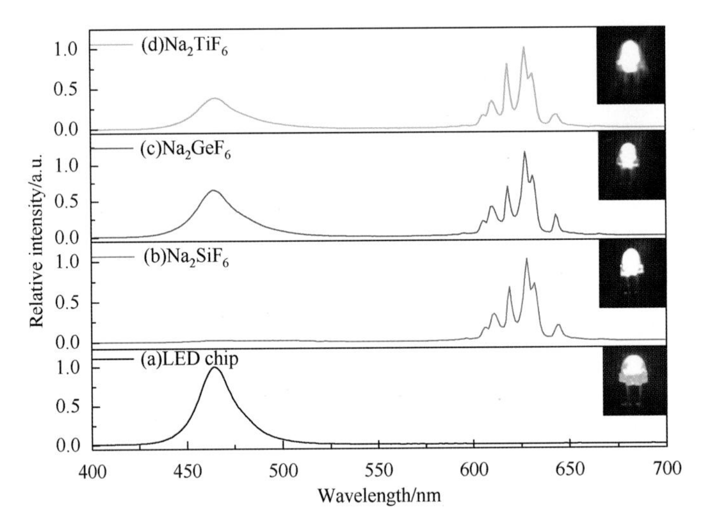


Fig.2-35 EL spectra of (a) the GaN-LED chip, the red LEDs based on (b) $Na_2SiF_6:Mn^{4+}$, (c) $Na_2GeF_6:Mn^{4+}$ and (d) $Na_2TiF_6:Mn^{4+}$ under 20 mA current excitation. The inserted figures are the images of the red LEDs

Fig.2-36 is the EL spectra of w-LEDs fabricated with or without NXFM. The white light is obtained by combining the blue emission of the LED chip and the yellow emission of YAG. Since the emission of YAG in the red regions is very weak, the as-obtained w-LED shows a high CCT (8440 K) and a low $R_a$ (78.4). With the introduction of NXFM, red emission peaks can be observed obviously at 609 nm, 617 nm, 627 nm, 631 nm and 642 nm respectively. These peaks are due to the $^2E_g \rightarrow {}^4A_2$ of $Mn^{4+}$ in NXFM. The emission of the LED chip gets weaker with the introduction of the red-emitting phosphor. Among the four curves, the red emission intensity of the w-LED based on $Na_2SiF_6:Mn^{4+}$ is the strongest. The intense white light can be observed by naked eyes from these w-LEDs when they are excited with a 20 mA current. The relative performances of the w-LEDs are summarized in Table 2-5. The warm w-LED based on $Na_2SiF_6:Mn^{4+}$ shows excellent optical performance, CCT = 3930 K, $R_a$ = 92.3, LE = 69.2 lm/W. These results demonstrate that the red phosphor has potential applications in warm w-LEDs.

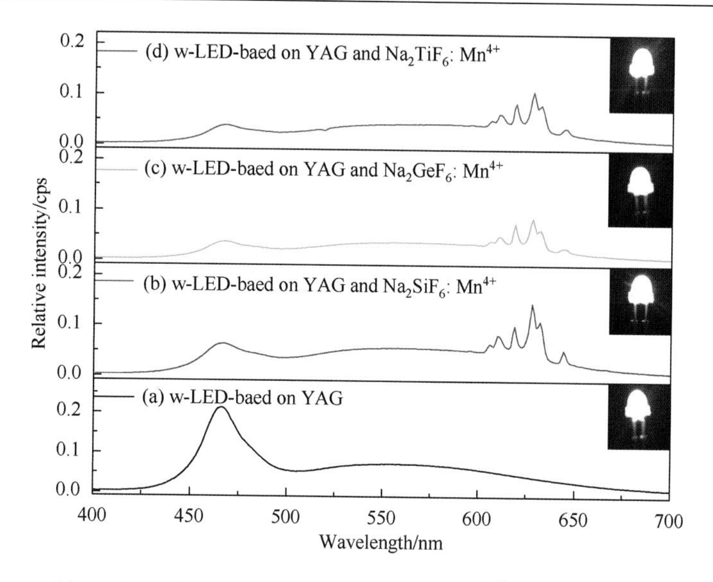


Fig.2-36 EL spectra of the w-LEDs based on (a) YAG, (b) $Na_2SiF_6:Mn^{4+}$ and YAG, (c) $Na_2GeF_6:Mn^{4+}$ and YAG, (d) $Na_2TiF_6:Mn^{4+}$ and YAG under 20 mA current excitation. The inserted figures are the images of the w-LEDs

**Table 2-5 Performance of the w-LEDs coated with (a) YAG, (b) YAG and $Na_2SiF_6$: $Mn^{4+}$, (c) YAG and $Na_2GeF_6$: $Mn^{4+}$, (d) YAG and $Na_2TiF_6$: $Mn^{4+}$ at 20 mA forward current**

| w-LEDs | CCT/K | $R_a$ | LE/(lm/W) | CIE $(x, y)$ |
|---|---|---|---|---|
| a | 8440 | 78.4 | 78.31 | (0.286, 0.314) |
| b | 3930 | 92.3 | 63.21 | (0.389, 0.397) |
| c | 3554 | 92.1 | 46.71 | (0.410, 0.412) |
| d | 3841 | 90.7 | 34.21 | (0.396, 0.409) |

In summary, $Na_2XF_6:Mn^{4+}$ red phosphors were prepared by the co-precipitation method, and their structure, morphology and luminescent properties were investigated. Among all the as-prepared NXFM samples, red phosphor $Na_2SiF_6:Mn^{4+}$ shows the strongest absorption band in the blue-light region and the strongest red emission with appropriate CIE coordinates ($x = 0.684$, $y = 0.315$). The w-LED based on $Na_2SiF_6:Mn^{4+}$ and commercial YAG emits intense white light with lower CCT and higher $R_a$. These results suggest that $Na_2SiF_6:Mn^{4+}$ phosphor is a promising red phosphor for w-LEDs.

#### 2.1.3.5 $K_2XF_6$: $Mn^{4+}$ (X = Ti, Si, Ge)

In this section, we propose an optimized precipitation method to prepare a series of red phosphors $K_2XF_6:Mn^{4+}$ (X = Ti, Si, Ge) with a low concentration of HF at mild temperature. All the obtained samples exhibit intensive red light with high color-purity under the blue light excitation. w-LEDs fabricated with $K_2XF_6:Mn^{4+}$ (marked as KXFM) red phosphors display warm white light with lower CCT and higher $R_a$.

Red phosphors KXFM have been synthesized by the co-precipitation approach with $H_2O_2$ as a reductant in the HF solution. In the current work, KXFM phosphors were prepared via an optimized co-precipitation method without $H_2O_2$. Pan et al. thought high HF concentrations might help to form a high concentration of $Mn^{4+}$ in such hexafluoride phosphors. In the lower content of the HF solution, a large amount of $H_2O$ may improve the hydrolyzation of $MnF_6^{2-}$. In this preparation process, acetic acid glacial is used to reduce HF and $H_2O$ consumption. $MnF_6^{2-}$ is obtained by the decomposition of $KMnO_4$ in the HF solution without $H_2O_2$. The chemical reaction is represented as follows:

$$(1-x)\,XO_2 + (4+x)\,HF + (2-x)\,KF + xKMnO_4 \longrightarrow K_2X_{1-x}Mn_xF_6 + 0.75xO_2 + (2+0.5x)\,H_2O \tag{2-2}$$

Fig.2-37(a) shows the XRD patterns of these red phosphors, accompanied by their corresponding standard diffraction cards. The XRD pattern of $K_2TiF_6$:$Mn^{4+}$ (KTFM) is similar to the standard diffraction card of $K_2TiF_6$ with a hexagonal structure (JCPDS card No. 08-0488, space group $P$-3$m$1, $a = b = 5.727$ nm, $c = 4.662$ nm) without any impurity peaks, which suggests that the as-prepared KTFM is of a single pure phase. With the doping of $Mn^{4+}$, a slight shift in the diffraction peaks toward higher angles can be found in Fig.2-37(b). This shift is due to the difference in ionic radius between $Mn^{4+}$(0.53 Å) and $Ti^{4+}$(0.61 Å). Furthermore, all the diffraction peaks of KTFM are very sharp, indicating that the obtained phosphor is well crystallized. Similarly, the XRD curves of $K_2SiF_6$:$Mn^{4+}$ (KSFM) and $K_2GeF_6$:$Mn^{4+}$ (KGFM) are also in agreement with their correspondingly standard diffraction cards of $K_2SiF_6$ (JCPDS card No. 07-0217, space group $Fm$-3$m$, $a = b = c = 8.133$ Å) and $K_2GeF_6$ (JCPDS card No. 07-0241, space group $P$-3m1, $a = b = 5.632$, $c = 4.668$ Å), respectively. As illustrated above, $Mn^{4+}$ can occupy the lattice site of $Si^{4+}$ and $Ge^{4+}$, due to the similar ionic radii ($Mn^{4+}$: 0.53 Å, $Si^{4+}$: 0.40 Å, and $Ge^{4+}$: 0.53 Å) and the same valance state.

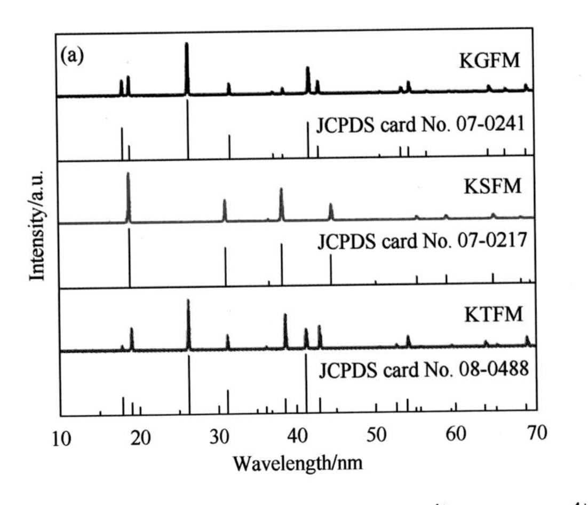


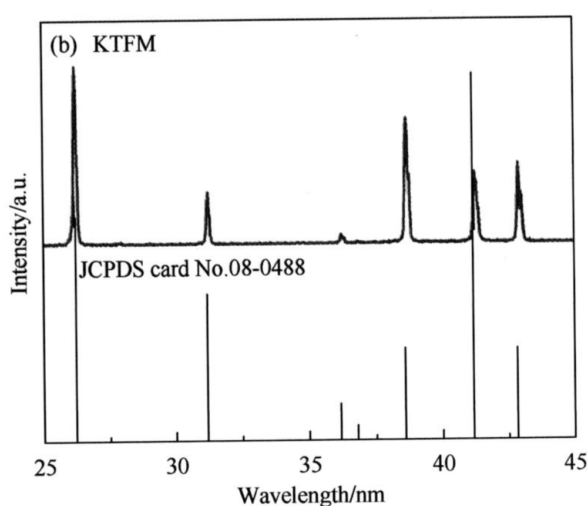


Fig.2-37 XRD patterns of $K_2TiF_6$:$Mn^{4+}$, $K_2SiF_6$:$Mn^{4+}$, $K_2GeF_6$:$Mn^{4+}$ (a) and enlarged diffraction peaks of $K_2TiF_6$:$Mn^{4+}$ in the regions from 25° to 45° (b)

The SEM images of the as-prepared yellowish products are shown in Fig.2-38. KTFM shares the hexagonal shape with a size range from 2 μm to 10 μm. Its EDS spectrum indicates that the obtained red phosphor is composed of K, Ti, F and Mn elements. The Si peak in the EDS spectrum is due to the silicon wafer which is used to measure the EDS. KSFM exhibits a smaller particle size than that of KTFM, which tends to form an octahedral shape with the particle size in the range of 2 μm to 6 μm. Moreover, the KGFM product in Fig.2-38(d) shows a hexagonal bipyramid shape with a particle size from 6 μm to 10 μm, whose size is much bigger than the former mentioned KTFM and KSFM products.

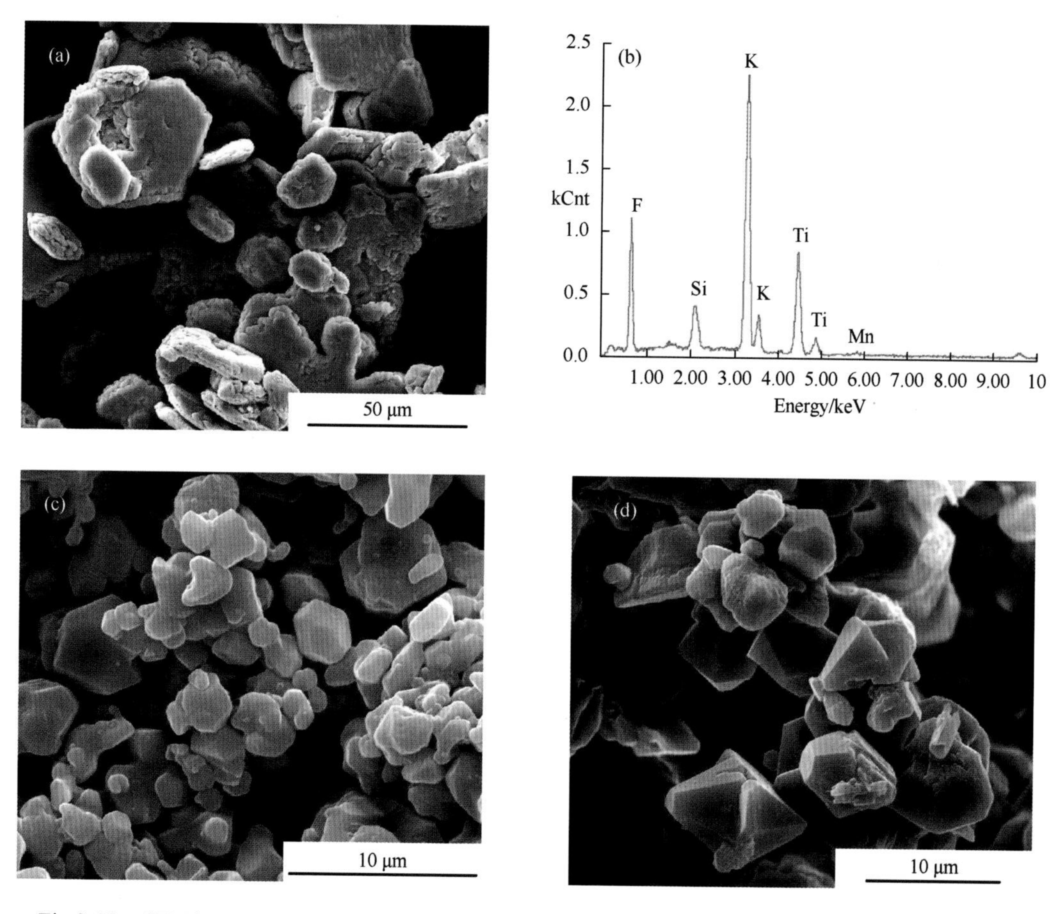


Fig.2-38 SEM images of the KTFM (a), KSFM (c) and KGFM (d) and the EDS spectrum of KTFM (b)

Fig.2-39 illustrates the excitation and emission spectra of the obtained KTFM, KSFM and KGFM red phosphors. These three phosphors are of similar excitation spectra with two intense broad absorption bands in the UV and blue regions from 300 nm to 550 nm. The former excitation band (350 nm) can be attributed to the spin-allowed between $^4A_{2g}$ and $^4T_{1g}$ transition of $Mn^{4+}$. The latter one (465 nm) is ascribed to the same transition of $^4A_{2g}$ to $^4T_{2g}$. The half peak width of the excitation band at 460 nm is about 50 nm, which is even broader than that of GaN-chip emission

(20 nm). According to the Tanabe–Sugano energy-level diagram, the excitation peak position of $Mn^{4+}$ is largely affected by different levels of the crystal field strength. The excitation spectrum of KTFM shows a little red-shift, comparing the excitation spectrum of KTFM with that of KSFM and KGFM. This result indicates that KTFM possesses a weaker crystal field strength. The excitation intensities of these phosphor increase in the order of KSFM<KGFM<KTFM, and this result is contrary to that reported by Paulusz. The excitation intensity of the phosphor is affected not only by its crystal structure but also by the content of the luminescent center. The AAS analysis is employed to confirm the concentration of $Mn^{4+}$, and the relevant results are listed in Table 2-6. The doping amounts of $Mn^{4+}$ in KSFM, KGFM and KTFM is 0.87% (mole fraction), 2.67% (mole fraction)and 3.85% (mole fraction) respectively. The sample KTFM shares the strongest excitation intensity due to the highest $Mn^{4+}$ concentration. Although the consumption of $KMnO_4$ is excessive, the concentrations of $Mn^{4+}$ in KXFM are still at a lower level. The reason may be as follows: ①$KMnO_4$ doesn't completely reduce to Mn(IV); ②$KMnO_4$ may be reduced further to Mn(II) or Mn(III), which is not incorporated into the KXFM crystal lattices.

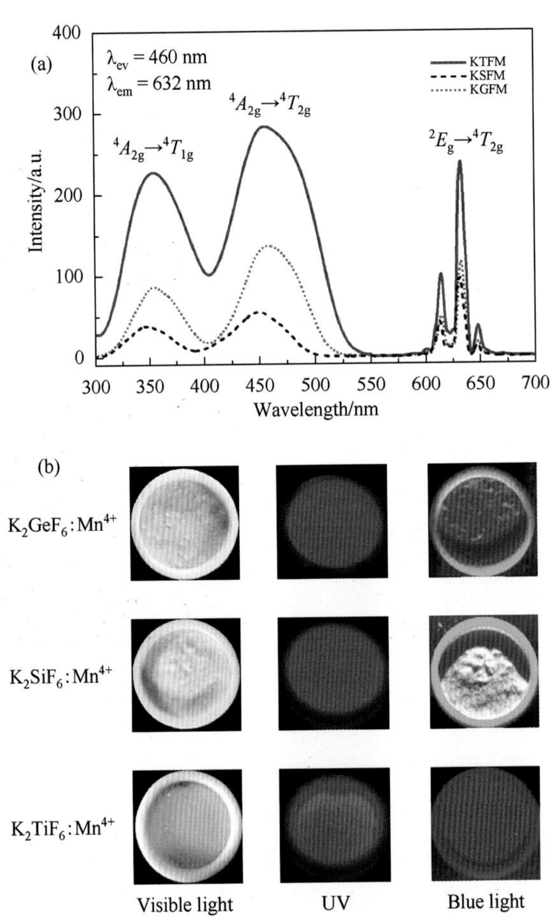


Fig.2-39　Excitation and emission spectra of KTFM, KSFM and KGFM (a) and the images of KTFM, KSFM and KGFM under visible, UV and blue light excitation (b)

Contrary to the excitation spectra, the red emissions of these phosphors between 600 nm and 650 nm belong to the spin-forbidden d-d transition from the $^2E_g \rightarrow ^4A_{2g}$ transition of $Mn^{4+}$, and the strongest emission peak is located at 632 nm. Since the energy of the $^2E_g$ state in the $d^3$ electronic configuration is independent of the crystal field, the energy of the emission peaks does not remarkably change with the different crystal field strength. The emission peaks of these phosphors located at 613 nm, 632 nm, 636 nm, 648 nm are ascribed as the transitions of the $\nu_6$ ($t_{2u}$), $\nu_6$ ($t_{2u}$), $\nu_4$ ($t_{1u}$) and $\nu_3$ ($t_{1u}$) vibronic modes respectively. The ZPL cannot be observed in current emission spectra at room temperature. Among these three phosphors, KTFM exhibits the strongest red emission with the appropriate CIE chromaticity coordinates ($x = 0.693$, $y = 0.307$), which is consistent with that of excitation spectra. The images of these red phosphors illuminated by the UV, blue and visible light are shown in Fig.2-39(b), intense red emission can be observed from these phosphors under the blue light excitation.

Fig.2-40 is the diffuse reflectance spectra of red-emitting phosphors $K_2TiF_6$ and $K_2TiF_6$:$Mn^{4+}$. Comparing with curve a, two obvious absorption bands at 350 nm and 465 nm in curve b are due to the absorption of $Mn^{4+}$ in KTFM. This result is in agreement with that of the excitation spectrum of KTFM.

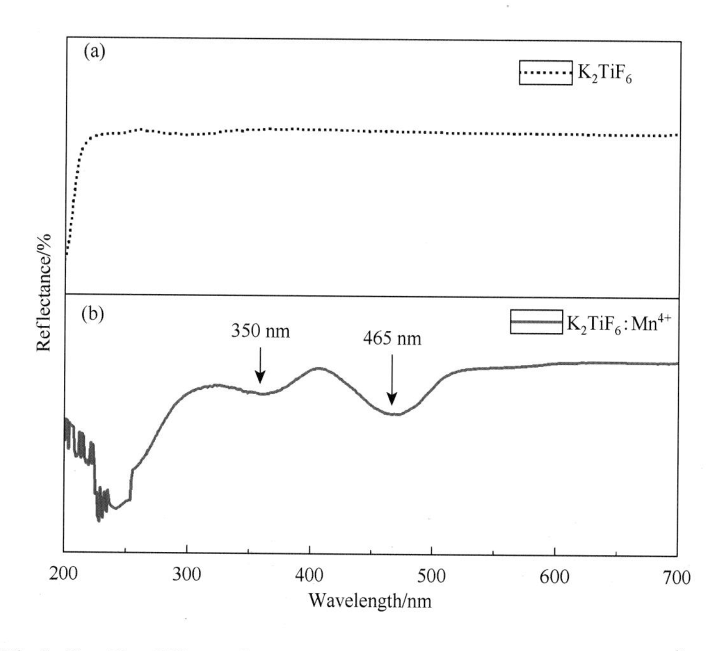


Fig.2-40 The diffuse reflectance spectra of $K_2TiF_6$ (a) and $K_2TiF_6$:$Mn^{4+}$ (b)

Fig.2-41 shows the decay curves for the $^2E_g \rightarrow ^4A_2$ transition (634 nm) of the $Mn^{4+}$ in KTFM, KSFM and KGFM red phosphors. These curves are fitted into a single-exponential function, and their lifetime $\tau$ values are 5.4 ms, 7.4 ms and 5.8 ms respectively.

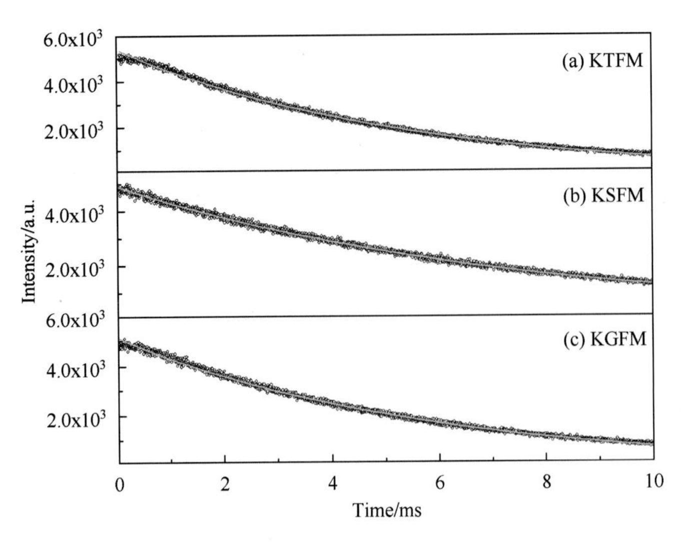


Fig.2-41　Decay curves of the $Mn^{4+}$ emission in KTFM (a), KSFM (b) and KGFM (c) ($\lambda_{ex}$ = 465 nm; $\lambda_{em}$ = 634 nm)

To investigate the potential application of KXFM red phosphors, several LED devices were fabricated by coating these phosphors with or without commercial YAG on the blue GaN chips. Fig.2-42(a) is the EL spectrum of the red LED based on KTFM phosphor under an excitation current of 20 mA. The intense red emission is ascribed to the emission of KTFM, and the emission of the GaN chip cannot be observed. This result indicates that KTFM is an excellent red phosphor for a blue LED chip. Fig.2-42(b) shows the EL spectra of the w-LEDs fabricated with different amounts of KTFM. With the introduction of KTFM, the red emission of KTFM in the red region can be found. The red emission intensity of the w-LED is increasing with the quantity improvement of KTFM, and the emission intensity of the GaN chip and YAG yellow phosphor is decreasing.

The performance of these w-LEDs is listed in Table 2-6. The CCT and $R_a$ of the w-LED without KTFM phosphor are 6509 K and 76.9 respectively. With the increase of the KTFM consumption, the examined CCT of the w-LEDs drops from 5865 K to 4050 K, and $R_a$ improves

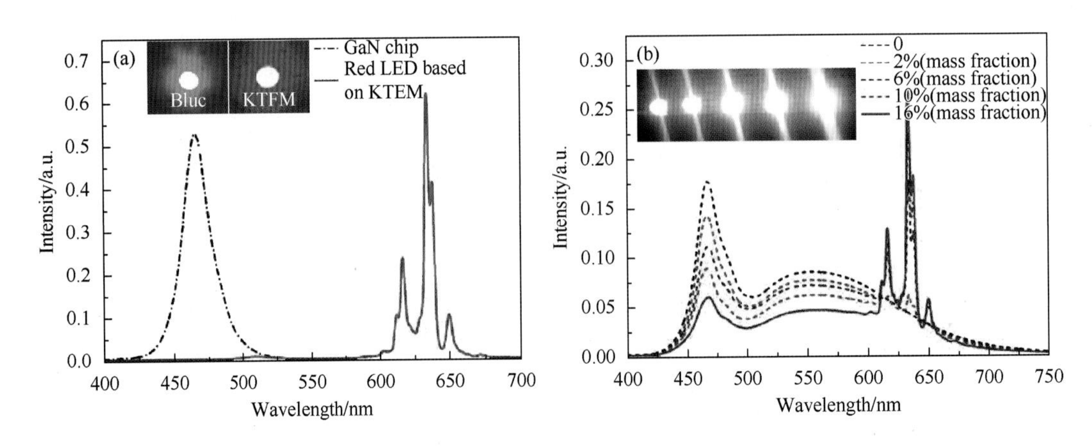

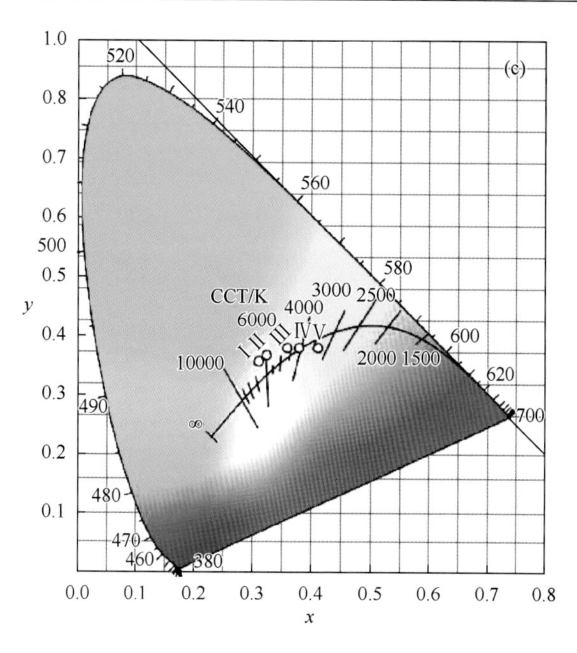


Fig.2-42 Electro-luminescent spectra of LEDs devices based on KTFM (a), the mixture of KTFM and YAG (b) under 20 mA current excitation, and their chromaticity coordinates in CIE 1931 (c)

from 79.1 to 93.4. This performance of the w-LEDs can satisfy the requirement of the indoor lighting. The corresponding CIE coordinates are shown in Fig.2-42(c), which reveals that the CCT moves to the warm white region with the increasing red color component. To obtain high $R_a$ and low CCT, the best mass ratio of epoxy resin, KTFM and YAG is 16∶1.6∶1. The efficiency of w-LED based on KTFM in this work is lower than that reported by Chen's group. This discrepancy is partially due to the small particle size of our synthesized material which results in an increase in scattering in the package and lower luminous efficacy.

**Table 2-6 AAS results of KTFM, KSFM and KGFM red phosphors**

| Names of samples | Quantity of samples/g | The molar ratio of $Mn^{4+}$/%(mole fraction) |
|---|---|---|
| KSFM | 0.0104 | 0.0087 |
| KGFM | 0.0115 | 0.0267 |
| KTFM | 0.0098 | 0.0385 |

**Table 2-7 Performance of the red LEDs and w-LEDs under 20 mA current excitation**

| Device | Phosphor | The ratio of resin, KXFM and YAG | CCT/K | $R_a$ | CIE $(x, y)$ | Efficacy/(lm/W) |
|---|---|---|---|---|---|---|
| GaN chip | | / | 10000 | −48.2 | (0.132, 0.066) | 12.8 |
| Red LED | KTFM | 1∶1∶0 | 1062 | 48.8 | (0.604, 0.311) | 29.2 |
| Red LED | KSFM | 1∶1∶0 | 1012 | 29.7 | (0.577, 0.281) | 25.1 |

continued

| Device | Phosphor | The ratio of resin, KXFM and YAG | CCT/K | $R_a$ | CIE ($x$, $y$) | Efficacy/(lm/W) |
|---|---|---|---|---|---|---|
| Red LED | KGFM | 1∶1∶0 | 2163 | −84 | (0.344, 0.180) | 17.5 |
| w-LED 1 | YAG | 16∶0∶1 | 6509 | 76.9 | (0.309, 0.355) | 89.8 |
| w-LED 2 | YAG + KTFM | 16∶0.32∶1 | 5865 | 79.1 | (0.323, 0.366) | 84.1 |
| w-LED 3 | YAG + KTFM | 16∶0.96∶1 | 4699 | 90.4 | (0.357, 0.377) | 79.4 |
| w-LED 4 | YAG + KTFM | 16∶1.6∶1 | 4050 | 93.4 | (0.377, 0.378) | 70.3 |
| w-LED 5 | YAG + KTFM | 16∶2.56∶1 | 3286 | 86.4 | (0.410, 0.378) | 56.1 |
| w-LED 6 | YAG + KSFM | 16∶3.2∶1 | 4205 | 82.7 | (0.382, 0.416) | 64.2 |
| w-LED 7 | YAG + KGFM | 16∶3.2∶1 | 4304 | 77.1 | (0.331, 0.23) | 32.1 |

Fig.2-43 exhibits the EL spectra of single red LEDs and w-LEDs based on KSFM or KGFM. The red emission at 634 nm in Fig.2-43(a) is due to the emission of the phosphor. Compared with Fig.2-42(a), the emission of GaN chip at 460 nm can be found. This result shows that the absorption of KSFM and KGFM is weaker than that of KTFM, which is in agreement with that of the excitation spectra in Fig.2-39(a). The corresponding w-LEDs based on KSFM and KGFM have been fabricated, and their performance is listed in Table 2-7.

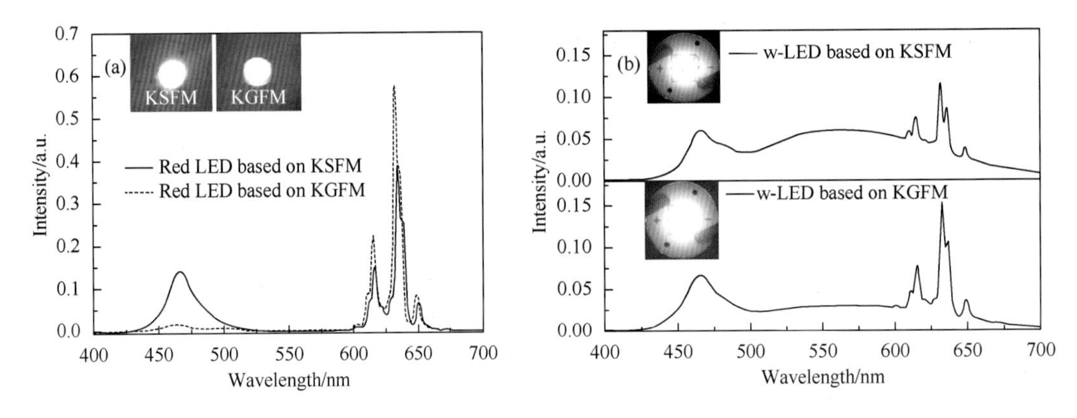


Fig.2-43 EL spectra of single red LEDs (a) and w-LEDs (b) based on $K_2SiF_6$:$Mn^{4+}$ or $K_2GeF_6$:$Mn^{4+}$ under 20 mA current excitation

In summary, a series of red phosphors KXFM were prepared by a one-step precipitation method with a low-concentration of HF. Among all the as-prepared KXFM samples, red phosphor KTFM shows the strongest absorption band in the blue-light region and the most intensive narrowband red emission around 632 nm. The w-LEDs based on the YAG-KTFM system exhibit intense white-light emission with lower CCT and higher $R_a$, compared with w-LED based on YAG.

### 2.1.3.6 $Cs_2TiF_6$:$Mn^{4+}$

Fig.2-44 shows the representative morphology and composition results of the obtained $Cs_2TiF_6$:$Mn^{4+}$ product examined from SEM and EDS measurements. In Fig.2-44(a), a close-up view shows that the sample is composed of many hexagonal prism-shaped crystals with clear edges and corners, as well as tiny particles on its smooth surfaces, indicating the product has been well crystallized. Closely view the tips of a typical hexagonal prism in Fig.2-44(b), its height is about 30 μm and two parallel symmetrical hexagons are base surfaces with a side length of about 6 μm. The corresponding EDS spectrum in Fig.2-44(c) confirms the existence of F, Ti,

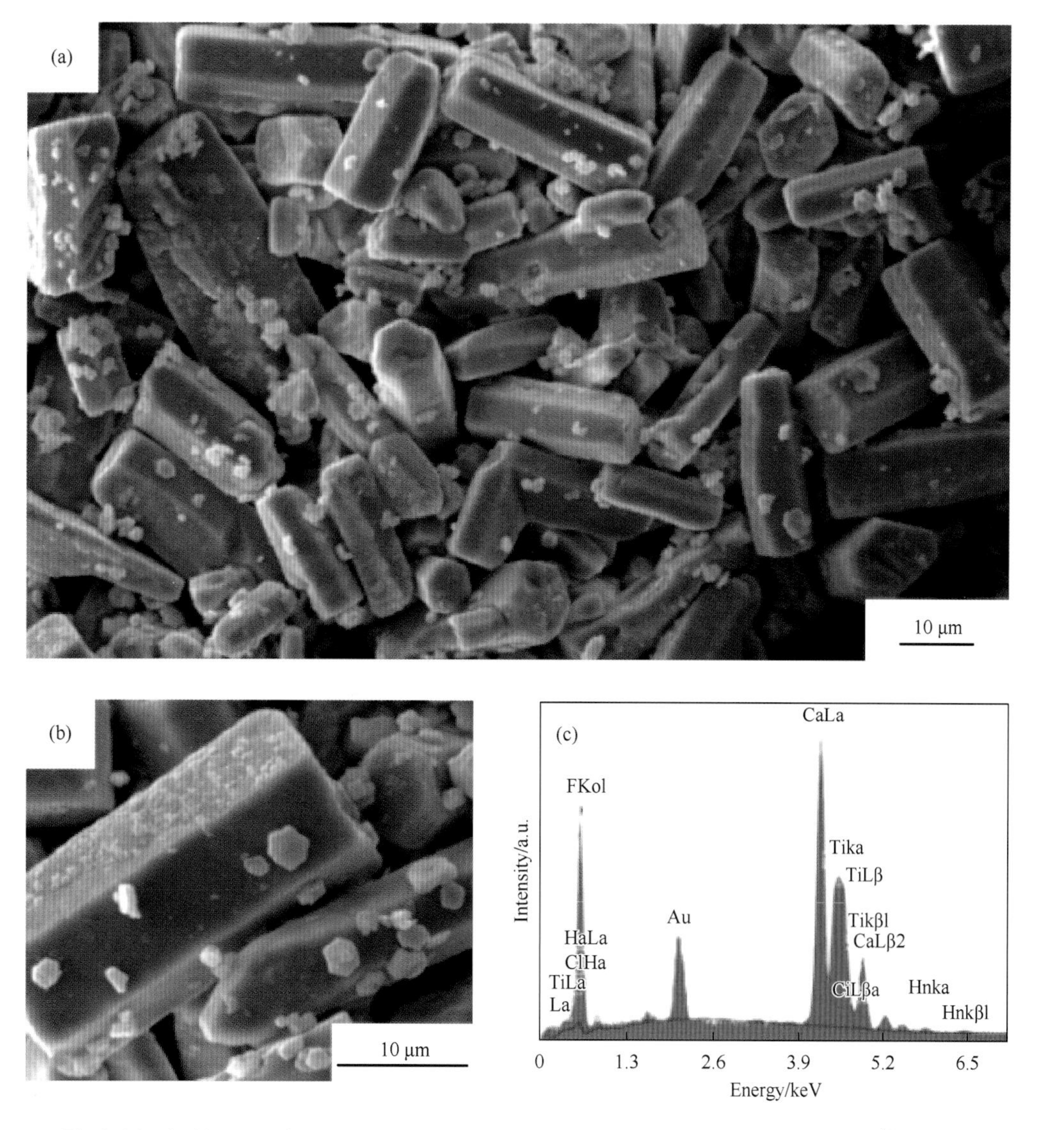


Fig.2-44 (a, b) SEM pictures and (c) EDS spectrum of the as-synthesized $Cs_2TiF_6$:$Mn^{4+}$ product

Cs and Mn elements, as well as the absence of O, suggesting that $Mn^{4+}$ has successfully occupied the lattice site of $Ti^{4+}$ and no $MnO_2$ is produced.

The doping amount of $Mn^{4+}$ in this sample, measured from the AAS system is 3.97%. Of course, with the various molar ratios between $TiO_2$ and $K_2MnF_6$, different doping amounts of $Mn^{4+}$ in the $Cs_2TiF_6$ matrix can be obtained. The results are displayed in Table 2-8. Moreover, the prepared $Cs_2TiF_6$:$Mn^{4+}$ powders are found to be a bright yellow color observed by naked eyes.

**Table 2-8 AAS results of $Cs_2TiF_6$:$Mn^{4+}$ red phosphors prepared from the different molar ratios between $TiO_2$ and $K_2MnF_6$**

| Samples | Molar ratios of $TiO_2$:$K_2MnF_6$ | Doping amount of $Mn^{4+}$/%（mole fraction） |
|---|---|---|
| 1 | 40：1 | 1.33 |
| 2 | 20：1 | 2.21 |
| 3 | 10：1 | 3.97 |
| 4 | 5：1 | 7.94 |
| 5 | 2.5：1 | 14.02 |

More detailed crystal structural studies on the fabricated $Cs_2TiF_6$:$Mn^{4+}$ product along with the pure $Cs_2TiF_6$ matrix and $K_2MnF_6$ source were carried out by XRD, and the results are exhibited in Fig.2-45. In curve a, all peaks belong to the hexagonal $K_2MnF_6$ crystal, implying $K_2MnF_6$ is pure single-phase. Comparing curves 2b with curve c, we can find all the pattern peaks can be well indexed to the space group *P*-3*m*1 of the hexagonal $Cs_2TiF_6$ crystal (JCPDS card No. 51-0612, $a = b = 6.167$ Å, $c = 4.999$ Å) and no traces of $K_2MnF_6$ residual and other secondary phase were identified. This indicates the cation exchange result of $Mn^{4+}$ substituting for $Ti^{4+}$ does not alter the crystal structure of the $Cs_2TiF_6$ matrix. That is because $Mn^{4+}$ not only has

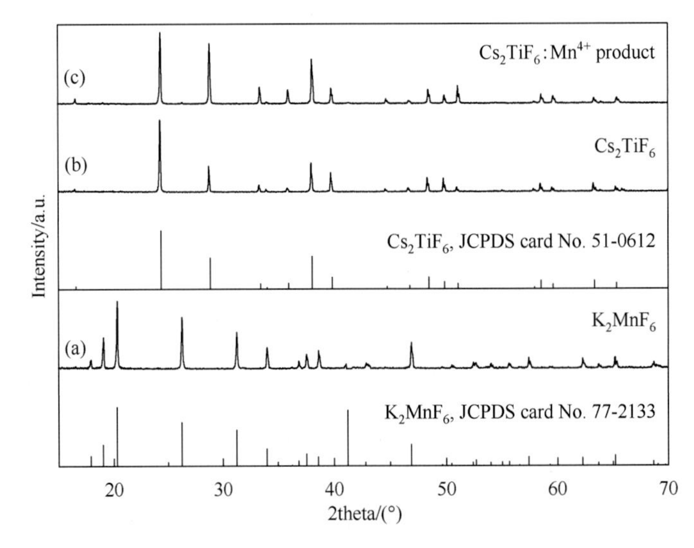


Fig.2-45 Representative XRD patterns of (a) $K_2MnF_6$, (b) $Cs_2TiF_6$ matrix and (c) $Cs_2TiF_6$:$Mn^{4+}$ product

the identical valence state but also a similar ionic radius as $Ti^{4+}$ (0.530 Å, CN = 6 vs. 0.605 Å, CN = 6), which is beneficial for $Mn^{4+}$ replacing the octahedral core site of $Ti^{4+}$ to coordinate with six $F^-$ forming stable $MnF_6^{2-}$ octahedron in the $Cs_2TiF_6$ crystal field. In this crystal structure, each $Ti^{4+}$ coordinates with six $F^-$ to form a regular $TiF_6^{2-}$ octahedron and each $Cs^+$ is at the center of twelve neighboring $F^-$.

It has been well accepted that the un-doped fluoride host absorbs little in the visible region whilst $Mn^{4+}$-doped fluoride phosphor presents broad absorption and intense emission in the blue and red regions, respectively. Fig.2-46(a) shows the excitation and DRS results of the above $Cs_2TiF_6$:$Mn^{4+}$ product. The excitation spectrum, monitored at 632 nm, is composed of two broad bands with peaks at 363 nm and 474 nm, which are located in a larger wavelength area than those of the previously reported $Na_2TiF_6$:$Mn^{4+}$ and $K_2TiF_6$:$Mn^{4+}$ phosphors. That is because the excitation peak position is highly dependent on the crystal field strength. $Cs_2TiF_6$:$Mn^{4+}$ has the same $TiF_6^{2-}$ ion group as $Na_2TiF_6$:$Mn^{4+}$ and $K_2TiF_6$:$Mn^{4+}$ while $Cs^+$ presents a much larger radius than those of $Na^+$ and $K^+$, which means $Cs_2TiF_6$ possesses larger unit cell volume than $Na_2TiF_6$ and $K_2TiF_6$, leading to its weaker crystal field strength and obvious red-shift excitation wavelength. Furthermore, the two broad peaks can be attributable to the spin-allowed transitions of $Mn^{4+}$ from the ground state $^4A_{2g}$ to excited states $^4T_{1g}$ and $^4T_{2g}$, respectively. The DRS result noticeably reveals a strong absorption in the same region with a maximum at 474 nm, which is perfectly in agreement with the above strongest excitation wavelength at 474 nm.

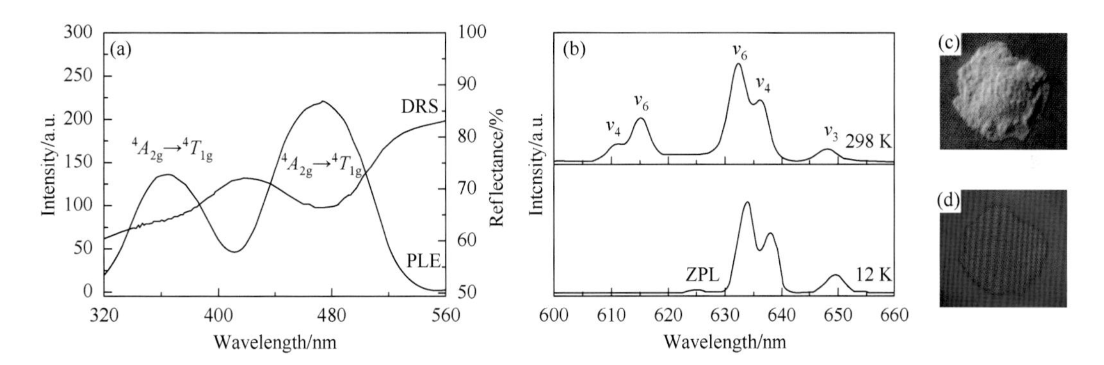


Fig.2-46 $Cs_2TiF_6$:$Mn^{4+}$ phosphor: (a) PLE and DRS results, (b) PL spectra examined at room and low temperature, respectively, photographs under (c) natural light and (d) 460 nm blue light

The emission spectra of $Cs_2TiF_6$:$Mn^{4+}$ phosphor were examined at different temperatures. Fig.2-46(b) shows two representative emission spectra recorded at room and low temperature, respectively. At $T$ = 25 ℃, the red emissions originate from the spin-forbidden $^2E_g\rightarrow^4A_{2g}$ transition. The five peaks are due to the transitions of anti-Stokes $v_4$, $v_6$ and Stokes $v_6$, $v_4$ and $v_3$ vibronic modes respectively, in which ZPL emission is hardly observed. At low temperatures, ZPL emission can be observed, and anti-Stokes lines of $v_4$ and $v_6$ disappear. These features are the same as those observed in $K_2GeF_6$:$Mn^{4+}$ phosphor. The characteristics of blue-excitation and red-emission for the $Cs_2TiF_6:Mn^{4+}$ phosphor meet well with the emission wavelength of the blue-chip and the red component need of YAG type w-LED. The sample shows a uniform yellow tint under natural light illumination [Fig.2-46(c)], which emits brilliant red light under the blue light (460 nm) excitation [Fig.2-46(d)].

The influence of the temperature on the PL properties is shown in Fig.2-47. Obviously, with the increasing temperature, it can be found that the emission peak position does not shift. Up to 100 ℃, over 100% of the integrated emission intensity can be preserved compared with that at 25 ℃.

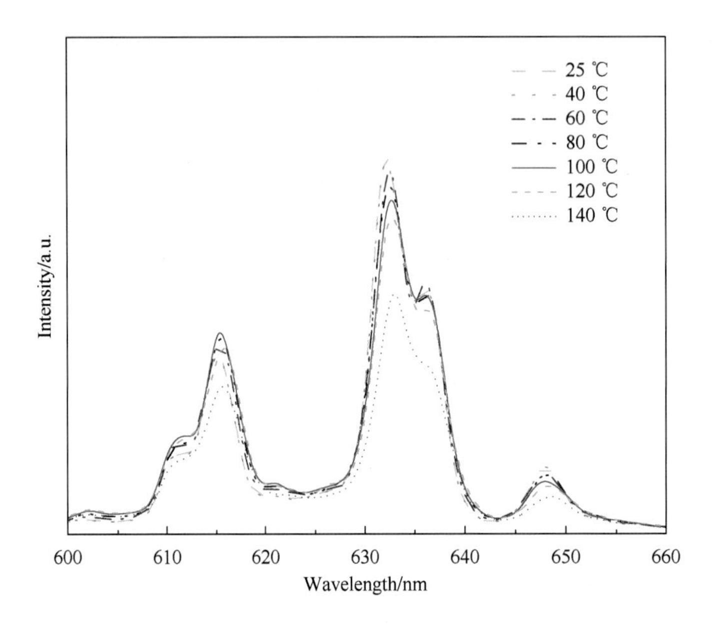


Fig.2-47　Temperature-dependent thermal luminescent spectra of $Cs_2TiF_6:Mn^{4+}$ red phosphor and relative intensity of the emission spectrum by integrating the spectral area

Furthermore, the influence of the doping amounts of $Mn^{4+}$ on the excitation and emission spectra has been investigated and the results are shown in Fig.2-48. $Cs_2TiF_6:Mn^{4+}$ samples doped with different contents of $Mn^{4+}$ are of similar excitation and emission spectra. When the content of $Mn^{4+}$ is about 4%, the as-prepared $Cs_2TiF_6:Mn^{4+}$ shares the strongest intensities of excitation and emission among these phosphors.

Due to the excellent properties of the new red phosphor, that is $Cs_2TiF_6:Mn^{4+}$, it is necessary to investigate the illumination performance of its devices. A series of w-LEDs were fabricated by merging YAG, various amounts of $Cs_2TiF_6:Mn^{4+}$, and epoxy resin on the blue GaN chips (450 nm). Their CIE coordinates are labeled in or close to the white light region in CIE 1931 color spaces as color points of i-vi in Fig.2-49(b). Photographs of the lighting w-LEDs are shown in the insert image of Fig.2-49(a). Consistent with the decrease of CCT, a noticeable warmer tone of the emitting light can be observed, owing to the increase of red light

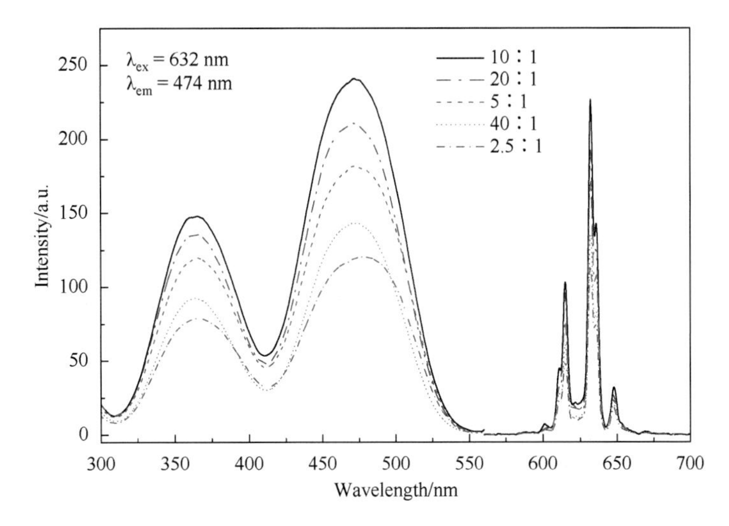


Fig.2-48 The excitation and emission spectra of $Cs_2TiF_6$:$Mn^{4+}$ red phosphors obtained from the different molar ratio between $TiO_2$ and $K_2MnF_6$

in the emission spectrum. Their EL spectra reconfirm the sharp emission lines of $Cs_2TiF_6$:$Mn^{4+}$ phosphor and more red components with the lower CCTs. Fig.2-49(c) shows the bright red light of a typical red LED based on $Cs_2TiF_6$:$Mn^{4+}$ phosphor.

Table 2-9 compares the CCT, $R_a$ and LE of these w-LEDs. Evidently, by adding $Cs_2TiF_6$:$Mn^{4+}$ phosphor into the YAG powders, warm white light with $R_a$>80 and CCT<4000 K can be readily achieved. The $R_a$ and CCT of the warmest w-LED are 80.1 and 3272 K with an excellent LE of 124.6 lm/W. Notably, the LE value is much higher than the previous reports.

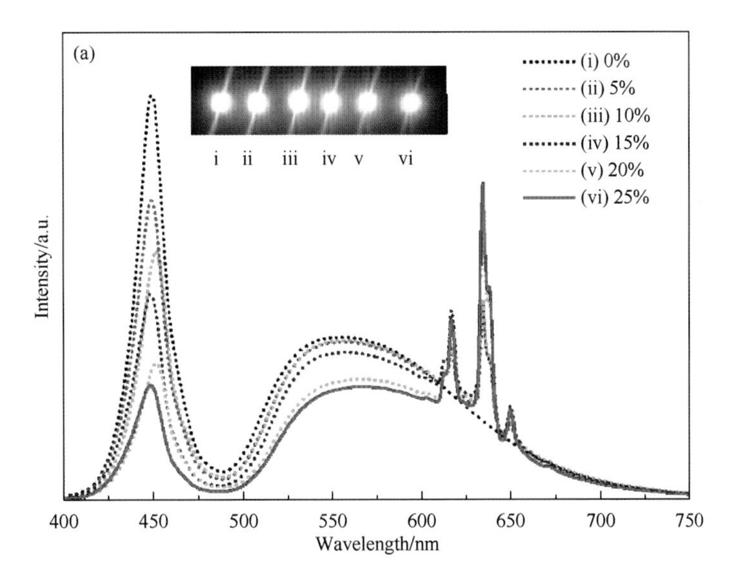

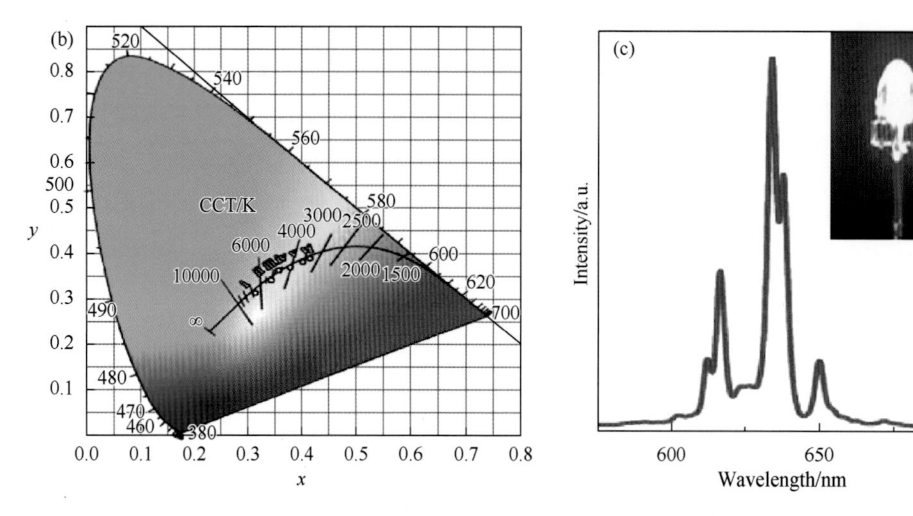


Fig.2-49 (a) EL spectra and photographs of five w-LEDs, (b) their CIE coordinates in CIE 1931 color spaces, (c) EL spectrum with a photograph of a red LED. All the LEDs were recorded under 20 mA drive current

**Table 2-9 Important LED photoelectric parameters with different amount of $Cs_2TiF_6$:$Mn^{4+}$ phosphor under a current of 20 mA**

| Device | Phosphor | CTFM/%(mass fraction) | CCT/K | $R_a$ | CIE (*x*, *y*) | LE/(lm/W) |
|---|---|---|---|---|---|---|
| i | YAG | 0 | 6620 | 73.9 | (0.313, 0.314) | 153.3 |
| ii | YAG + CTFM | 5 | 5052 | 74.7 | (0.343, 0.341) | 148.8 |
| iii | YAG + CTFM | 10 | 4650 | 75.8 | (0.356, 0.363) | 143.2 |
| iv | YAG + CTFM | 15 | 4020 | 77.5 | (0.378, 0.370) | 136.5 |
| v | YAG + CTFM | 20 | 3540 | 78.8 | (0.401, 0.382) | 130.4 |
| vi | YAG + CTFM | 25 | 3272 | 80.1 | (0.415, 0.389) | 124.6 |

In conclusion, a novel $Cs_2TiF_6$:$Mn^{4+}$ red phosphor, belonging to the hexagonal *P*-3*m*1 crystal structure, has been successfully prepared through a cation exchange route. The $Cs_2TiF_6$:$Mn^{4+}$ product is well crystallized into a single-phase with hexagonal micro-rod morphology. The $Cs_2TiF_6$:$Mn^{4+}$ phosphor can absorb the broadband blue light and emit intense bright narrowband red light efficiently. Therefore, with the addition of $Cs_2TiF_6$:$Mn^{4+}$ red phosphor, obvious improvement in $R_a$ and CCT data has been achieved, resulting in the $Cs_2TiF_6$:$Mn^{4+}$ phosphor can be a potential candidate for the indoor lighting w-LEDs.

#### 2.1.3.7 $Cs_2GeF_6$:$Mn^{4+}$

In this part, red phosphor $Cs_2GeF_6$:$Mn^{4+}$ was prepared via the cation exchange method using $K_2MnF_6$ as $Mn^{4+}$ source. The optical properties of this phosphor have been investigated in detail. The as-prepared $Cs_2GeF_6$:$Mn^{4+}$ shows intense red emission with high color-purity under the blue-light excitation. The optical performance of w-LEDs can be improved by introducing red phosphor $Cs_2GeF_6:Mn^{4+}$.

Fig.2-50 is the XRD pattern of as-prepared $Cs_2GeF_6:Mn^{4+}$, which is consistent with the corresponding JCPDS card of $Cs_2GeF_6$ (No. 76-1398). This result indicates that the obtained phosphor $Cs_2GeF_6:Mn^{4+}$ shares the single phase with the cubic structure (space group of *Fm*-3*m*, $a = b = c = 8.99$ Å) of $Cs_2GeF_6$. A little doping of $Mn^{4+}$ does not change the crystal structure of this $Cs_2GeF_6$ host. Fig.2-50(b) illustrates the crystal structure of $Cs_2GeF_6$ unit cell viewed in [110] direction. In the present crystal, each $Ge^{4+}$ is coordinated with six $F^-$ to form a regular $GeF_6^{2-}$ octahedron. Because $Mn^{4+}$ not only has the same valence state as $Ge^{4+}$ but also the identical ionic radius with $Ge^{4+}$ (0.53 Å, CN = 6 vs. 0.53 Å, CN = 6), $Mn^{4+}$ in $Cs_2GeF_6:Mn^{4+}$ will occupy the site of $Ge^{4+}$ in the center of an octahedron.

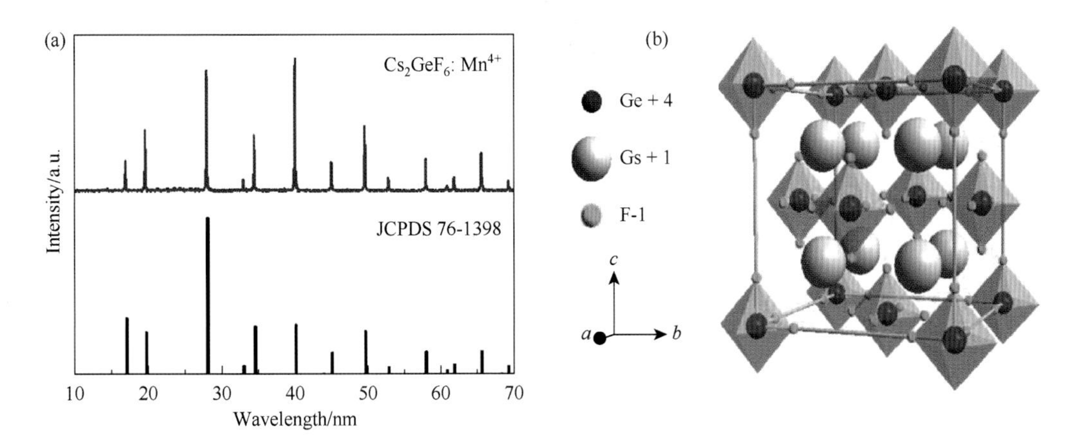


Fig.2-50 (a) XRD pattern of the $Cs_2GeF_6:Mn^{4+}$ and (b) crystal structure of $Cs_2GeF_6$ unit cell viewed in [110] direction

The morphology and composition of the obtained $Cs_2GeF_6:Mn^{4+}$ were examined by the SEM and EDS analysis, and the representative results are shown in Fig.2-51. The obtained products exhibit an irregular morphology with smooth surfaces. Closely inspecting the particle size distribution among them, we can find that the $Cs_2GeF_6:Mn^{4+}$ product displays an apparent larger size (30 μm). Compared with $Cs_2GeF_6$: $Mn^{4+}$ prepared by chemically etching Ge shots, obvious edges and corners can be found from our sample. This result shows that the sample has been well-crystallized during the cation exchange process. Fig.2-51(b) is the corresponding EDS spectrum. All peaks belong to F, Ge, Cs, Mn elements respectively. No other element can be observed. This result indicates that the Mn element has been indeed doped into the matrix lattice to occupy the lattice site of Ge. The peak of the Si element in Fig.2-51(b) is due to the silicon wafer used during the measurement of SEM and EDS. Moreover, the absence of oxygen peak in these EDS spectra implies that there is no $MnO_2$ produced during this preparation process.

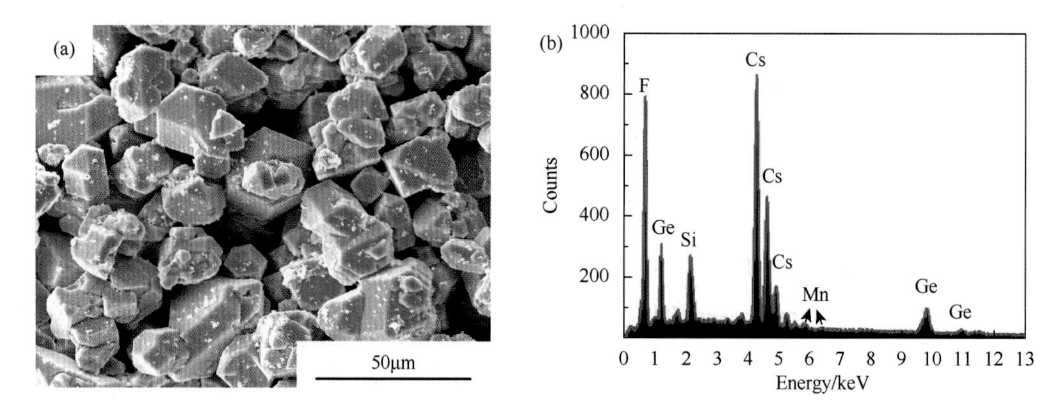


Fig.2-51 (a) SEM image and (b) EDS spectrum of the $Cs_2GeF_6:Mn^{4+}$

The DRS of $Cs_2GeF_6$ and $Cs_2GeF_6:Mn^{4+}$ are shown in Fig.2-52. Comparing with the $Cs_2GeF_6$ host, $Cs_2GeF_6:Mn^{4+}$ exhibits two obvious absorption bands at 350 nm and 455 nm in Fig.2-52(b), which are due to the absorption of $Mn^{4+}$ in $Cs_2GeF_6:Mn^{4+}$. Fig.2-52(c) is the excitation spectrum of $Cs_2GeF_6:Mn^{4+}$ by monitoring 633 nm emission. Two broad excitation bands peaking at 350 nm and 455 nm, which can be assigned to the spin-allowed transitions of $Mn^{4+}$ from the ground state $^4A_{2g}$ to the excited states $^4T_{1g}$ and $^4T_{2g}$, respectively. This result is in agreement with that of the DRS of $Cs_2GeF_6:Mn^{4+}$. The strongest excitation band of $Cs_2GeF_6:Mn^{4+}$ is located at 455 nm, which just meets with the emission wavelength ($\lambda = 450$ nm) of the GaN blue LED chip.

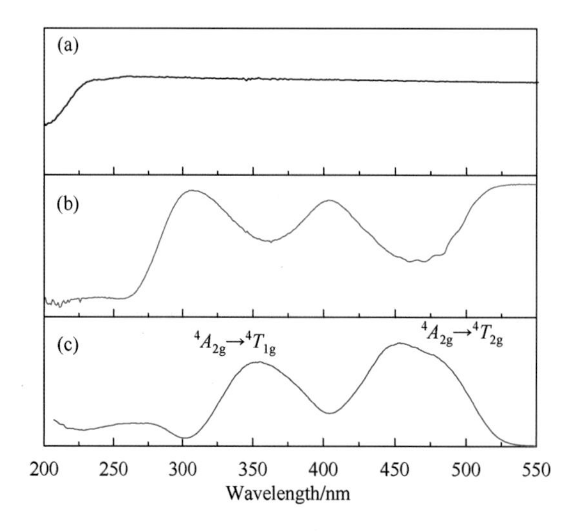


Fig.2-52 Diffuse reflectance spectra of (a) $Cs_2GeF_6$ and (b) $Cs_2GeF_6:Mn^{4+}$, (c) excitation spectrum of $Cs_2GeF_6:Mn^{4+}$

Fig.2-53 is the emission spectra of $Cs_2GeF_6:Mn^{4+}$ excited by 455 nm light at 293 K and 12 K. The as-prepared sample exhibited intense red emission at 12 K, which is due to the spin-forbidden ${}^2E_g \to {}^4A_{2g}$ transition of $Mn^{4+}$. The series of emission peaks located at 615 nm, 624 nm, 633 nm, 636 nm, 649 nm are ascribed as the transitions of the $v_6$, zero phonon line (ZPL), $v_6$, $v_4$, $v_3$ vibronic modes, respectively. Weaker ZPL emission of the phosphor can be found from the emission spectrum at 12 K. At 293 K, the ZPL emission of $Cs_2GeF_6:Mn^{4+}$ disappears, and a new emission peak ($v_4$) at 611 nm appears in the emission spectrum. This result is in agreement with that reported by S. Adachi. The bright red light can be observed from this sample excited by the 460 nm blue light. The CIE chromaticity coordinates according to the emission spectrum of $Cs_2GeF_6:Mn^{4+}$ are calculated to be $x = 0.69$, $y = 0.31$, which are very close to the NTSC standard values for red. To ensure the LE of the as-prepared phosphor, the luminescent property of commercial $K_2TiF_6:Mn^{4+}$ red phosphor was investigated, compared with that of $Cs_2GeF_6:Mn^{4+}$. Commercial $K_2TiF_6:Mn^{4+}$ shares one similar emission spectrum with that of $Cs_2GeF_6:Mn^{4+}$, except for a little blue-shift of emission positions. The emission intensity of as-prepared $Cs_2GeF_6:Mn^{4+}$ is about 1.14 times higher than that of commercial $K_2TiF_6:Mn^{4+}$.

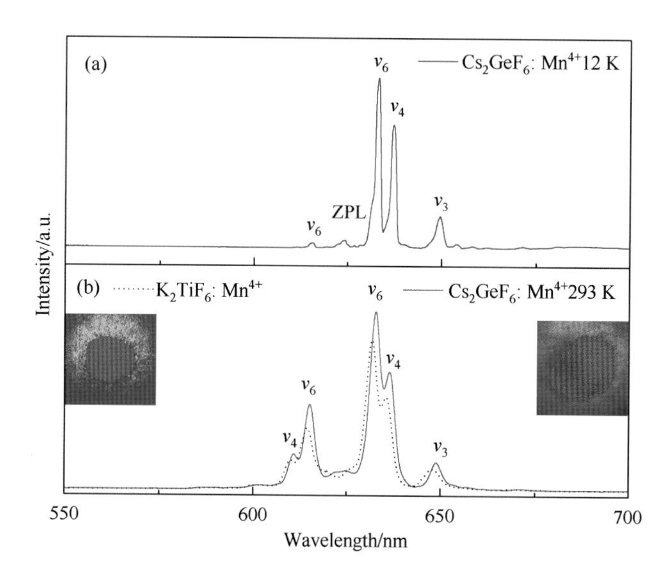


Fig.2-53 Emission spectra of the $Cs_2GeF_6:Mn^{4+}$ and $K_2TiF_6:Mn^{4+}$ (a) at 12 K and (b) 293 K

To investigate the influence of the $Mn^{4+}$ content on the PL properties, a series of $Cs_2GeF_6:Mn^{4+}$ prepared with the different molar ratios of $GeO_2$ and $K_2MnF_6$ were synthesized, and their emission spectra are shown in Fig.2-54. All the emission spectra are of similar shapes with five main emission peaks from 600 nm to 650 nm. With the increase of the $K_2MnF_6$ consumption, the emission intensity of $Cs_2GeF_6:Mn^{4+}$ is increasing. When the molar ratio of $GeO_2$ and $K_2MnF_6$ is 10 ∶ 1, the as-obtained sample is of the strongest emission intensity.

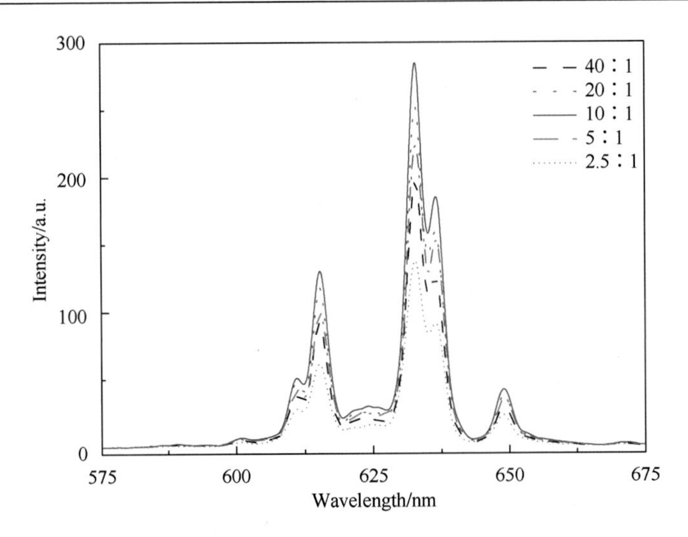


Fig.2-54 Emission spectra of $Cs_2GeF_6:Mn^{4+}$ prepared with the different molar ratios of $GeO_2$ and $K_2MnF_6$

To confirm the concentration quenching phenomenon of $Mn^{4+}$ in $Cs_2GeF_6:Mn^{4+}$, atomic absorption spectrophotometer is adopted to measure the relative concentration of $Mn^{4+}$, and the results are presented in Table 2-10. With the increase of the $K_2MnF_6$ consumption, the concentration of $Mn^{4+}$ in $Cs_2GeF_6:Mn^{4+}$ also increases.

**Table 2-10 Relative concentration of $Mn^{4+}$ in $Cs_2GeF_6:Mn^{4+}$ prepared with the different molar ratios of $GeO_2$ and $K_2MnF_6$**

| Sample | Molar ratios of $GeO_2$ and $K_2MnF_6$ | $x$ |
|---|---|---|
| 1 | 50 : 1 | 0.0220 |
| 2 | 25 : 1 | 0.0468 |
| 3 | 10 : 1 | 0.0878 |
| 4 | 5 : 1 | 0.1449 |
| 5 | 2.5 : 1 | 0.2104 |

Fig.2-55 is the concentration dependence of the relative emission intensity of $Mn^{4+}$ $^4A_{2g}\rightarrow^4T_{2g}$ transition in $Cs_2GeF_6:Mn^{4+}$. When the concentration of $Mn^{4+}$ is 8.78%, the emission intensity is the strongest among these phosphors.

Since LEDs work usually at a temperature below 150 ℃, the thermal stability and the temperature dependence on the luminescent efficiency are important parameters for phosphors. As shown in Fig.2-56, TG and DSC curves of $Cs_2GeF_6:Mn^{4+}$ show that its thermal decomposition temperature ($T_d$) is about 450 ℃. Such high decomposition temperature suggests that this phosphor has high thermal stability and it is enough to be applied in w-LED devices.

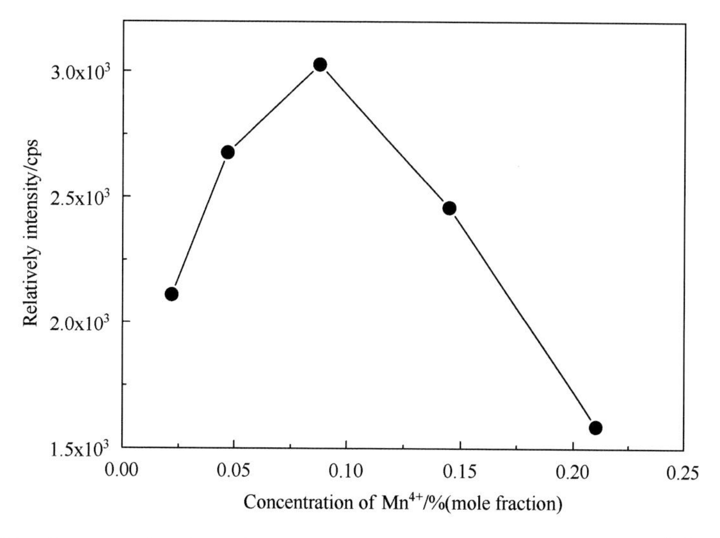


Fig.2-55 Concentration dependence of the relative emission intensity of $Mn^{4+}$ ${}^4A_{2g}\rightarrow{}^4T_{2g}$ transition in $Cs_2GeF_6:Mn^{4+}$

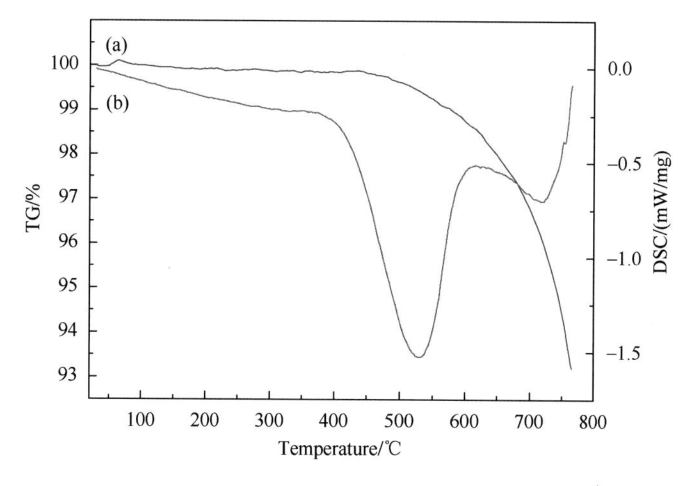


Fig.2-56 (a) TG and (b) DSC curves of $Cs_2GeF_6:Mn^{4+}$

Fig.2-57 exhibits the emission spectra of $Cs_2GeF_6:Mn^{4+}$ under different temperatures. All the emission peaks are in the same positions with the strongest emission peak at 633 nm. No obvious emission peak position shift can be found. The inserted figure in Fig.2-57 is the temperature dependence of the relative emission intensity of $Cs_2GeF_6:Mn^{4+}$. The integrated intensity of the sample at 140 ℃ is still higher than that of the sample at room temperature, which indicates that this red phosphor shares high thermal stability, and this result is in agreement with that of $Rb_2SiF_6:Mn^{4+}$.

Fig.2-58 shows the decay curve for the ${}^2E_g\rightarrow{}^4A_{2g}$ transition (633 nm) of the $Mn^{4+}$ in $Cs_2GeF_6:Mn^{4+}$ red phosphor. This decay curve is well fitted into a single-exponential function, and the lifetime $\tau$ value of $Cs_2GeF_6:Mn^{4+}$ is 8.8 ms. This result complements the experimental data of previous reports prepared by other methods.

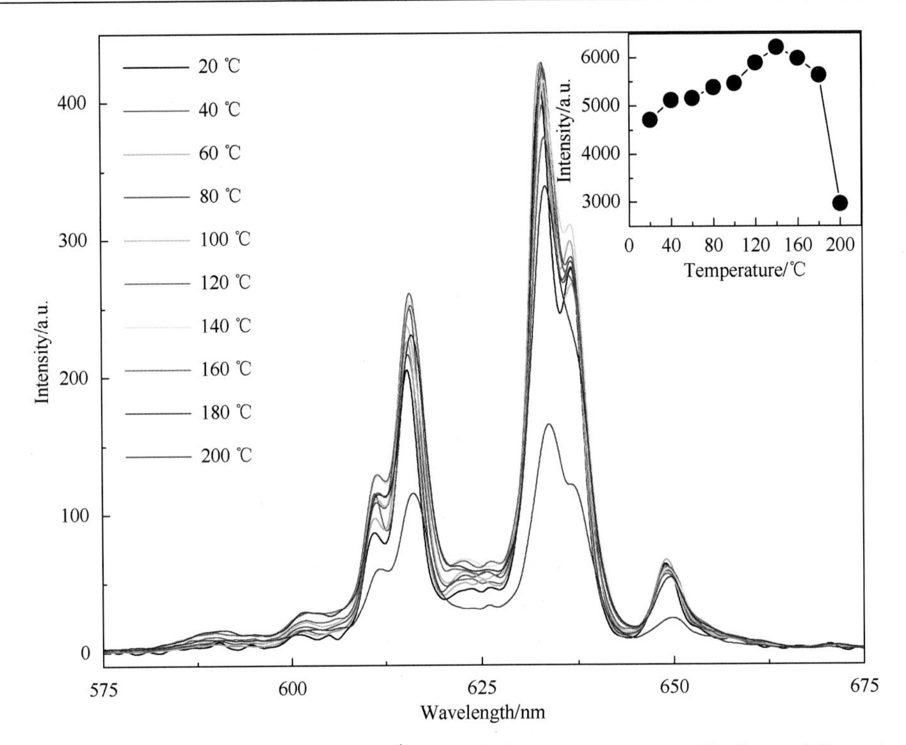


Fig.2-57　Emission spectra of $Cs_2GeF_6:Mn^{4+}$ under different temperatures. The inserted figure is the relationship between the temperature and the relative emission intensity

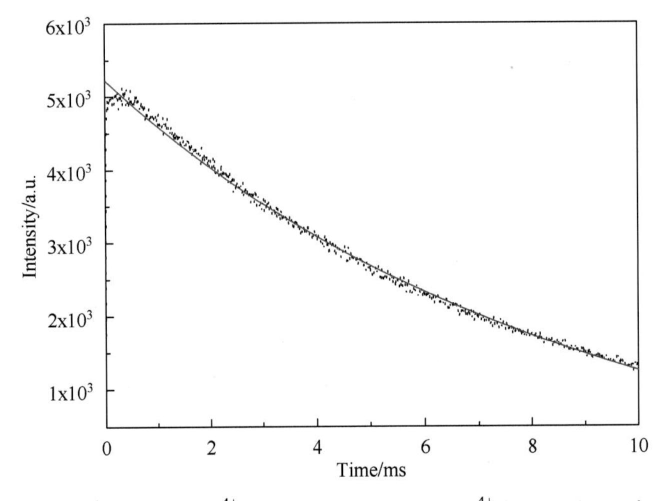


Fig.2-58　Decay curve of the $Mn^{4+}$ emission in $Cs_2GeF_6:Mn^{4+}$($\lambda_{ex}$ = 460 nm; $\lambda_{em}$ = 633 nm)

Fig.2-59 is the EL spectra of the LED chip and red LED based on $Cs_2GeF_6:Mn^{4+}$ under 20 mA current excitation. The emission of the GaN LED chip is at 450 nm, and its half-peak width is about 20 nm. Compared with the spectrum a, the emission of the LED chip gets weak in the spectrum b, indicating that $Cs_2GeF_6:Mn^{4+}$ can efficiently absorb the emission of the LED chip. The red emission between 600 nm and 650 nm is due to the $^2E_g \rightarrow ^4A_{2g}$ transition of $Mn^{4+}$. The bright red light can be observed from this red LED based on $Cs_2GeF_6:Mn^{4+}$.

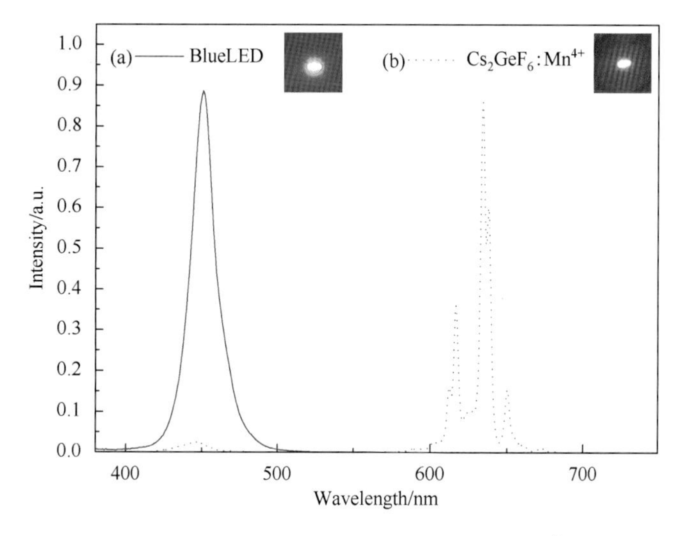


Fig.2-59 EL spectra of (a) LED chip and (b) red LED based on $Cs_2GeF_6$:$Mn^{4+}$ under 20 mA current excitation

Fig.2-60 is the EL spectra of w-LEDs with YAG. The broadband in the blue region is due to the emission of the GaN chip, and the greenish-yellow emission is due to the emission of YAG and different amounts of $Cs_2GeF_6$:$Mn^{4+}$. Since the emission of YAG in the red regions is very weak, this w-LED based on YAG exhibits high CCT (6090 K) and low $R_a$ (72.7). With the introduction of $Cs_2GeF_6$:$Mn^{4+}$, red emission peaks can be observed obviously at 615 nm, 624 nm, 633 nm, 636 nm, 649 nm, which are due to the $^2E_g \rightarrow ^4A_{2g}$ transition of $Mn^{4+}$. And the emission of the LED chip turns weaker. The related parameters of these w-LEDs are listed in Table 2-11. When the amount of $Cs_2GeF_6$:$Mn^{4+}$ is adjusted from 5% to 20%, the examined CCT of the w-LEDs drops from 5193 K to 3673 K, and $R_a$ is improved from 75.2 to 84.9.

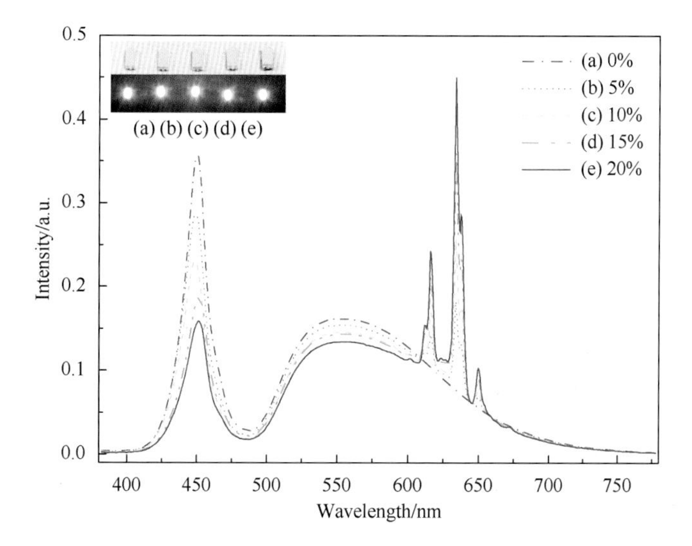


Fig.2-60 EL spectra of w-LEDs based on YAG and different amount of $Cs_2GeF_6$:$Mn^{4+}$ under 20 mA current excitation

**Table 2-11 Performance of the w-LEDs with different amount of $Cs_2GeF_6:Mn^{4+}$ at 20 mA forward current**

| Devices | $Cs_2GeF_6$: $Mn^{4+}$/%(mass fraction)* | CCT/K | $R_a$ | CIE (*x*, *y*) | LE/(lm/W) |
|---|---|---|---|---|---|
| a | 0 | 6090 | 72.7 | (0.321, 0.329) | 158.9 |
| b | 5 | 5193 | 75.2 | (0.340, 0.344) | 153.9 |
| c | 10 | 4567 | 77.2 | (0.358, 0.359) | 147.2 |
| d | 15 | 4081 | 82.0 | (0.377, 0.375) | 144.3 |
| e | 20 | 3673 | 84.9 | (0.395, 0.383) | 141.5 |

$$*\%(\text{mass fraction}) = \frac{m_{(Cs_2GeF_6:Mn^{4+})}}{m_{(\text{epoxy resin})}} \times 100\%.$$

This performance of the w-LEDs can satisfy the requirement of the indoor lighting. The corresponding CIE coordinates are shown in Fig.2-61, which reveals that the CCT moves to the warm white region with the increase of the red color component. According to the survey of *x*, *y*. warm w-LEDs based on red phosphors doped $Mn^{4+}$ with a luminous efficacy higher than

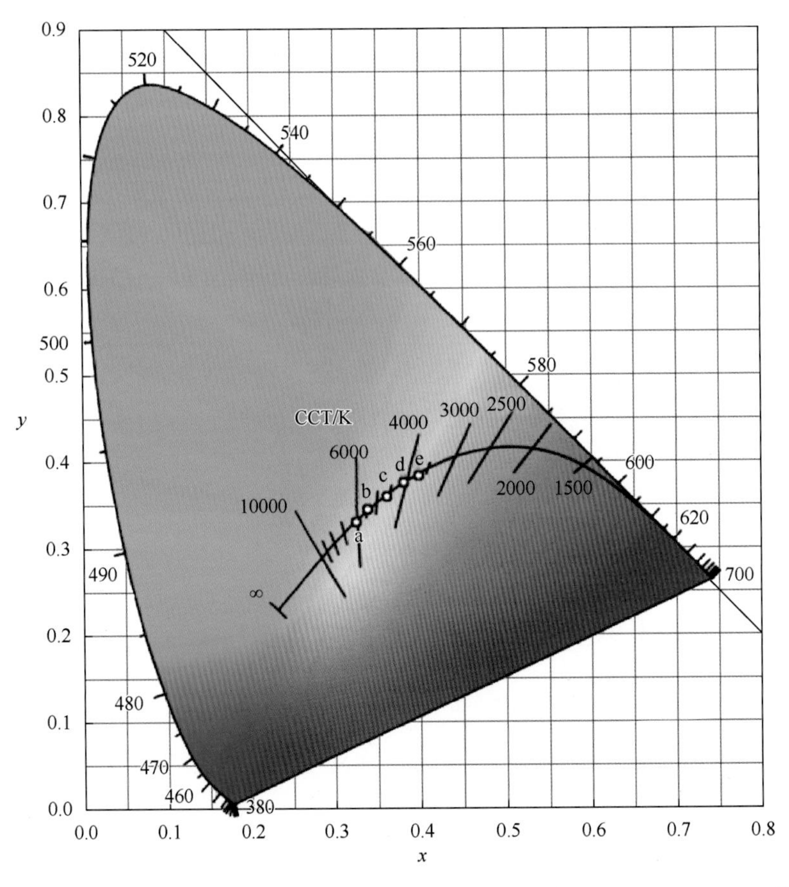


Fig.2-61 Chromaticity coordinates of w-LEDs fabricated with different amounts of $Cs_2GeF_6:Mn^{4+}$ in CIE 1931

90 lm/W had never been achieved before they fabricated the warm w-LED based on $K_2TiF_6:Mn^{4+}$ (LE = 124 lm/W) under 20 mA current excitation. In this work, we fabricated high- performance warm w-LED based on $Cs_2GeF_6:Mn^{4+}$ with low CCT (3673 K), high $R_a$ (84.9) and high LE (141.5 lm/W). The bright white light can be observed by naked eyes from this w-LED when it is excited with 20 mA current. These results demonstrate that this red phosphor shares excellent PL properties, which has potential application in warm w-LEDs.

In conclusion, red phosphor $Cs_2GeF_6:Mn^{4+}$ was prepared by the cation exchange method, and their structure, morphology and optical properties were investigated. The as-prepared phosphor $Cs_2GeF_6:Mn^{4+}$ with high thermal stability shows the intense and broad excitation band in the blue-light region, and red emission with appropriate CIE coordinates ($x = 0.69$, $y = 0.31$). The w-LEDs fabricated with $Cs_2GeF_6:Mn^{4+}$ and commercial YAG exhibit intense white light with good optical performances (CCT = 3673 K, $R_a = 84.9$, LE = 141.5 lm/W). Hence, $Cs_2GeF_6:Mn^{4+}$ is a promising red phosphor for warm w-LEDs.

#### 2.1.3.8 $Cs_2ZrF_6$: $Mn^{4+}$

In this part, we design a new $Mn^{4+}$-activated red phosphor, $Cs_2ZrF_6:Mn^{4+}$ for the first time, with its synthesis route and optical properties. The product presents sharp red line emission, excellent thermal stability and high color purity under the blue light illumination. The influence of the doping amount on luminescence properties has been studied comprehensively. With the introduction of $Cs_2ZrF_6:Mn^{4+}$ into YAG yellow phosphor, the obtained w-LED exhibits warmer white light than that made from only one YAG component.

Crystal structural studies on the pre-set $Mn^{4+}$ source, together with the as-prepared $Cs_2ZrF_6:Mn^{4+}$ product are carried out by XRD, and the representative patterns along with the corresponding standard diffraction cards of $K_2MnF_6$ and $Cs_2ZrF_6$ are displayed in Fig.2-62. In curve a, all the diffraction peaks can be well indexed to the hexagonal $K_2MnF_6$ crystal, indicating itself pure single-phase. Curve b is the XRD pattern of the cation exchange reaction product of $Cs_2ZrF_6:Mn^{4+}$, in which all the diffraction peaks are in agreement with the standard JCPDS card of $Cs_2ZrF_6$ (No.74-0173, space group $P\text{-}3m1$, $a = b = 6.41$ Å, $c = 5.01$ Å), indicating this product is of a single pure phase. In its crystal structure, each $Zr^{4+}$ coordinates with six $F^-$ to form a regular $ZrF_6^{2-}$ octahedron and one $Cs^+$ ion locates at the center of twelve neighboring $F^-$ anions. Furthermore, it is easy to find that no traces of $K_2MnF_6$ residual and other secondary phases are identified from curve b, indicating the introduction of $Mn^{4+}$ into the $Cs_2ZrF_6$ matrix does not change the crystal structure of $Cs_2ZrF_6$. This is because $Mn^{4+}$ and $Zr^{4+}$ not only have the same valence state but also share a similar ionic radius (0.53 Å, CN = 6 vs. 0.72 Å, CN = 6), which results in the replacement of $Mn^{4+}$ for the octahedral core site of $Zr^{4+}$, to coordinate with six $F^-$ forming stable $MnF_6^{2-}$ octahedron in the crystal field of $Cs_2ZrF_6$.

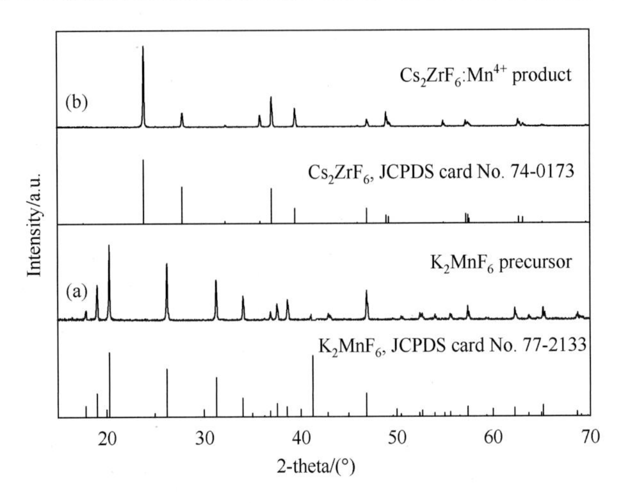


Fig.2-62　Representative XRD patterns of (a) $K_2MnF_6$ precursor and (b) $Cs_2ZrF_6$:$Mn^{4+}$ product

It is well known that the fluoride matrix doped with little $Mn^{4+}$ exhibits broad absorption bands in the visible region. To further verify $Mn^{4+}$ ions have been doped into the $Cs_2ZrF_6$ matrix, the DRS of the $Cs_2ZrF_6$ host and $Cs_2ZrF_6$:$Mn^{4+}$ product are investigated and the results are shown in Fig.2-63. Comparing curve a with curve b, it can be found that the $Cs_2ZrF_6$:$Mn^{4+}$ product reveals two strong absorption bands peaking at 366 nm and 475 nm in the UV and blue regions, respectively. This phenomenon perfectly accords with the typical property of $Mn^{4+}$ ion activated red phosphor. Therefore, we can conclude that the cation exchange reaction for the replacement of $Mn^{4+}$ for $Zr^{4+}$ in the $Cs_2ZrF_6$ matrix is successful.

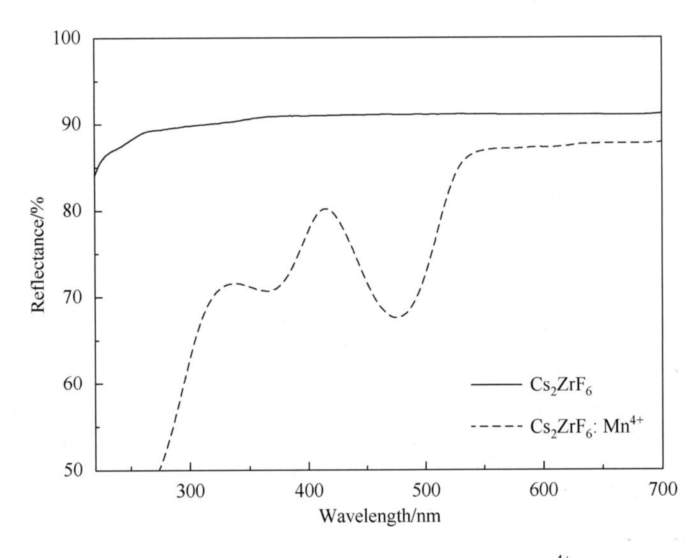


Fig.2-63　Diffuse reflectance spectra of $Cs_2ZrF_6$ matrix and $Cs_2ZrF_6$:$Mn^{4+}$ product examined at 298 K

More detailed morphological and compositional results of the obtained $Cs_2ZrF_6$:$Mn^{4+}$ product examined from SEM and EDS measurements are presented in Fig.2-64. This product is composed of a lot of irregular particulates with smooth surfaces, clear edges and corners, which implies that our sample has been well crystallized. Closely viewing the particle size distribution among them, it can be found that the $Cs_2ZrF_6$:$Mn^{4+}$ product displays an apparent large size (30 μm). From Fig.2-64(b), the corresponding EDS spectrum, we can recognize the existence of F, Zr, Cs and Mn elements as well as the absence of O. This result suggests that the Mn element has indeed successfully occupied the lattice site of Zr and no $MnO_2$ is produced during the entire preparation process. Moreover, observed by naked eyes, the sample shows a soft khaki tint under the natural light illumination [Fig.2-64(c)] whist emits brilliant red light under the blue-light (460 nm) excitation [Fig.2-64(d)].

Fig.2-64 $Cs_2ZrF_6$:$Mn^{4+}$ red phosphor: (a) SEM image, (b) EDS spectrum with typical photos illuminated under (c) the natural light and (d) 460 nm blue light, (e) crystal structure

As illustrated above, it has been well accepted that the un-doped fluoride host absorbs little in the visible region, whilst $Mn^{4+}$-doped fluoride phosphor presents broad absorption and intense emission in the blue and red regions respectively. Fig.2-65 shows the PL properties of the $Cs_2ZrF_6:Mn^{4+}$ phosphor. Noticeably, the excitation spectrum monitored at 631 nm presents two broad excitation bands placing in the same absorption region as the DRS illustrated above with peaks at 364 nm and 475 nm respectively. From the Tanabe-Sugano diagram, we know that they belong to the spin-allowed and parity-forbidden transitions of $Mn^{4+}$ from the ground state $^4A_{2g}$ to excited states $^4T_{1g}$ and $^4T_{2g}$. The stronger absorption peak locates in a larger wavelength area than those of the previously reported $Cs_2TiF_6:Mn^{4+}$ and $Cs_2GeF_6:Mn^{4+}$ red phosphors. This can be attributed to the different crystal field strength among the three phosphors. $Zr^{4+}$ ion has a much larger radius than those of $Ti^{4+}$ and $Ge^{4+}$ (0.72 Å vs.0.605 Å vs.0.53 Å), which results in $Cs_2ZrF_6$ possessing the largest unit cell volume, the weakest crystal field strength, the lowest excitation energy, and hence the largest excitation wavelength among them.

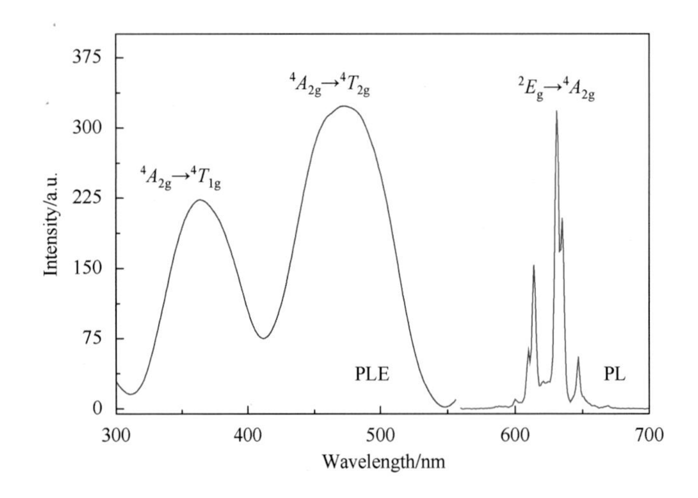


Fig.2-65 Excitation and emission spectra of the $Cs_2ZrF_6:Mn^{4+}$ product recorded at 293 K ($\lambda_{em}$ = 631 nm, $\lambda_{ex}$ = 475 nm)

Under the blue light illumination, the PL spectrum is found to be positioned in the red region. This is a typical property of $Mn^{4+}$ activated fluoride red phosphor, and the red emissions originate from the spin-forbidden $^2E_g$→$^4A_{2g}$ transition. By close inspection of the tips on this spectrum, we can find that it is composed of five strong peaks, which are attributable to the anti-Stokes $\nu_4$, $\nu_6$ and Stokes $\nu_6$, $\nu_4$ and $\nu_3$ vibronic emissions respectively, and the ZPL emission is hardly observed. Therefore, the properties of wide blue-excitation and sharp red-emission for the as-prepared $Cs_2TiF_6:Mn^{4+}$ phosphor meet well with the emission wavelength of the blue GaN chip and the red component need of YAG type w-LED.

For comparison, the emission spectra of the $Cs_2ZrF_6:Mn^{4+}$ phosphors synthesized with different mole ratios of $H_2ZrF_6$ to $K_2MnF_6$: (i) 80 ∶ 1, (ii) 60 ∶ 1, (iii) 40 ∶ 1 and (iv) 20 ∶ 1 are recorded under the 460 nm light excitation and the results are shown in Fig.2-66(a). Five emission peaks located at 610 nm, 614 nm, 631 nm, 635 nm and 647 nm strongly indicate that the emitted light is red, which are identified as the anti-Stokes and Stokes transitions between $^2E_g$ and $^4A_{2g}$ levels of $Mn^{4+}$ respectively. Among all the emission peaks, the strongest peak locates at 631 nm. This result furtherly verifies that the $Mn^{4+}$ ions are active centers in $Cs_2ZrF_6$ to emit red light under the blue light excitation. The doping amount of $Mn^{4+}$ in this sample is 3.30%, and the quantum efficiency (QY) for the as-prepared sample with 3.30% is 56.9%. Of course, with the various molar ratios between $H_2ZrF_6$ and $K_2MnF_6$, different amounts of $Mn^{4+}$ are doped into the $Cs_2ZrF_6$ matrix. The results are displayed in Table 2-12.

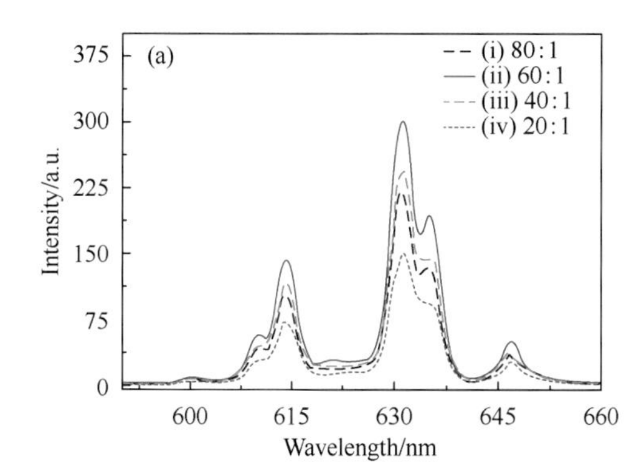


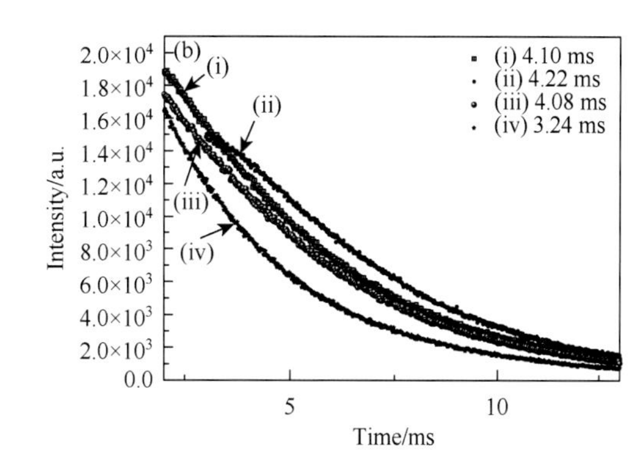


Fig.2-66 (a) PL spectra and (b) decay curves for the $Cs_2ZrF_6:Mn^{4+}$ products prepared from different molar ratios of $H_2ZrF_6$ to $K_2MnF_6$: (i) 80 ∶ 1, (ii) 60 ∶ 1, (iii) 40 ∶ 1 and (iv) 20 ∶ 1($\lambda_{em}$ = 631 nm, $\lambda_{ex}$ = 475 nm)

**Table2-12 AAS results of $Cs_2ZrF_6:Mn^{4+}$ red phosphors prepared from the different molar ratios of $H_2ZrF_6$ to $K_2MnF_6$**

| Samples | Molar ratios of $H_2ZrF_6$: $K_2MnF_6$ | Doping amounts of $Mn^{4+}$ /%(mole fraction) |
|---|---|---|
| i | 80:1 | 2.25 |
| ii | 60:1 | 3.30 |
| iii | 40:1 | 4.68 |
| iv | 20:1 | 6.65 |

Fig.2-66(b) presents the PL decay properties of the emitting state $^2E_g$ of $Mn^{4+}$ doping in the $Cs_2ZrF_6$ matrix with different molar ratios of 2.25%, 3.30%, 4.68% and 6.65% examined at room temperature. The curves are all well fitted into the single-exponential function with lifetime $\tau$ values of 4.10 ms, 4.22 ms, 4.08 ms and 3.24 ms. The increasing lifetime of these samples should be owing to the enhanced radiative transition probability. As the concentration of $Mn^{4+}$ furtherly increases in the $Cs_2ZrF_6$ matrix, the increasing exchange interaction probability between $Mn^{4+}$ pairs induces the decrease of the decay time.

According to the results obtained above, the optimized molar ratios of $H_2ZrF_6$ to $K_2MnF_6$ is 60 ∶ 1. Consequently, this optimized sample is chosen for the following thermal behavior, temperature-dependence and w-LED performance investigations. Fig.2-67 displays the thermal behavior of this sample measured by a TG measurement. It is obvious that the $Cs_2TiF_6:Mn^{4+}$ red-emitting phosphor is chemically stable and begins to decompose at a temperature of 564 ℃.

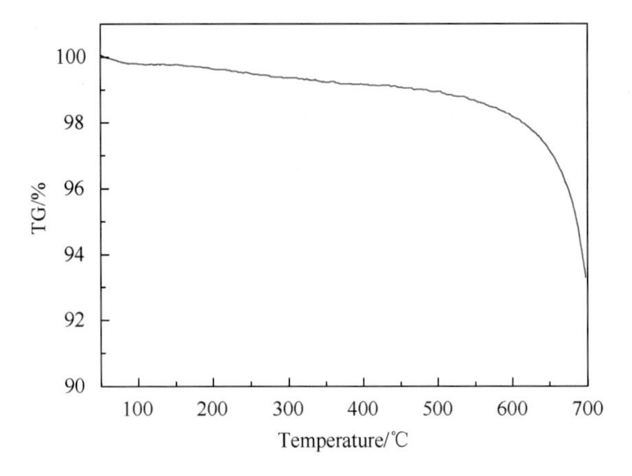


Fig.2-67 TG curve of the $Cs_2ZrF_6:Mn^{4+}$ red phosphor under $N_2$ atmosphere. The thermal stability is investigated from Perkin Elmer STA 8000 at a heating rate of 10 K/min

The influence of temperature on the PL properties of the obtained $Cs_2ZrF_6$: $Mn^{4+}$ (3.30% mole fraction) red phosphor at 20-140 ℃ under the 475 nm blue light excitation is investigated and the representative results are presented in Fig.2-68. It can be observed that with the temperature increasing from 20 ℃ to 140 ℃, the emission intensity exhibits a lower tendency because of the increasing non-radiative transition processes. Furthermore, with the temperature increasing, not only

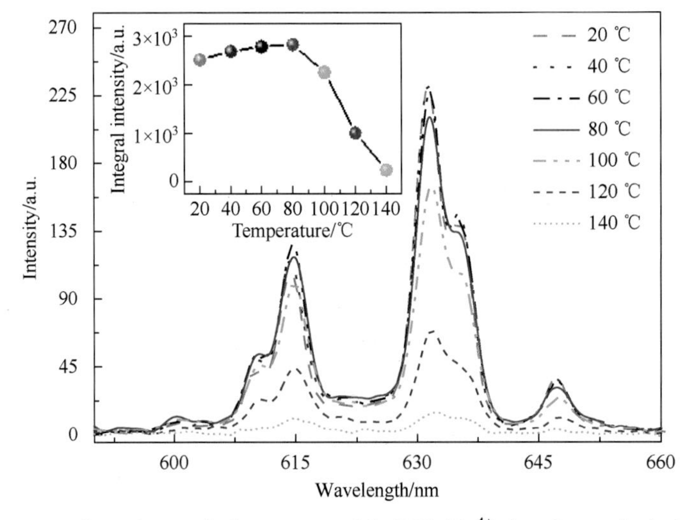


Fig.2-68 Temperature-dependent emission spectra of $Cs_2ZrF_6:Mn^{4+}$ phosphor and relative intensity of the emission spectrum by integrating the spectral area. The inserted figure is the relationship between the temperature and the relative emission intensity

the emission spectrum gradually becomes broader but also the main emission peak positions exhibit a swift red-shift. This is due to the expansion of the unit cell and the enhancement of vibration modes of $MnF_6^{2-}$ octahedral in a hot environment. This result is in agreement with Liu's report. Furthermore, the relationship between temperature and integral emission intensity is inserted in Fig.2-68, which displays considerable stability for $Cs_2ZrF_6:Mn^{4+}$ red phosphor with the temperature increase from 20 ℃ to 100 ℃. When the temperature is 80 ℃, nearly 113% of the integrated emission intensity can be achieved, compared with that at 20 ℃. Up to 100 ℃, the relative intensity still preserves about 90%.

As illustrated above, full width at half maximum for the excitation spectrum of $Cs_2ZrF_6:Mn^{4+}$ phosphor in the blue-light region is much wider than the emission band of the GaN chip (20 nm), and their emission peak locations are very close. This is why $Cs_2ZrF_6:Mn^{4+}$ phosphor can absorb the emission of GaN chip to illuminate intense red light and be used in the YAG-type w-LED possessing warm white light. From a practical point of view, it is necessary to investigate the illumination performance of the YAG-type w-LED fabricated with $Cs_2ZrF_6:Mn^{4+}$ phosphor. Fig.2-69 is the EL spectra of a series of w-LEDs fabricated by merging YAG, $Cs_2TiF_6:Mn^{4+}$, and epoxy resin on blue GaN chips recorded at 20 mA drive current. Only with a single YAG component, the obtained white light exhibits weak emissions in the red region with high CCT (5485 K) and low $R_a$ (71.1). After the addition of $Cs_2ZrF_6:Mn^{4+}$, several obvious red-emitting peaks from the as-selected phosphor in the spectra can be observed. Increasing its addition amount of $Cs_2ZrF_6:Mn^{4+}$ from 5% to 20%, a noticeable stronger tone of the emitting red light can be observed.

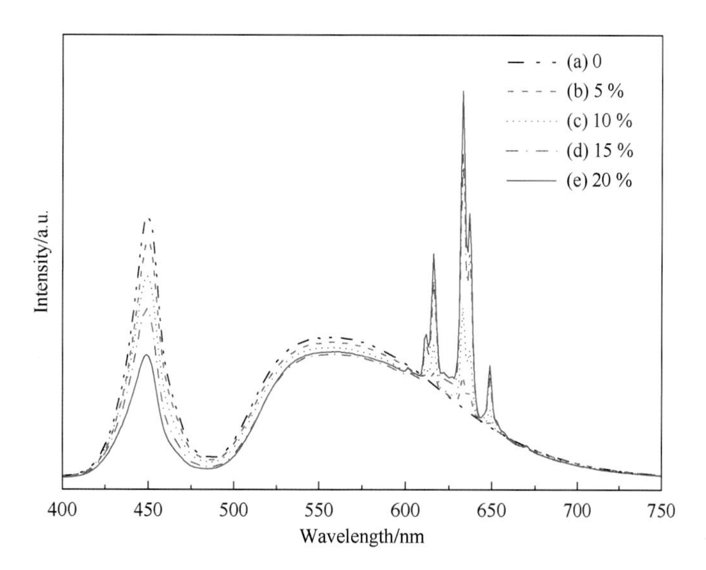


Fig.2-69 Electro-luminescent spectra and photographs of five LED devices recorded with 20 mA drive current

The CIE coordinates of these w-LED devices are labeled in or close to the white light region in CIE 1931 color spaces as color points (from a to e) in Fig.2-70. Photographs of the lighting w-LEDs are shown in the insert image of Fig.2-70. Their corresponding CCT, $R_a$ and LE data are compared, as listed in Table 2-13. The EL spectra reconfirm the sharp emission lines of $Cs_2ZrF_6:Mn^{4+}$ phosphor and more red components with lower CCTs. Furthermore, by adding the $Cs_2ZrF_6:Mn^{4+}$ phosphor into the YAG powders with its addition amount from 5% to 20%, the CCT data drops from 5060 K to 3469 K and $R_a$ improves from 71.7 to 82.4. In a word, the $R_a$ and CCT of the warmest w-LEDs are 82.4 and 3469 K with an excellent LE of 132.2 lm/W.

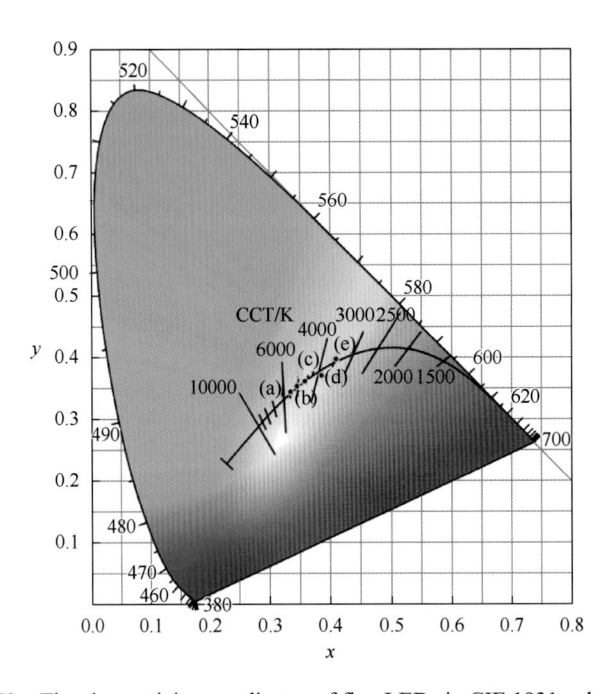


Fig.2-70　The chromaticity coordinates of five LEDs in CIE 1931 color spaces

**Table 2-13　Typical LED photoelectric parameters with different amount of $Cs_2TiF_6:Mn^{4+}$ phosphor under a current of 20 mA**

| Devices | Phosphor | CTFM/%(mass fraction) | CCT/K | $R_a$ | CIE ($x$, $y$) | LE/(lm/W) |
|---|---|---|---|---|---|---|
| a | YAG | 0 | 5485 | 71.1 | (0.333, 0.344) | 140.9 |
| b | YAG + CZFM | 5 | 5060 | 71.7 | (0.344, 0.353) | 136.3 |
| c | YAG + CZFM | 10 | 4609 | 74.3 | (0.357, 0.362) | 132.4 |
| d | YAG + CZFM | 15 | 3818 | 79.9 | (0.386, 0.371) | 129.1 |
| e | YAG + CZFM | 20 | 3459 | 82.4 | (0.410, 0.398) | 132.2 |

In conclusion, a new red phosphor $Cs_2ZrF_6:Mn^{4+}$ was prepared using a cation exchange method, and its structure, morphology and optical properties were investigated. The

$Cs_2ZrF_6$:$Mn^{4+}$ product is well crystallized into single-phase and shows broadband excitation in the blue-light region and narrowband emission in the red-light region. With the addition of $Cs_2ZrF_6$:$Mn^{4+}$ phosphor, the fabricated w-LED devices exhibit warm white light with good illumination performance (CCT = 3469 K, $R_a$ = 82.4, LE = 132.2 lm/W), implying its potential application in the indoor lighting w-LEDs.

2.1.3.9 $Rb_2TiF_6$:$Mn^{4+}$

In this section, some red phosphors $Rb_2TiF_6$:$Mn^{4+}$ (RTFM) were prepared through the ion exchange method. The PL properties of $Mn^{4+}$ in $Rb_2TiF_6$ have been investigated in detail. The obtained phosphor emits intensive red emission under the blue light excitation with high thermal-quenching resistance, implying this fluoride phosphor is a potential red component for warm w-LEDs.

Fig.2-71(a) is the XRD patterns of the obtained RTFM samples prepared with different molar ratios of $TiO_2$ to $K_2MnF_6$ (40 ∶ 1, 20 ∶ 1, 10 ∶ 1, 5 ∶ 1). They are in good agreement with the ICDD PDF-2 standard card 510611 of $Rb_2TiF_6$ with the hexagonal structure ($a = b = 5.892$ Å, $c = 4.796$ Å). No other phase can be detected, indicating that each sample is phase-pure. The Rietveld refinement of RTFM prepared with the $TiO_2/K_2MnF_6$ molar ratio of 10 ∶ 1 is performed on the Topas Academic software to further confirm the crystal structure. Fig.2-71(b) shows the corresponding powder X-ray diffraction (XRD) patterns. The finally refined structural parameters are summarized in Table 2-14 and the reliability factors are $R_{wp}$ = 9.42%, $R_p$ = 7.01%, $R_{Bragg}$ = 1.09%, which indicates that the diffraction patterns satisfy the reflection condition. According to the refinement results, RTFM is crystallized in the $P$-3$m$1 space group

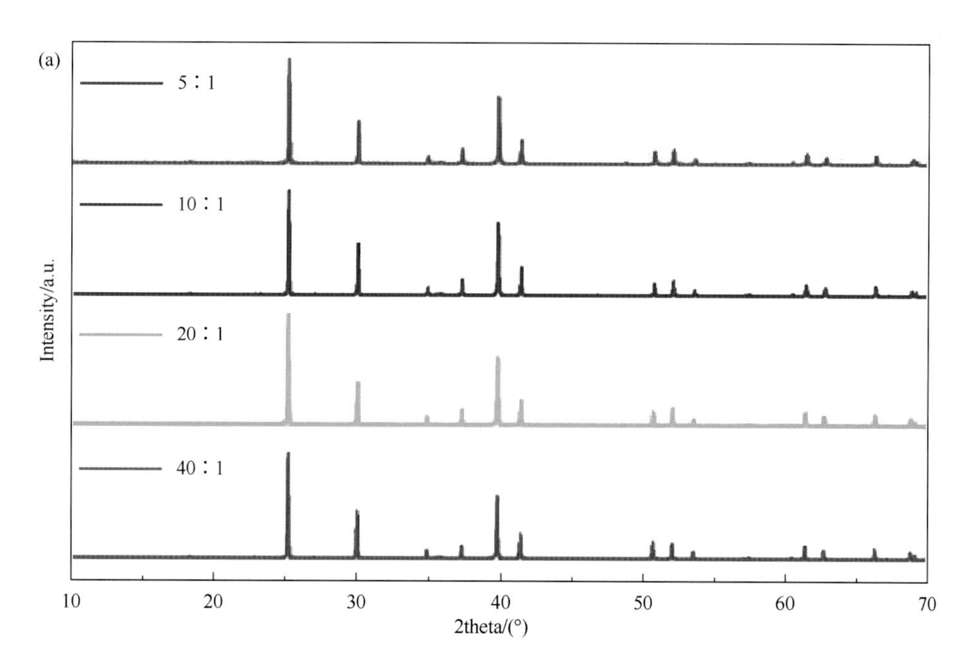

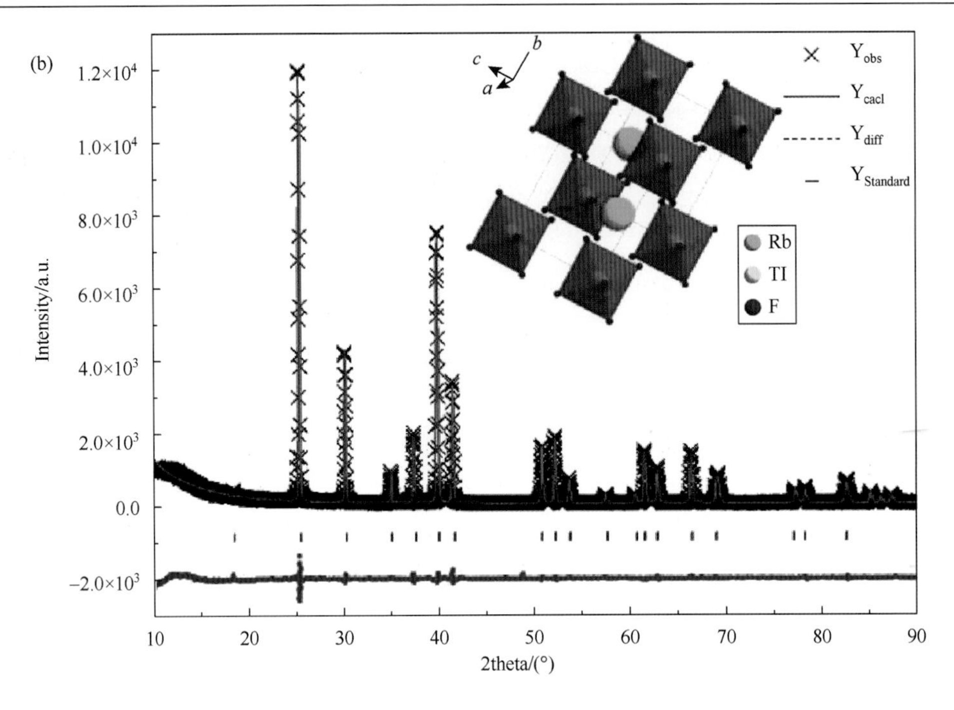


Fig.2.71 XRD patterns of RTFM prepared with different $TiO_2/K_2MnF_6$ molar ratios 40∶1, 20∶1, 10∶1, 5∶1 (a); experimental, calculated XRD patterns of RTFM and their difference at room temperature (b)

with the cell parameters of $a = b = 5.9033$ (7) Å and $c = 4.8004$ (0) Å. In this structure, each $Ti^{4+}$ is coordinated with six $F^-$ to form a regular $TiF_6^{2-}$ octahedron. Because of the identical valence state and similar ionic radii of $Ti^{4+}$ and $Mn^{4+}$, the dopants $Mn^{4+}$ would be incorporated into the $Ti^{4+}$ sites with six-fold coordination.

**Table 2-14 Crystal data of RTFM from the Rietveld refinement**

| Atom | Site | *x* | *y* | *z* | Occupancy |
|---|---|---|---|---|---|
| Rb | 3*m* | 0.3333 (0) | 0.6667 (0) | 0.6925 (4) | 1 |
| Ti | 3*m* | 0.0000 (0) | 0.0000 (0) | 0.0000 (0) | 0.904 |
| Mn | 3*m* | 0.0000 (0) | 0.0000 (0) | 0.0000 (0) | 0.096 |
| F | *m* | 0.1554 (9) | 0.3109 (8) | 0.2223 (7) | 1 |

Space group: *P*-3*m*1, $a = b = 5.9033$ (7) Å (0), $c = 4.8004$ (0) Å, $\alpha = \beta = 90°$, $\gamma = 120°$, cell volume: 144.88 (1) $Å^3$. $R_{wp} = 9.42\%$, $R_p = 7.01\%$, $R_{Bragg} = 1.09\%$

The surface morphology and composition of the RTFM with the $TiO_2/K_2MnF_6$ molar ratio of 20∶1 are analyzed, as shown in Fig.2-72. It can be observed that this sample exhibits a stone shape with a particle size of 30-40 μm. The peaks of F, Ti, Rb, and Mn elements can be easily found in the corresponding EDS spectrum [Fig.2-72(b)], but the peaks of the K element cannot be observed, indicating that as-prepared RTFM does not contain the impurity of $K_2MnF_6$.

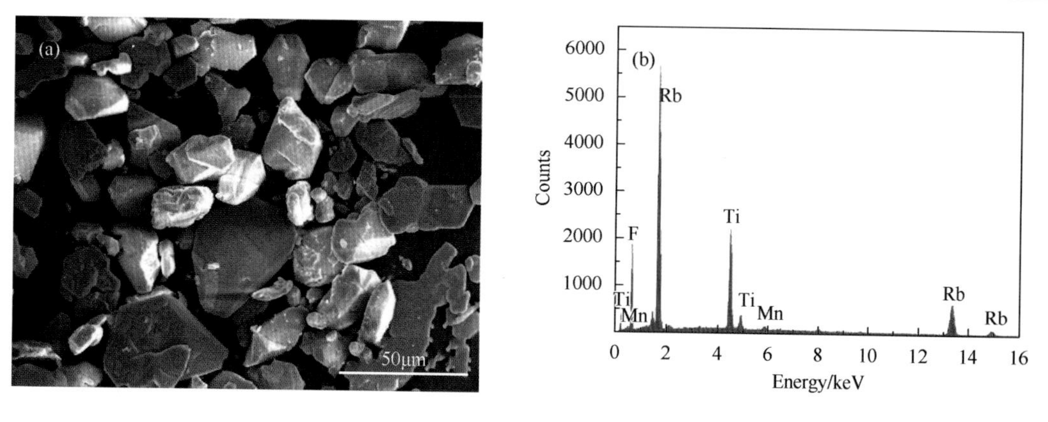


Fig.2-72 SEM image (a) and EDS spectrum (b) of the RTFM sample

Fig.2-73 displays the TEM and high-resolution TEM (HRTEM) images of the typical RTFM sample, demonstrating the single-crystalline nature of this sample. The smooth surface with clear edges and corners can be found from this TEM image. The HRTEM image shows a clear lattice fringe, which corresponds well to the (110), (101), (211) crystal faces of $Rb_2TiF_6$.

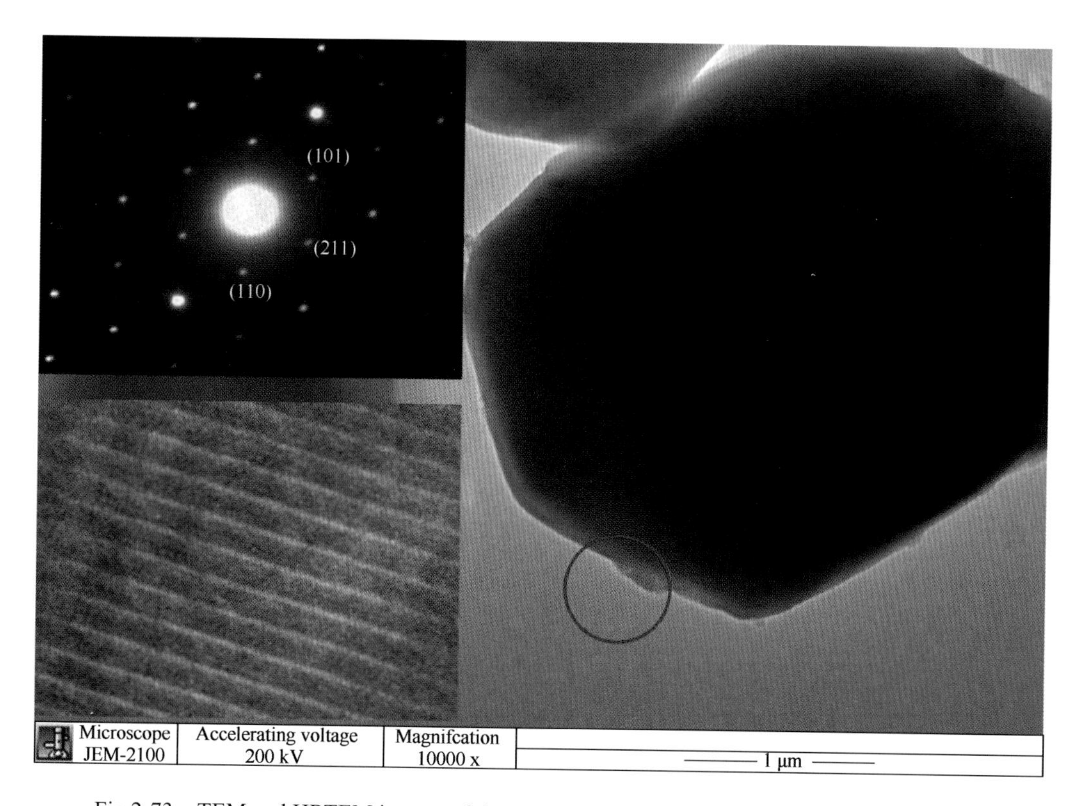


Fig.2-73 TEM and HRTEM images of the RTFM with the $TiO_2/K_2MnF_6$ molar ratio of 20 : 1

The diffuse reflectance spectra of $Rb_2TiF_6$ and RTFM are shown in Fig.2-74. The $Rb_2TiF_6$

host shows very weak absorption in the UV and blue regions. In contrast, RTFM exhibits three strong absorption bands at 238 nm, 353 nm and 460 nm, which can be attributed to the charge transfer band (CTB), spin-allowed transitions $^4A_{2g}\rightarrow^4T_{1g}$ and $^4A_{2g}\rightarrow^4T_{2g}$ of $Mn^{4+}$ respectively. The strongest absorption peak of RTFM is located at 460 nm, which overlaps very well with the emission band of the GaN chip (460 nm).

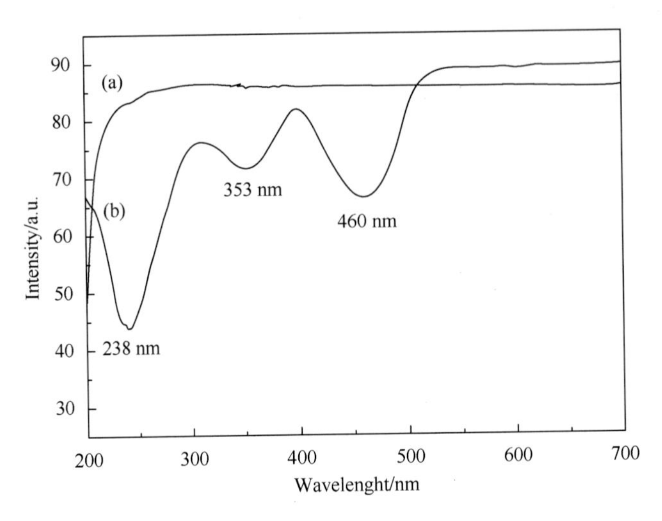


Fig.2-74 Diffuse reflectance spectra of $Rb_2TiF_6$ (a) and RTFM (b)

Fig.2-75 is the excitation and emission spectra of the typical RTFM. Three broad excitation bands in the spectral ranges of 210-300 nm, 300-400 nm and 400-550 nm can be found in the excitation spectrum. Those are due to the charge transfer band, the spin-allowed and parity-forbidden transitions of $Mn^{4+}$ from the ground state $^4A_{2g}$ to the excited states $^4T_{1g}$ and

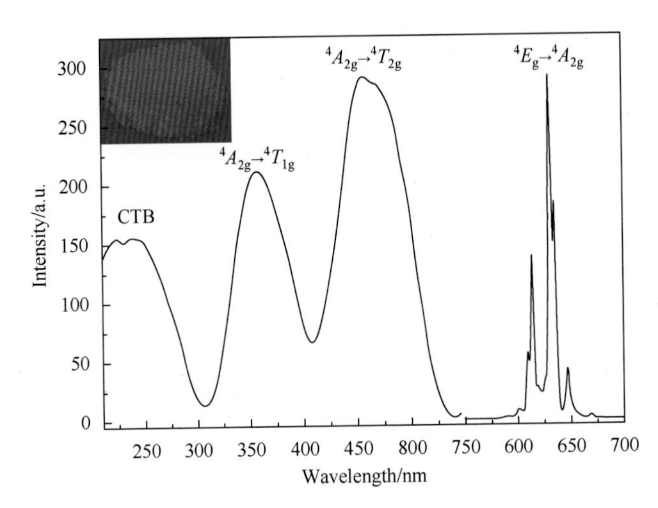


Fig.2-75 Excitation ($\lambda_{em}$ = 632 nm) and emission ($\lambda_{ex}$ = 460 nm) spectra of RTFM with the $TiO_2/K_2MnF_6$ molar ratio of 20：1. The inserted figure is the photograph of RTFM under 460 nm blue light excitation

${}^4T_{2g}$, respectively, as mentioned in Fig.2-74. The strongest excitation band at 460 nm with a full width at half maximum (FWHM) of 70 nm completely covers the emission of the blue GaN chip (460 nm). Intense red emission can be observed from this phosphor under the 460 light irradiation as shown in Fig.2-75, which is due to the ${}^2E_g \rightarrow {}^4A_{2g}$ spin-forbidden transition of $Mn^{4+}$ and contains five emission peaks with maxima at 609 nm, 614 nm, 631 nm, 635 nm and 647 nm, corresponding to the anti-Stokes transitions with the $v_4$, $v_6$ vibronic modes and Stokes transitions with the $v_6$, $v_4$ and $v_3$ vibronic modes, correspondingly. No ZPL of $Mn^{4+}$ can be observed from this emission spectrum at room temperature.

It's well known that the spectroscopic properties of the $Mn^{4+}$ ions in an octahedral environment can be interpreted by the Tanabe–Sugano energy-level diagram, as shown in Fig.2-76. The crystal field strength of $Mn^{4+}$ in RTFM is about 2174 $cm^{-1}$, which is roughly estimated by the ${}^4A_{2g} \rightarrow {}^4T_{2g}$ transition energy gap (21739 $cm^{-1}$). The ${}^4A_{2g} \rightarrow {}^4T_{1g}$ transition energy for $Mn^{4+}$ in RTFM is about 28329 $cm^{-1}$, and the calculated value of $B$ is 616 $cm^{-1}$. Thus the value of $D_q/B$ is 3.53, indicating that $Mn^{4+}$ experiences a strong crystal field in such fluoride matrix crystal field.

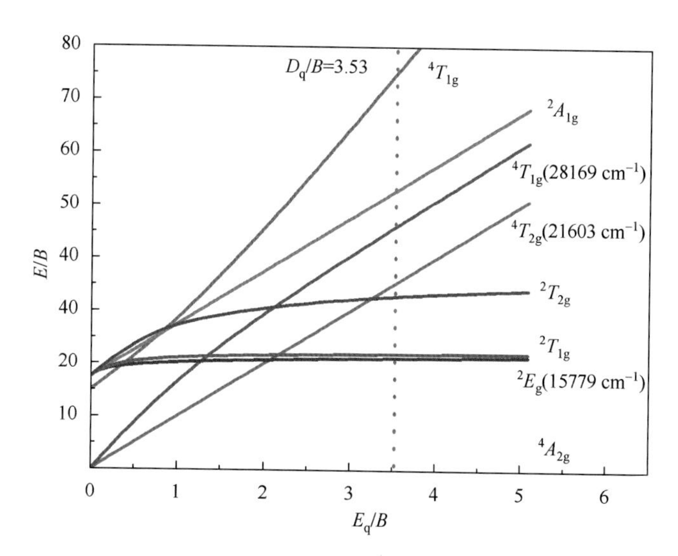


Fig.2-76 Tanabe-Sugano energy-level diagram of $Mn^{4+}$ in an octahedral crystal field

The Racah parameter $C$ also can be calculated as follows:

$$E\,({}^2E_g)/B = 3.05C/B + 7.9 - 1.8B/D_q \tag{2-3}$$

According to the reference, the ${}^2E_g \rightarrow {}^4A_{2g}$ transition energy is about 16051 $cm^{-1}$. So the Racah parameter $C$ is calculated to be 3770 $cm^{-1}$.

The energy of the ${}^2E_g \rightarrow {}^4A_{2g}$ transition of $Mn^{4+}$ is determined mainly by the nephelauxetic effect, which is correlated with the overlap of the wave functions between the $Mn^{4+}$ ions and the ligands. A new parameter $\beta_1$ can be used to describe the nephelauxetic effect in the spectroscopy of $Mn^{4+}$. The value of $\beta_1$ can be obtained as follows:

$$\beta_1 = \sqrt{\left(\frac{B}{B_0}\right)^2 + \left(\frac{C}{C_0}\right)^2} \tag{2-4}$$

where $B_0$ (1160 $cm^{-1}$) and $C_0$ (4303 $cm^{-1}$) are the Racah parameters for free $Mn^{4+}$. The $\beta_1$ value for $Mn^{4+}$ in the RTFM is calculated to be 1.02. Then we can estimate the $E$ ($^2E_g$) state energy based on the following linear relationship:

$$E\,(^2E_g) = -142.83 + 15544.02\beta_1 + \sigma \tag{2-5}$$

where the value of $\sigma$ is 365 $cm^{-1}$. The calculated value $E$ ($^2E_g$) of 16077 $cm^{-1}$ is close to the experimental value (16051 $cm^{-1}$).

For a further analysis of the spectroscopic properties of RTFM with $Mn^{4+}$ ions, the exchange charge model (ECM) of the crystal field is used to calculate the energy levels of the $Mn^{4+}$ ions in $Rb_2TiF_6$. The main advantage of the model is that it allows taking into account the covalent effects caused by the overlap of the wave functions of impurity ions and ligands. The crystal field parameters (CFPs) $B_p^k$ are represented as a sum of two contributions $B_{p,q}^k$ (point charge contributions) and $B_{p,S}^k$ (exchange charge contributions). All relevant equations and explanations can be found in the references and are not given here for the sake of brevity. The structural data for the prepared $Rb_2TiF_6$: $Mn^{4+}$ samples are given in Table 2-14. A large cluster consisting of more than 80000 ions is generated to ensure convergence of the crystal lattice sums, which are needed to calculate the CFPs acting on the $Mn^{4+}$ optical electrons. The calculated CFPs and $Mn^{4+}$ energy levels are given in Table 2-15 and Table 2-16, respectively. The structure of the crystal field Hamiltonian confirms the trigonal symmetry of the $Mn^{4+}$ position (only three CFPs, $B_2^0, B_4^0, B_4^3$) are not zero. Besides, all orbital triplets are split into a singlet and orbital doublet.

**Table 2-15 Calculated values of CFPs (Stevens normalization, in $cm^{-1}$) for $Rb_2TiF_6$:$Mn^{4+}$. The ECM parameter $G$= 13.485**

| CFP | $B_{p,q}^k$ | $B_{p,S}^k$ | Total value $B_{p,q}^k + B_{p,S}^k$ |
|---|---|---|---|
| $B_2^0$ | −191.2 | −1118.5 | −1309.7 |
| $B_4^0$ | −232.0 | −3364.4 | −3596.4 |
| $B_4^3$ | 7304.7 | 102128.4 | 109433.1 |

**Table 2-16 Calculated and experimental energy levels (in $cm^{-1}$) for $Mn^{4+}$ in $Rb_2TiF_6$. The orbital doublet states are denoted with an asterisk. The Racah parameters (in $cm^{-1}$) are given in the text**

| $O_h$ group notation and “parent” LS term | Calc. | Exp. |
|---|---|---|
| $^4A_{2g}$ ($^4$F) | 0 | 0 |
| $^2E_g$ ($^2$G) | 15779* | 16051 |

continued

| $O_h$ group notation and "parent" LS term | Calc. | Exp. |
|---|---|---|
| $^2T_{1g}$ ($^2$G) | 16111, 16337* | |
| $^4T_{2g}$ ($^4$F) | 21603, 21803* | 21739 |
| $^2T_{2g}$ ($^2$G) | 23974*, 24450 | |
| $^4T_{1g}$ ($^4$F) | 28169*, 28624 | 28329 |
| … | | |
| $^4T_{1g}$ ($^4$P) | 45425, 46499* | |

Inspection of Table 2-16 confirms the good agreement between the calculated and experimentally observed $Mn^{4+}$ energy levels.

$Mn^{4+}$ content plays an important role in the luminescent efficiency of RTFM red phosphor, so it is necessary to investigate its optimal doping amount in the $Rb_2TiF_6$ host. Fig.2-77 shows the excitation and emission spectra of RTFM prepared with different molar ratios of $TiO_2$ to $K_2MnF_6$. All these excitation and emission spectra are of similar shapes with a broad excitation band in the blue light region and red-emitting peaks from 600 nm to 650 nm. When the molar ratio $TiO_2$ to $K_2MnF_6$ is 20∶1, the obtained RTFM sample has the strongest red emission with high color purity ($x$ = 0.692, $y$ = 0.308).

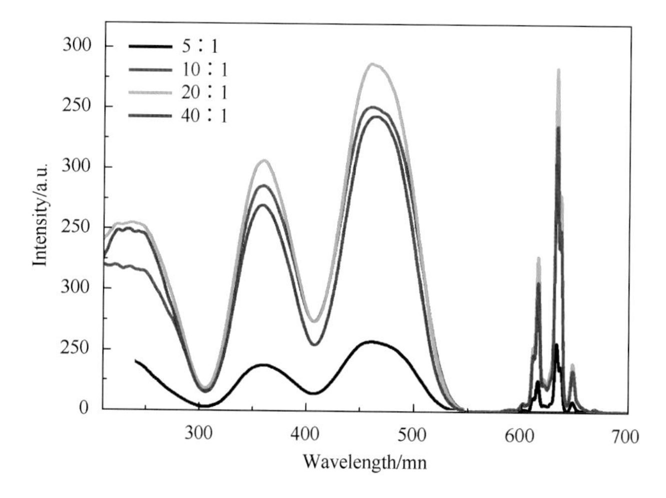


Fig.2-77 Excitation ($\lambda_{em}$ = 632 nm) and emission ($\lambda_{ex}$ = 460 nm) spectra of RTFM prepared with different molar ratios of $TiO_2$ to $K_2MnF_6$

The AAS is used to detect the doping concentration of $Mn^{4+}$ in these phosphors, and the corresponding results are listed in Table 2-17. The replacement concentration of $Mn^{4+}$ for $Ti^{4+}$ in the $Rb_2TiF_6$ host increases with the increasing consumption of $K_2MnF_6$. When the molar ratio of $TiO_2$ to $K_2MnF_6$ is 20∶1, the doping amount of $Mn^{4+}$ is measured to be 6.51% (mole fraction). With the further increase of $K_2MnF_6$ consumption, the concentration quenching for $Mn^{4+}$ can be observed from the emission spectra in Fig.2-77.

**Table 2-17　AAS results of RTFM prepared with different molar ratios of $TiO_2$ to $K_2MnF_6$**

| Samples | Molar ratios of $TiO_2$ to $K_2MnF_6$ | Doping amounts of $Mn^{4+}$/% (mole fraction) |
|---|---|---|
| 1 | 40:1 | 3.06 |
| 2 | 20:1 | 6.51 |
| 3 | 10:1 | 9.60 |
| 4 | 5:1 | 12.91 |

Thermal-quenching resistance is an important parameter for phosphors to be used for high-power w-LEDs since the joint-temperature of LED chips often reaches over 150 ℃. Fig.2-78(a) displays the temperature-dependent emission spectra of the optimal RTFM upon the 460 nm light excitation. All the emission spectra exhibit similar shapes with the strongest emission peak at 631 nm. No obvious shift of emission peaks can be found. The CIE coordinates according to the corresponding emission spectrum at 160 ℃ are calculated to be $x = 0.681$, $y = 0.319$, which are close to those at room temperature. This result shows that RTFM has high color stability.

With the increasing temperature, the intensity of anti-Stokes transitions ($W_a$) shows a rising trend [Fig.2-78(b)]. In contrast, the intensity of Stokes transitions ($W_s$) slightly increases. They reach the top when the temperature is at 140 ℃. This phenomenon is in according to the following equations:

$$W_a(T) = D \cdot \frac{1}{\exp\left(\frac{h\omega}{kT}\right) - 1} \tag{2-6}$$

$$W_s(T) = D \cdot \frac{\exp(h\omega / kT)}{\exp\left(\frac{h\omega}{kT}\right) - 1} \tag{2-7}$$

where $D$ is the proportional coefficient, $\hbar\omega$ is the energy of the coupled vibronic mode, $T$ is temperature and $k$ is the Boltzmann constant. So it is easy to deduce that the intensity of anti-Stokes transitions ($W_a$) and Stokes transitions ($W_s$) would increase with the rise of temperature. The intensity of $W_a$ will rise more rapidly than that of $W_s$ with the increasing temperature. The integrated emission intensity ($I_e$) of RTFM also reaches the top at the same temperature, which is about 1.2 times higher than that at 20 ℃. Meanwhile, the emission

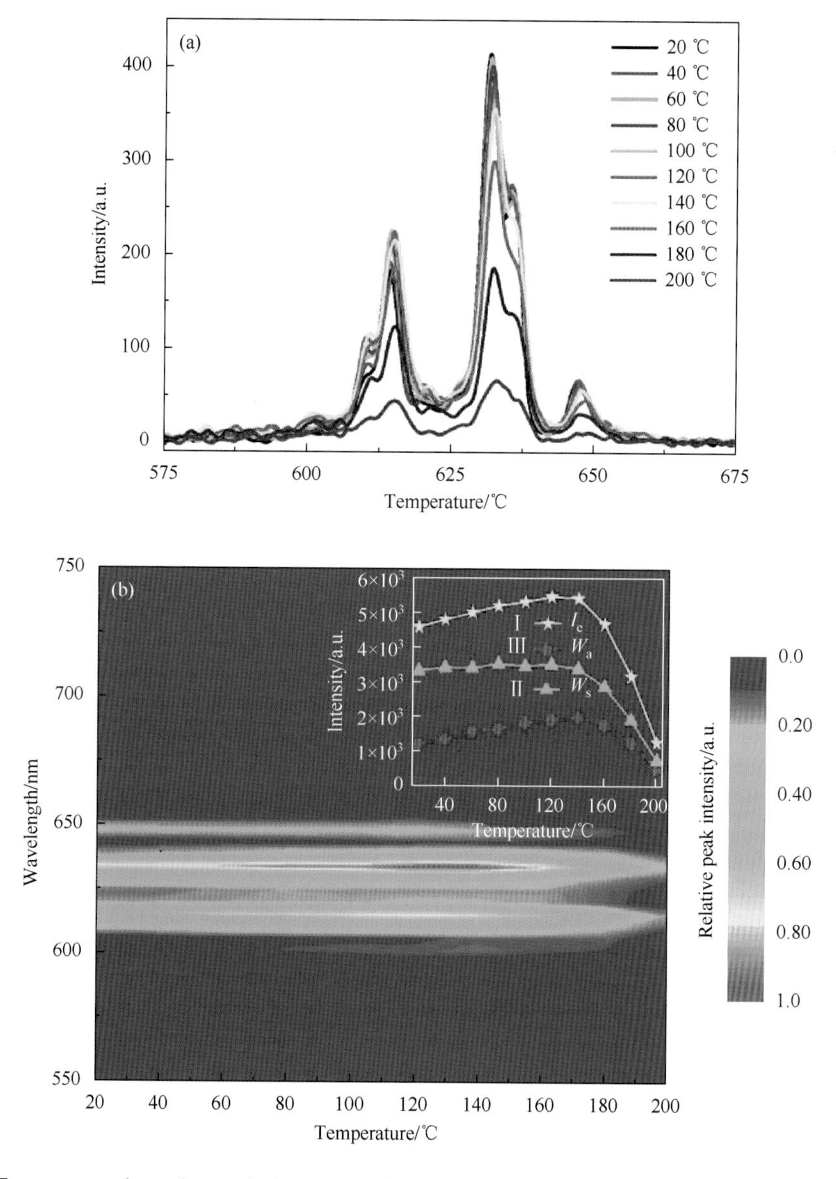


Fig.2-78 Temperature-dependent emission spectra (a) and the temperature-dependent emission intensity of the optimal RTFM upon 460 nm light excitation (b)

intensity of RTFM at 160 ℃ is still higher than that at 20 ℃. These results indicate that RTFM shares excellently thermal quenching resistance with high color stability, which may be applied in high power w-LEDs.

To further investigate the potential application of RTFM, a series of the w-LED devices were fabricated by coating the as-prepared sample on blue LED chips (460 nm). The EL spectra of w-LEDs-based on commercial YAG and RTFM are shown in Fig.2-79(a). Without RTFM (the black curve), the w-LED exhibits cold bright white light with the $R_a$ of 72.1 and CCT of 5070 K (Table 2-18). After the introduction of RTFM (other five curves), the sharp red emission can be observed from the w-LEDs, and the related parameters of these w-LEDs are listed in Table 2-18. With the addition of RTFM, the examined CCT of the w-LEDs drops from 5070 K to 3123 K, and $R_a$ is improved from 72.1 to 91.5, showing that the addition of RTFM is favorable for the improvement of $R_a$ and CCT levels of the YAG type w-LED. However, the LE slightly comes down from 202.1 lm/W to 187.9 lm/W. The corresponding CIE coordinates are shown in Fig.2-79(b), which reveal that the CCT moves to the warm white region with the increase of the red color component.

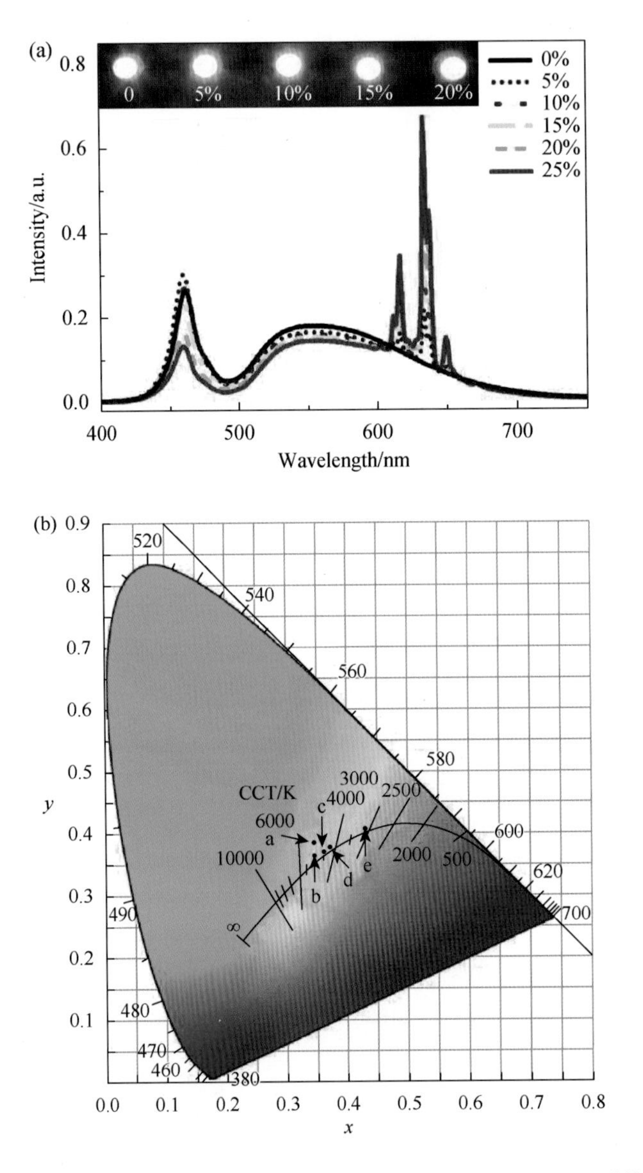


Fig.2.79 EL spectra (a) and Chromaticity coordinates (b) of w-LEDs based on different amounts of commercial YAG and RTFM under 20 mA drive current

**Table 2-18 Important photoelectric parameters for w-LEDs with different amounts of RTFM phosphor under a current of 20 mA**

| w-LEDs | Phosphors | RTFM/%(mass fraction)* | CCT/K | $R_a$ | LE/(lm/W) | CIE $(x, y)$ |
|---|---|---|---|---|---|---|
| a | YAG | 0 | 5070 | 72.1 | 202.1 | (0.346, 0.386) |
| b | YAG + RTFM | 5 | 4998 | 79.7 | 200.0 | (0.346, 0.364) |
| c | YAG + RTFM | 10 | 4512 | 84.5 | 197.0 | (0.362, 0.371) |
| d | YAG + RTFM | 15 | 4258 | 84.6 | 189.9 | (0.372, 0.378) |
| e | YAG + RTFM | 20 | 3123 | 91.5 | 187.9 | (0.431, 0.406) |

* %(mass fraction) $= \frac{m_{(\text{RTFM})}}{m_{(\text{epoxy resin})}} \times 100\%$.

As shown in Fig.2-80, the EL spectra of the selected w-LED (the mass ratio of YAG, epoxy resin, and RTFM is 1 ∶ 10 ∶ 2) were further measured under different currents to investigate the high power w-LED application. These EL spectra keep similar shapes with one broadband in the blue region and narrow bands in the red region, but the EL intensity is improved as the current increases until the current reaches 140 mA. With the increasing current, the luminous flux (LF) gradually rises and the LE gradually falls. When the current is 120 mA, the LE value still keeps at 115.3 lm/W. Moreover, as the current is improved, the $R_a$ value shows a slightly downward trend, and CCT does not change significantly. This performance of the w-LEDs can satisfy the requirement of high power w-LEDs. Hence, all of these results demonstrate that RTFM is a good red-emitting supplement for warm w-LEDs.

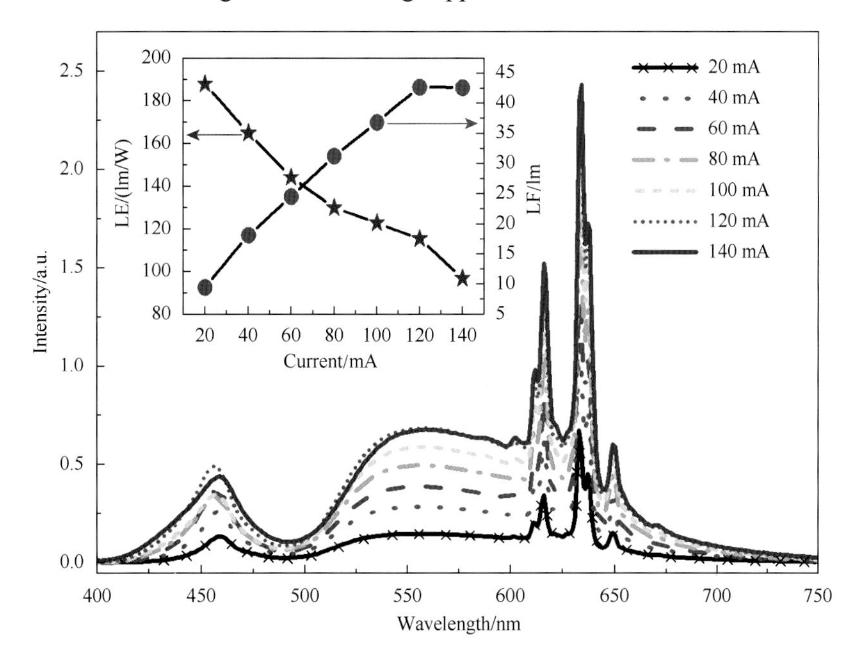


Fig.2-80 EL spectra of w-LED based on YAG and RTFM under different drive currents. The inserted figure is the current dependence of LE and LF

In summary, red phosphor RTFM with the hexagonal structure, which exhibits a broad excitation band in the blue-light region and sharp line-like emission in the red-light range due to d-d transitions of $Mn^{4+}$, was prepared by the ion exchange method. The luminescence of $Mn^{4+}$ at the octahedral $Ti^{4+}$ sites of $Rb_2TiF_6$ has been discussed. The calculated energy levels of $Mn^{4+}$ are in good agreement with the experimentally observed ones. The as-prepared phosphor presents high thermal stability and thermal quenching resistance with excellent color stability. w-LEDs based on this red-emitting phosphor show nice photoelectric performance with a low color temperature, a high color rendering index and a high LE under different drive currents. Hence, this red phosphor can find application in high power w-LEDs.

#### 2.1.3.10 $A_2HfF_6:Mn^{4+}$ ($A = Rb^+, Cs^+$)

In this work, red-emitting phosphors $A_2HfF_6:Mn^{4+}$ ($A = Rb^+, Cs^+$) were synthesized by the ion exchange method at room temperature. These phosphors show sharply red emission and broad absorption in the blue light region, which can improve the optical performance of w-LEDs based on YAG yellowish phosphor.

Fig.2-81(a) reveals the XRD patterns of as-prepared $Rb_2HfF_6:Mn^{4+}$ (RHFM) and $Cs_2HfF_6:Mn^{4+}$ (CHFM), along with their corresponding standard diffraction cards (No. 13-0443 for $Rb_2HfF_6$ and No. 74-0175 for $Cs_2HfF_6$). The diffraction peaks of these two samples can be well indexed to the corresponding JCPDS cards of $Rb_2HfF_6$ ($a = b = 6.14$ Å; $c = 4.82$ Å; $V = 157.65$ Å$^3$) and $Cs_2HfF_6$ ($a = b = 6.39$ Å; $c = 5.0$ Å; $V = 176.81$ Å$^3$). No second phase can be detected in these patterns. This result indicates that RHFM and CHFM are all of a single phase with the same hexagonal structure of the space group of $P$-3$m$1 and the doping of $Mn^{4+}$ does not change the phase of $Rb_2HfF_6$ and $Cs_2HfF_6$. Although RHFM and CHFM are of the iso-structure, they have different crystal cell parameters. Compared with CHFM, RHFM has a smaller crystal volume since the ionic radius of $Rb^+$ is smaller than that

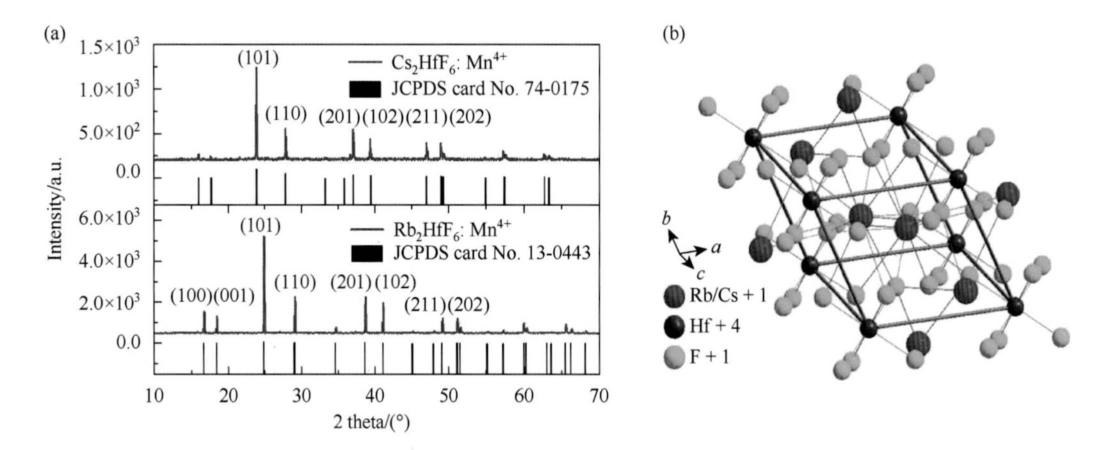


Fig.2-81 (a) XRD patterns of RHFM and CHFM (b) crystal structure of $A_2HfF_6$ ($A = Cs^+, Rb^+$)

of $Cs^{+}$, and the diffraction peaks of RHFM show a red-shift. According to the crystal structure of $A_2HfF_6$ ($A = Cs^{+}$, $Rb^{+}$) in Fig.2-81(b), each $Hf^{4+}$ in the host of $Rb_2HfF_6$ or $Cs_2HfF_6$ is coordinated with six $F^{-}$ to form a regular octahedron. Since $Mn^{4+}$ not only has the same valence state with $Hf^{4+}$ but also has a similar ionic radius with $Hf^{4+}$. $Mn^{4+}$ in RHFM or CHFM will occupy the site of $Hf^{4+}$ in the center of a regular octahedron.

The electronic structure of $Rb_2HfF_6$ is calculated with the CASTEP module of the Materials Studio package, and the calculated band structure of $Rb_2HfF_6$ is shown in Fig.2-82(a). The calculated bandgap is approximately 6.2 eV, which shows the $Rb_2HfF_6$ is a great luminescent host for doping $Mn^{4+}$ due to its wide bandgap. To analyze the top of the valence band maximum compare with that the conduction band minimum, the top point on the valence band is found at the K-point of G and the bottom point on the conduction band is located also at the K-point of G, which indicates that it is a direct bandgap semiconductor. To further analyze the composition of the calculated energy band, the partial density of states (PDOS) and total density of states (TDOS) are shown in Fig.2-82(b). The upper part of the valence band is composed mainly of F 2p states and the bottom of the conduction band is composed mainly of Hf 5d states. That is to say, the F-Hf group determines the optical properties of the $Rb_2HfF_6$ crystal.

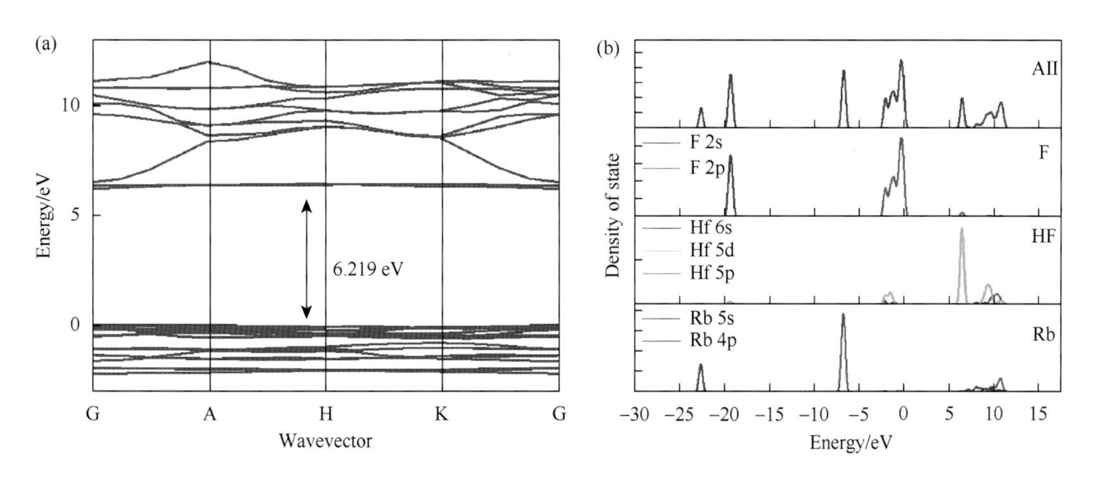


Fig.2-82 (a) calculated energy band structure of $Rb_2HfF_6$ (b) total and partial density of states for $Rb_2HfF_6$

Fig.2-83 presents the electronic structure of $Cs_2HfF_6$ which is similar to that of $Rb_2HfF_6$. The calculated band gap is about 6.3 eV, showing that $Cs_2HfF_6$ is also a pretty luminescent host for dopants of $Mn^{4+}$.

The composition and morphology of the as-prepared RHFM and CHFM were investigated, and the corresponding results are shown in Fig.2-84. The elements of Rb or Cs, F, Hf, Mn can be easily recognized from the corresponding EDS spectra, and the element O cannot be

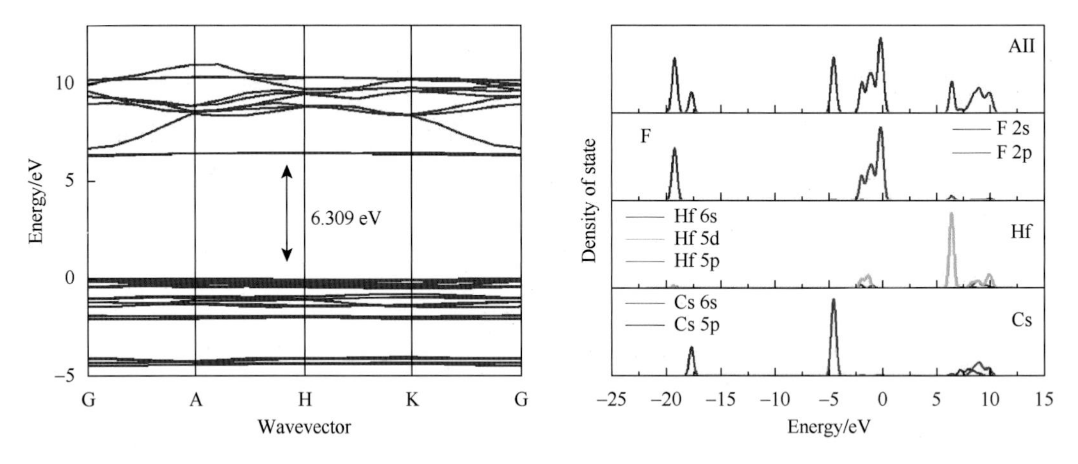


Fig.2-83 (a) calculated energy band structure of $Cs_2HfF_6$ (b) total and partial density of states for $Cs_2HfF_6$

found. This result proves that no manganese oxides are produced during the entire preparation process and manganese ions have already been doped into the crystal lattices of RHFM and CHFM. The particles of these two samples exhibit heavy agglomeration from their SEM images in the inserted figures of Fig.2-84. Compared with RHFM, CHFM shares a bigger particle size (20-30 μm) with clear edges and corners, which implies that this sample has been well crystallized.

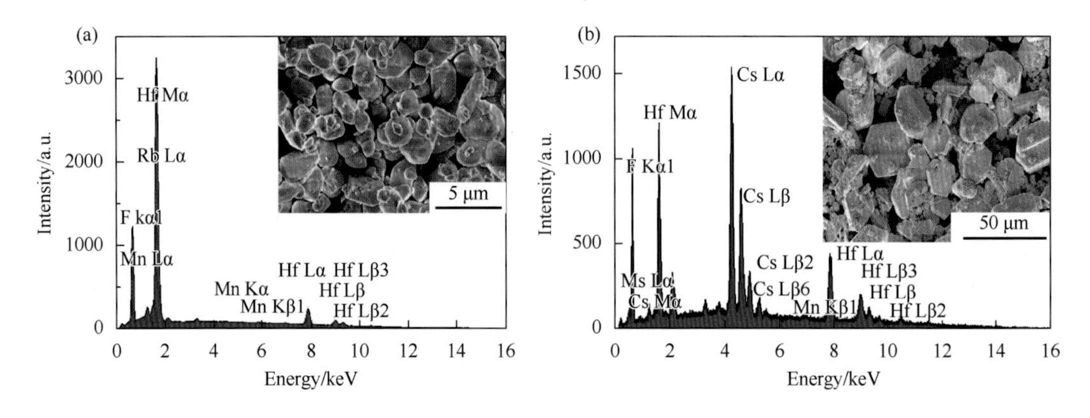


Fig.2-84 (a) EDS spectra of RHFM and (b) CHFM. The inserted figures are the corresponding SEM images

Besides, the chemical thermal stabilities of RHFM and CHFM have been investigated through the TG curves, as shown in Fig.2-85. $A_2HfF_6$:$Mn^{4+}$ shows high thermal decomposition temperatures that exceed 500 ℃. Such high decomposition temperatures suggest that they have high thermal stability for applying in LED devices.

Fig.2-86 exhibits the DRS of the $Rb_2HfF_6$ host and $Rb_2HfF_6$:$Mn^{4+}$. In contrast with the pure host of $Rb_2HfF_6$, red phosphor $Rb_2HfF_6$:$Mn^{4+}$ has two intense absorption bands in the range of wavelength from 300 nm to 530 nm, which are due to the spin-allowed transitions of

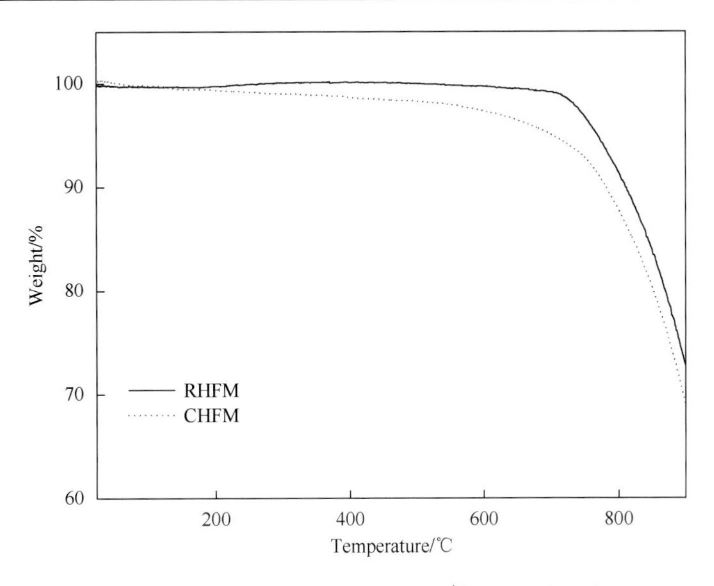


Fig.2-85 TG curves of $A_2HfF_6$:$Mn^{4+}$ (A = $Rb^+$, $Cs^+$)

${}^4A_{2g}\rightarrow{}^4T_{1g}$ and ${}^4A_{2g}\rightarrow{}^4T_{2g}$ of $Mn^{4+}$ respectively. This result means that this phosphor with the dopant of $Mn^{4+}$ can be effectively excited by the blue emission of the GaN chip.

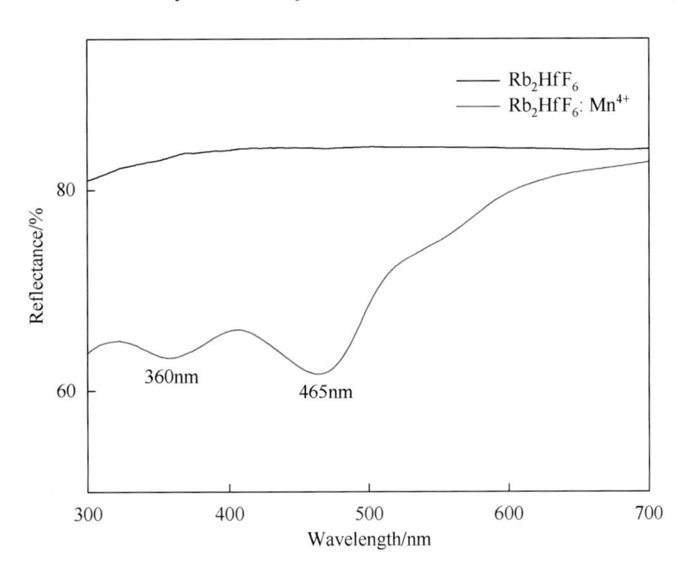


Fig.2-86 DRS results of (a) $Rb_2HfF_6$ and (b) $Rb_2HfF_6$:$Mn^{4+}$

Fig.2-87 is the PL spectra of the as-prepared RHFM and CHFM. These two samples show similar excitation spectra with two broad absorption bands between 300 nm and 530 nm which are consistent with their DRS results. The strongest excitation band at 480 nm with a full width at half maximum (FWHM) of about 60 nm completely covers the emission of the blue GaN chip (460 nm), so that these red phosphors can be efficiently excited by the blue light from one GaN chip. Moreover, the excitation band of the CHFM sample shows a slight red-shift, compared with that of RHFM. This is due to the different crystal field strength around $Mn^{4+}$ in such two different hosts.

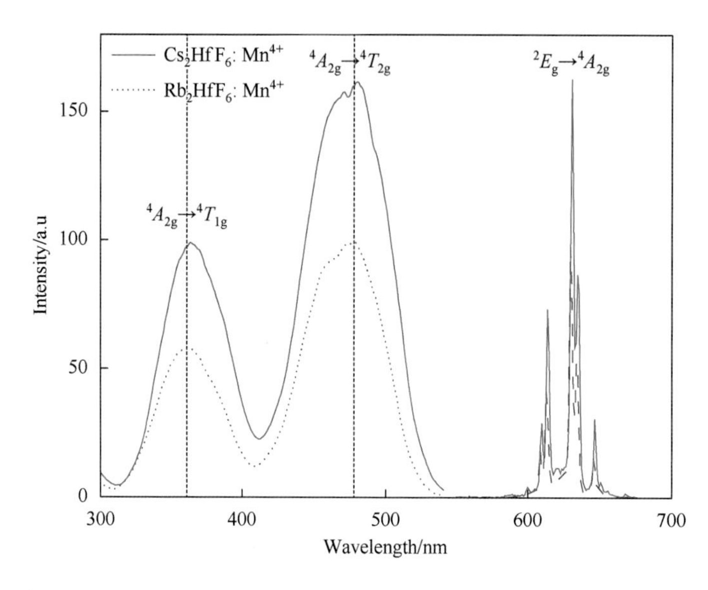


Fig.2-87 PL spectra of RHFM and CHFM ($\lambda_{em}$ = 630 nm, $\lambda_{ex}$ = 476 nm)

According to the emission spectra in Fig.2-87, it can be found that these two samples are of the same spectral shapes with the strongest emission peak located at 630 nm. As we know, the energy of the $^2E_g$ state of $Mn^{4+}$ is independent of the crystal strength. So the emission peaks of $Mn^{4+}$ in $A_2HfF_6$ are located at the same positions of 599 nm, 609 nm, 614 nm, 630 nm, 634 nm and 646 nm. The first three groups of emission peaks are due to the anti-Stokes transitions with the $v_3$, $v_4$ and $v_6$ vibronic modes, and the last three ones are ascribed to Stokes transitions with the $v_6$, $v_4$, and $v_3$ vibronic modes. Since $Mn^{4+}$ in $A_2HfF_6$ occupies the center of one regular octahedron, no obvious ZPL can be found in the emission spectra at room temperature. Besides, the internal quantum yields of RHFM and CHFM are obtained to be 0.556 and 0.652, respectively.

To obtain highly efficient phosphors, we investigated the influence of the doping amount of $Mn^{4+}$ on the PL properties. Fig.2-88 presents the PL spectra of CHFM with different molar ratios of $HfO_2$ to $K_2MnF_6$. All these samples exhibit similar PL spectra except for their intensities. Two broad excitation bands in the UV light and the blue light regions are attributed to the $^4A_{2g}\rightarrow{}^4T_{1g}$ and $^4A_{2g}\rightarrow{}^4T_{2g}$ transitions of $Mn^{4+}$. A couple of red-emitting peaks are all due to the transitions from $^2E_g$ to $^4A_{2g}$. Initially, the intensities of excitation and emission exhibit a rising tendency with the increasing consumption of $K_2MnF_6$. When the molar ratio of $HfO_2$ to $K_2MnF_6$ is 15：1, the as-obtained sample is of the strongest emission intensity which is about 2.23 times higher than that of CHFM prepared with the molar ratio of $HfO_2$ to $K_2MnF_6$ of 40：1. With the further increase of $K_2MnF_6$, the concentration quenching effect for $Mn^{4+}$ can be observed.

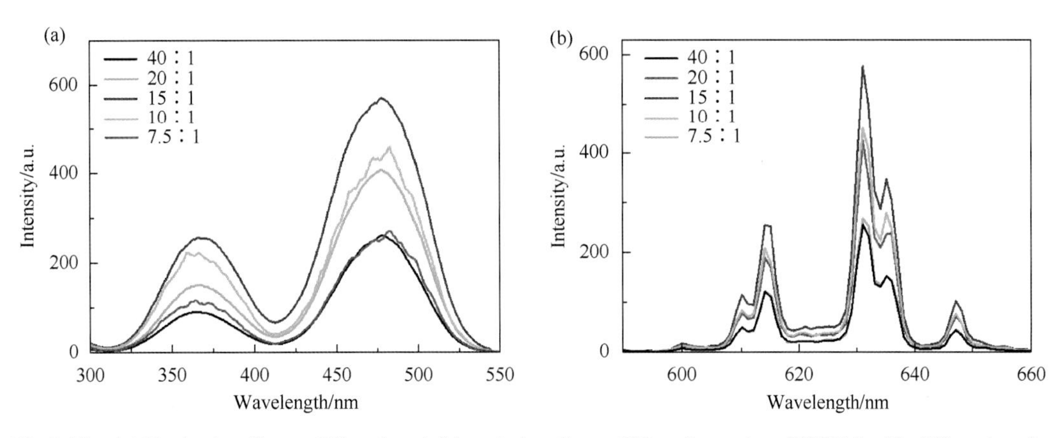


Fig.2-88 (a) Excitation ($\lambda_{em}$ = 630 nm) and (b) emission ($\lambda_{ex}$ = 476 nm) spectra of CHFM with different molar ratios of $HfO_2$ to $K_2MnF_6$ (40 : 1, 20 : 1, 15 : 1,10 : 1, 7.5 : 1)

To confirm the quenching concentration of $Mn^{4+}$ in this host, the AAS is used to measure the relative concentration of $Mn^{4+}$ in these samples, and the results are listed in Table 2-19. The relative concentration of $Mn^{4+}$ in these samples increases with the rising consumption of $K_2MnF_6$, but the quenching content of $Mn^{4+}$ in $Cs_2HfF_6$ is about 2.70% (mole fraction).

**Table 2-19 AAS results of CHFM prepared with different molar ratios of $H_2HfF_6$ to $K_2MnF_6$**

| Samples | Molar ratios of $HfO_2$ to $K_2MnF_6$ | Doping amounts of $Mn^{4+}$ /% (mole fraction) |
|---|---|---|
| i | 40 : 1 | 0.43 |
| ii | 20 : 1 | 1.85 |
| iii | 15 : 1 | 2.70 |
| iv | 10 : 1 | 6.05 |
| v | 7.5 : 1 | 6.42 |

Fig.2-89 is the PL spectra of RHFM prepared with different contents of $Mn^{4+}$. $Mn^{4+}$ in $Rb_2HfF_6$ exhibits a similarly luminescent behavior with $Cs_2HfF_6$. The as-obtained sample with the molar ratio of $HfO_2$ to $K_2MnF_6$ is 30 : 1 shows the strongest excitation and emission among

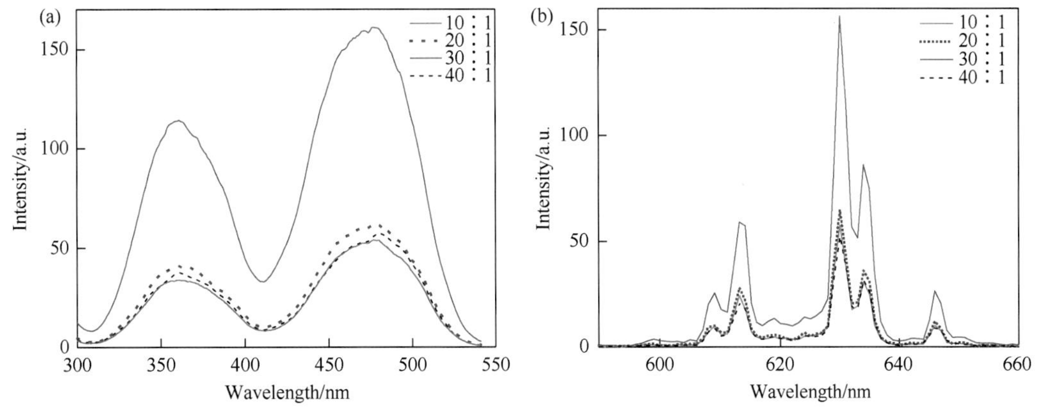


Fig.2-89 (a) Excitation and (b) emission spectra of the $Rb_2HfF_6$: $Mn^{4+}$ samples with different molar ratios of $HfO_2$ to $K_2MnF_6$ (40 : 1, 30 : 1, 20 : 1, 10 : 1) ($\lambda_{em}$ = 630 nm, $\lambda_{ex}$ = 476 nm)

these phosphors. Further improving the consumption of $K_2MnF_6$, the red emission of this phosphor is quenched by the excessive $Mn^{4+}$.

We also study the thermal stability of RHFM and CHFM. Fig.2-90 displays the emission

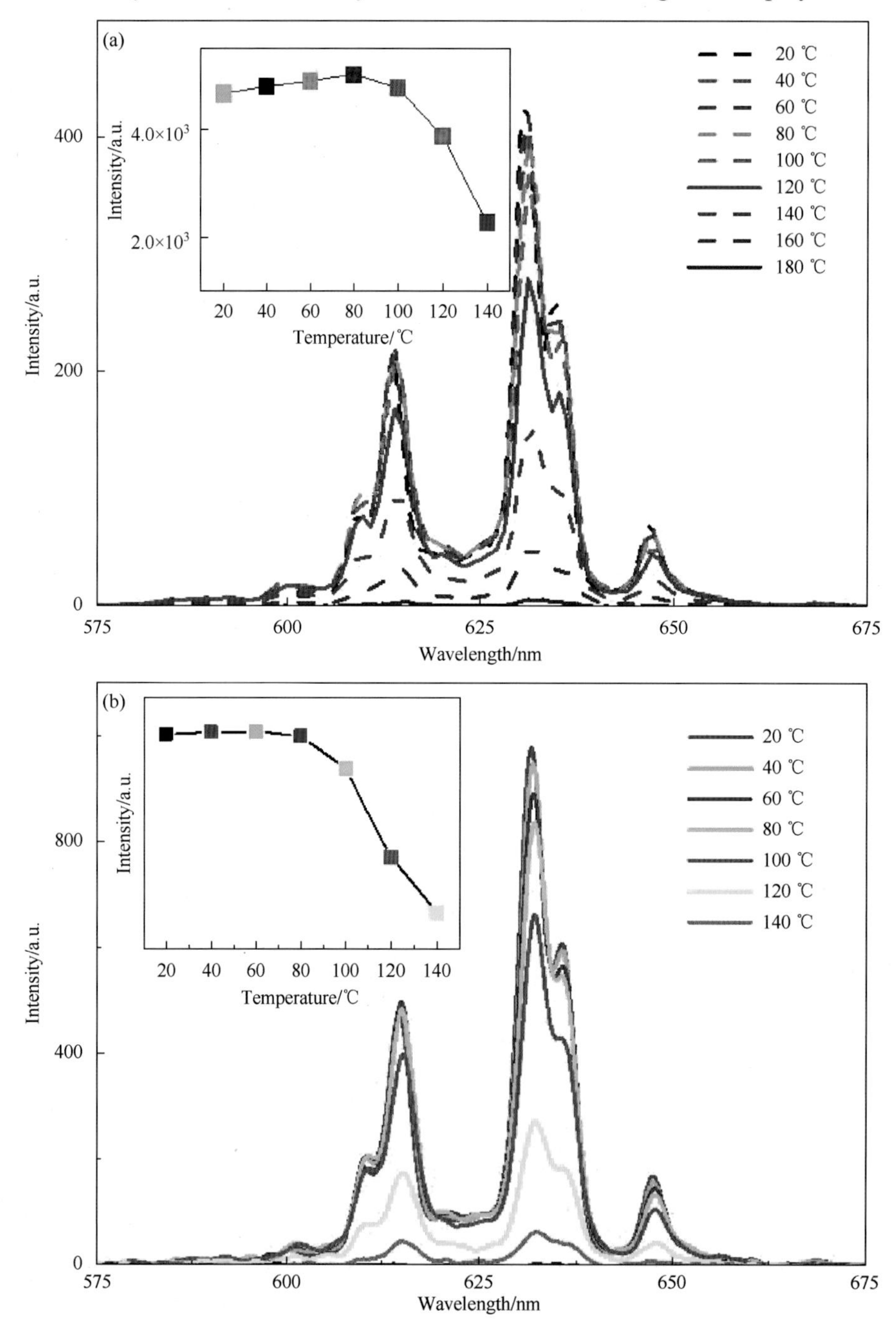


Fig.2-90　Emission spectra of (a) $Rb_2HfF_6$:$Mn^{4+}$ and (b) $Cs_2HfF_6$:$Mn^{4+}$ at different temperatures. The inserted figures are the relationships between the temperature and the relative emission intensity of $Rb_2HfF_6$: $Mn^{4+}$ and $Cs_2HfF_6$: $Mn^{4+}$

spectra of these two samples at different temperatures. Six groups of emission peaks can be found from these spectra, and the positions of these peaks have no obvious shift, indicating they share high color stability. The integrated emission intensity (IEI) of RHFM exhibits a slight uptrend until the temperature is up to 80 ℃, and the IEI at 100 ℃ is still higher than that at room temperature. This means that the RHFM sample shares excellent thermal stability. Moreover, $Mn^{4+}$ in $Cs_2HfF_6$ also exhibits a similar luminescent behavior at different temperatures.

The decay curves for the $^2E_g \rightarrow ^4A_{2g}$ transition (630 nm) of the $Mn^{4+}$ in RHFM and CHFM have been shown in Fig.2-91, which are well fitted into a single-exponential function. Their lifetimes are 5.18 ms and 3.44 ms, respectively.

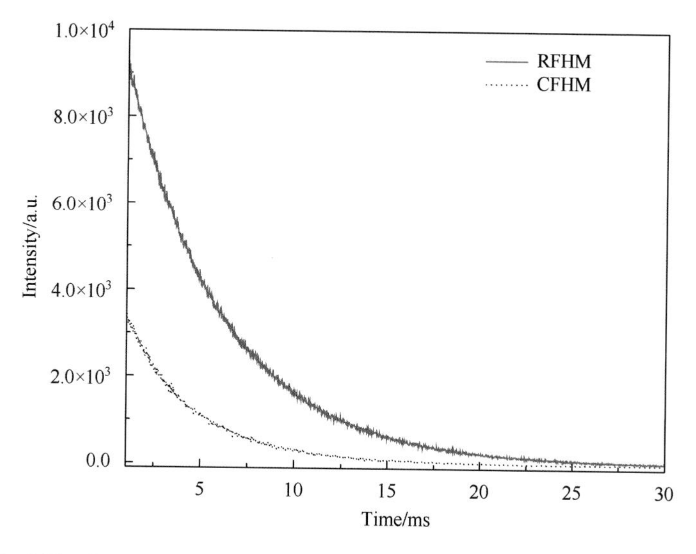


Fig.2-91 Decay curves of RHFM and CHFM ($\lambda_{em}$ = 630 nm, $\lambda_{ex}$ = 476 nm)

According to the above sections, we can find that RHFM and CHFM are of excellent PL properties, high thermal stability, and excellent thermal quenching resistance, so it is necessary to investigate their application on the LED devices. Fig.2-92 is the EL spectra of these red LEDs based on RHFM and CHFM. The blue emission band is due to the emission of the blue LED chip, and a set of red peaks are ascribed to the $^2E_g \rightarrow ^4A_{2g}$ transitions of $Mn^{4+}$ which are excited by the emission of the LED chip. Undoubtedly, the intensity of red emission is higher than that of the blue emission and the intense red light can be observed by naked eyes, which further indicates that RHFM and CHFM are excellent red-emitting phosphors.

Therefore, we studied the EL properties of w-LEDs based on RHFM and CHFM. Fig.2-93 exhibits the EL spectra of a series of w-LEDs based on red-emitting phosphor CHFM and yellow phosphor YAG. With the introduction of CHFM, the blue and yellowish-green peaks get gradually weak. In contrast, red emission from CHFM gradually strengthens, indicating that the white light turns from cold to warm. This result also can be proved by the series of w-LED images in Fig.2-93.

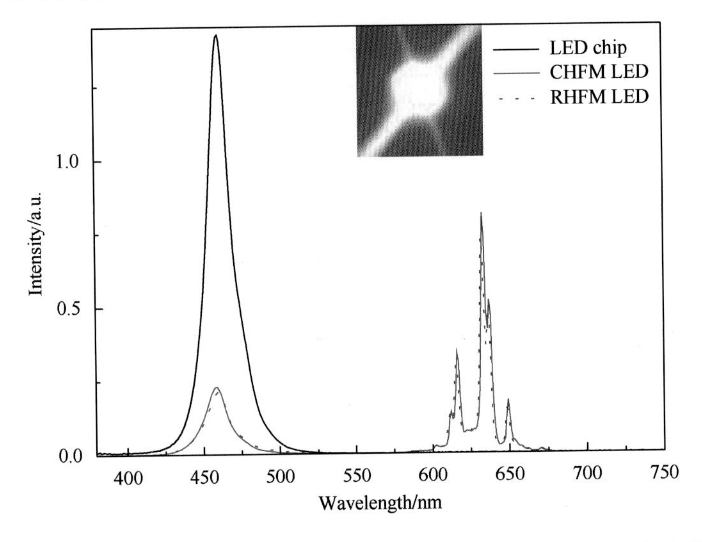


Fig.2-92 EL spectra of LED chip and red LEDs based on CHFM and RHFM

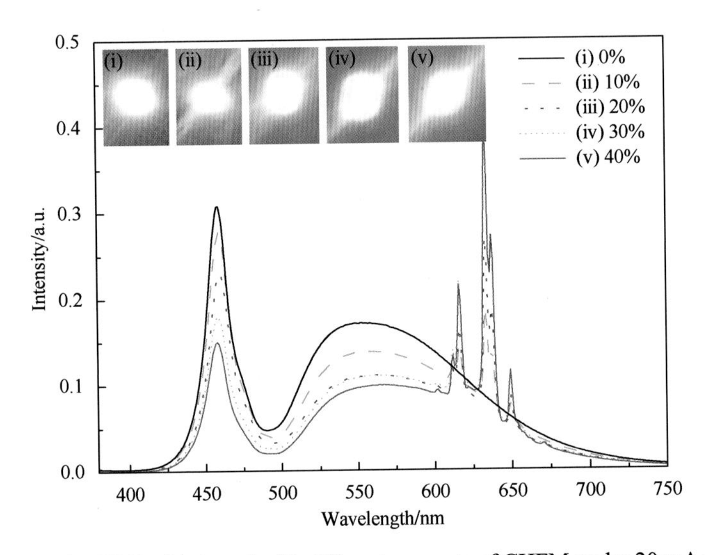


Fig.2-93 EL spectra of w-LEDs fabricated with different amounts of CHFM under 20 mA current excitation. The inserted figures are their corresponding w-LED images

The photoelectric parameters of these w-LEDs are listed in Table 2-20. CHFM can decrease the CCT value of these w-LEDs from 5390 K to 3337 K, and improve the $R_a$ value from 73.5 to 93.0. Meanwhile, the LE drops from 204.2 lm/W to 107.1 lm/W.

**Table2-20 Typical LED photoelectric parameters with different amount of CHFM phosphor under a current of 20 mA**

| Devices | CHFM/%(mass fraction) | CCT/K | $R_a$ | CIE $(x, y)$ | LE/(lm/W) |
|---|---|---|---|---|---|
| i | 0 | 5390 | 73.5 | (0.335, 0.367) | 204.2 |
| ii | 5 | 5071 | 82.0 | (0.343, 0.352) | 169.4 |

continued

| Devices | CHFM/%(mass fraction) | CCT/K | $R_a$ | CIE ($x$, $y$) | LE/(lm/W) |
|---|---|---|---|---|---|
| iii | 10 | 4560 | 89.7 | (0.356, 0.346) | 140.5 |
| iv | 15 | 3657 | 92.6 | (0.389, 0.363) | 118.5 |
| v | 20 | 3377 | 93.0 | (0.402, 0.369) | 107.1 |

Moreover, the luminescent characters of RHFM on w-LEDs were investigated, as shown in Fig.2-94. Being similar to CHFM, RHFM also can supply the red emission of the emission spectra of w-LEDs and optimize the optical performances of w-LEDs.

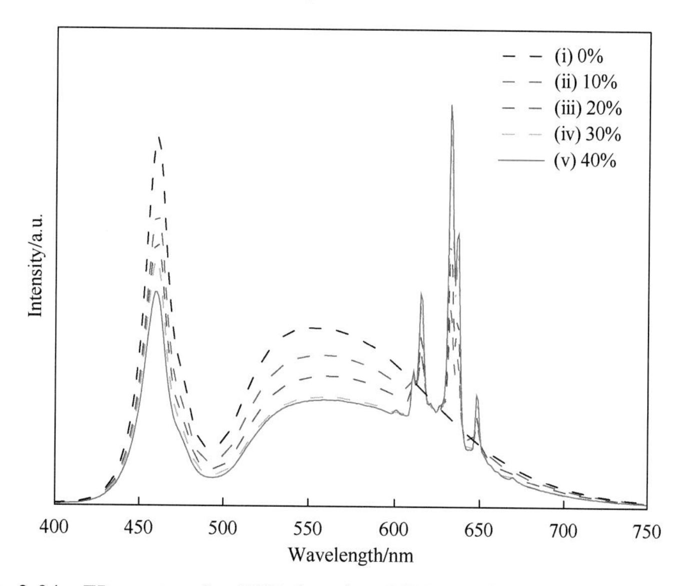


Fig.2-94 EL spectra of w-LEDs based on RHFM under 20 mA drive current

In conclusion, novel red-emitting phosphors $A_2HfF_6:Mn^{4+}$ ($A = Rb^+$, $Cs^+$) have been prepared through the ion exchange method. The as-obtained samples share a single- phase with the hexagonal structure and appropriate band gaps for doping $Mn^{4+}$. These phosphors can emit intense red light efficiently with one broad excitation band which can perfectly overlap the blue emission of the GaN chip. Meanwhile, $A_2HfF_6:Mn^{4+}$ phosphors are of high thermal stabilities and better quantum yields. With the introduction of these red phosphors, warm w-LEDs with excellent optical performances have been achieved successfully. Hence, these samples share potential applications for w-LEDs as red components.

## 2.2 $Mn^{4+}$ activated oxides red phosphors for blue LED chips

### 2.2.1 $SrGe_4O_9:Mn^{4+}$

$Mn^{4+}$ ions doped phosphors have been used in various applications, due to the efficiently red emission of $Mn^{4+}$ with one broadband in the UV and blue light regions. $Mn^{4+}$ ions show different PL properties in different hosts since their PL spectra are due to the d-d transition, which is highly influenced by the crystal field. According to the doped host, red phosphors doped with $Mn^{4+}$ ions can be divided into two categories: fluoride hosts and composite oxide hosts. $Mn^{4+}$ doped fluoride red phosphors exhibit intense red emission with one broadband at 460 nm, which could be used on w-LEDs. The PL properties of some composite oxides doped with $Mn^{4+}$ have been investigated in detail. Such as, Wang prepared a series of red phosphors $SrAl_{12}O_{19}$:$Mn^{4+}$, M (M = $Li^+$, $Na^+$, $K^+$, $Mg^{2+}$), which could be demonstrated to be potential candidates for high color rendering w-LEDs.

$SrGe_4O_9$ doped with $Mn^{4+}$ has been prepared by the hydrothermal method, and its PL property has been investigated. As we know, the concentration of $Mn^{4+}$ plays a crucial role in the PL properties of red phosphors doped with $Mn^{4+}$, and it is complicated to be determined, especially when the phosphors were synthesized from the hydrothermal route. To the best of our knowledge, no future research on this kind of phosphor has been reported.

In this work, red phosphors $SrGe_4O_9$ with different contents of $Mn^{4+}$ have been prepared by the solid-state reaction at a high temperature, and their structure and PL properties were investigated. At last, the EL properties of LED devices based on $SrGe_4O_9$ were investigated.

Fig.2-95 is the XRD pattern of $SrGe_4O_9$:0.024$Mn^{4+}$, which is consistent with the ICSD

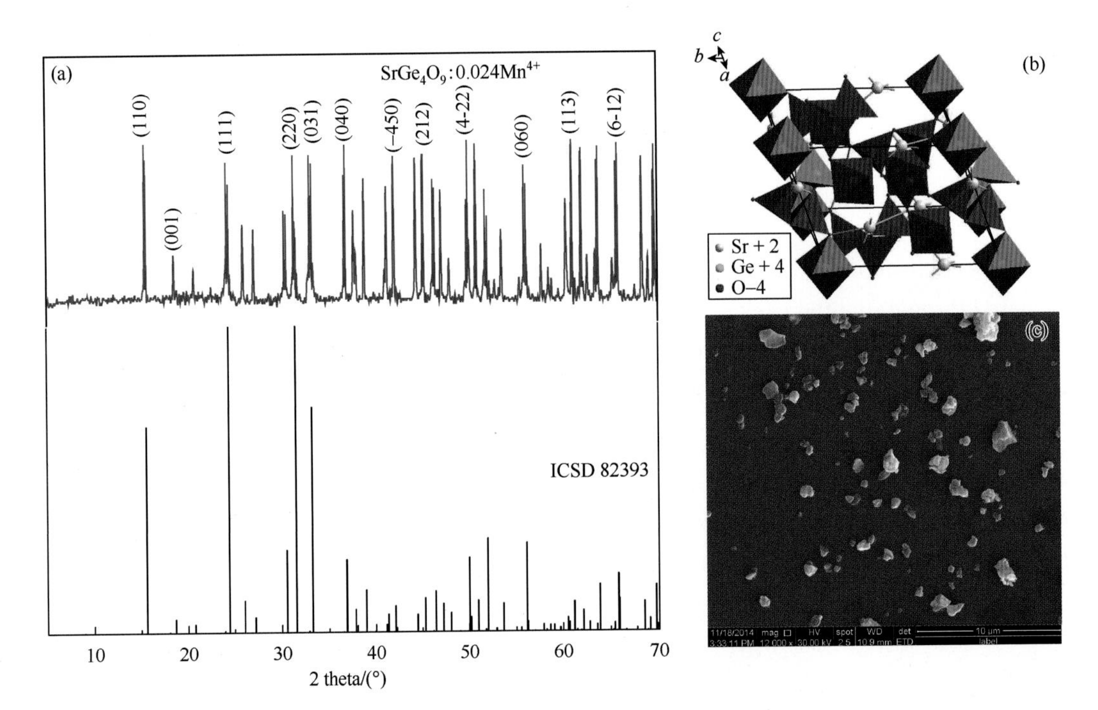


Fig.2-95　XRD pattern of $SrGe_4O_9$:0.024$Mn^{4+}$ (a), crystal structure of $SrGe_4O_9$ (b) and SEM image of $SrGe_{3.976}O_9$:0.024$Mn^{4+}$ (c)

card of $SrGe_4O_9$ (No. 82393, space group *P*321). This result indicates the phosphor is of a single-phase with the trigonal structure. A little doping of $Mn^{4+}$ does not change the crystal structure of this $SrGe_4O_9$ host. The crystal structure of $SrGe_4O_9$ is shown in Fig.2-95(b) There are two kinds of $Ge^{4+}$ sites in this crystal structure. One kind of $Ge^{4+}$ is coordinated with six $O^{2-}$ to form a regular octahedron. Another kind of $Ge^{4+}$ is in the center of one tetrahedron with four $O^{2-}$. Since $Mn^{4+}$ has the same charge and radius as $Ge^{4+}$, $Mn^{4+}$ maybe occupy the sites of $Ge^{4+}$ in the $SrGe_4O_9:Mn^{4+}$ crystal structure. The corresponding SEM image is shown in Fig.2-95(c). The as-prepared sample synthesized at high temperature is of irregular morphology with a particle size of 1-3 μm.

Fig.2-96 is the excitation and emission spectra of $SrGe_4O_9:4xMn^{4+}$ ($x = 0.002, 0.004, 0.006, 0.008, 0.010$). These samples share similar excitation spectra with two broad excitation bands in the UV (280 nm) and blue light (430 nm) regions, which are attributed to the spin-allowed transitions of $Mn^{4+}$ from the ground state $^4A_{2g}$ to the excited states $^4T_{1g}$ and $^4T_{2g}$, respectively. Fig.2-96(b) is the emission spectra of these phosphors $SrGe_4O_9:4xMn^{4+}$. The strongest emission peak is located at 655 nm, which is ascribed to the electric-dipole-forbidden transition $^2E_g \rightarrow {}^4A_{2g}$ of $Mn^{4+}$. The inserted figure in Fig.2-96(b) is the concentration dependence of the relative emission intensity of $Mn^{4+}$ $^4A_{2g} \rightarrow {}^4T_{2g}$ transition in $SrGe_4O_9:4xMn^{4+}$. When the concentration of $Mn^{4+}$ is 0.6%, the emission intensity of $SrGe_4O_9:0.024Mn^{4+}$ is the strongest among these phosphors.

The critical concentration of the concentration quenching can be used as a measure of the critical distance ($R_c$) of energy transfer. The $R_c$ value can be calculated using the following equation:

$$R_c = 2\left(\frac{3V}{4\pi x_c N}\right)^{\frac{1}{3}} \tag{2-8}$$

where $x_c$ is critical concentration, $N$ is the number of $Mn^{4+}$ ions in the unit cell and $V$ is the volume of the unit cell. In this case, the values for $x_c$, $N$, $V$ are 0.024, 3, 529.37 $Å^3$ respectively. The calculated $R_c$ value is about 24.12 Å for substitution of $Mn^{4+}$ at the $Ge^{4+}$ site. From this value, the $R_c$ of energy transfer is larger than the distance ($R$) between the $Mn^{4+}$ ions (the value of $R$ is about 1.80 Å, see ICSD 82393). As it can be seen that $R<R_c$, it is presumed that energy transfer between $Mn^{4+}$ ions dominates in the case of these red phosphors.

Fig.2-97 shows the decay curve for the $^2E_g \rightarrow {}^4A_{2g}$ transition (633 nm) of the $Mn^{4+}$ in $SrGe_4O_9:0.024Mn^{4+}$ red phosphor. This decay curve is well fitted into one single-exponential function, and the lifetime $\tau$ value of $SrGe_4O_9:0.024Mn^{4+}$ is about 0.99 ms. This result complements the experimental data of previous reports prepared by other methods.

Several LED devices were fabricated by coating the phosphor $SrGe_4O_9:0.024Mn^{4+}$ with or without YAG on the blue GaN chips. Curve a (Fig.2-98) is the EL spectrum of the GaN chip. The emission peak of the LED chip is located at 460 nm. The single red LED fabricated with

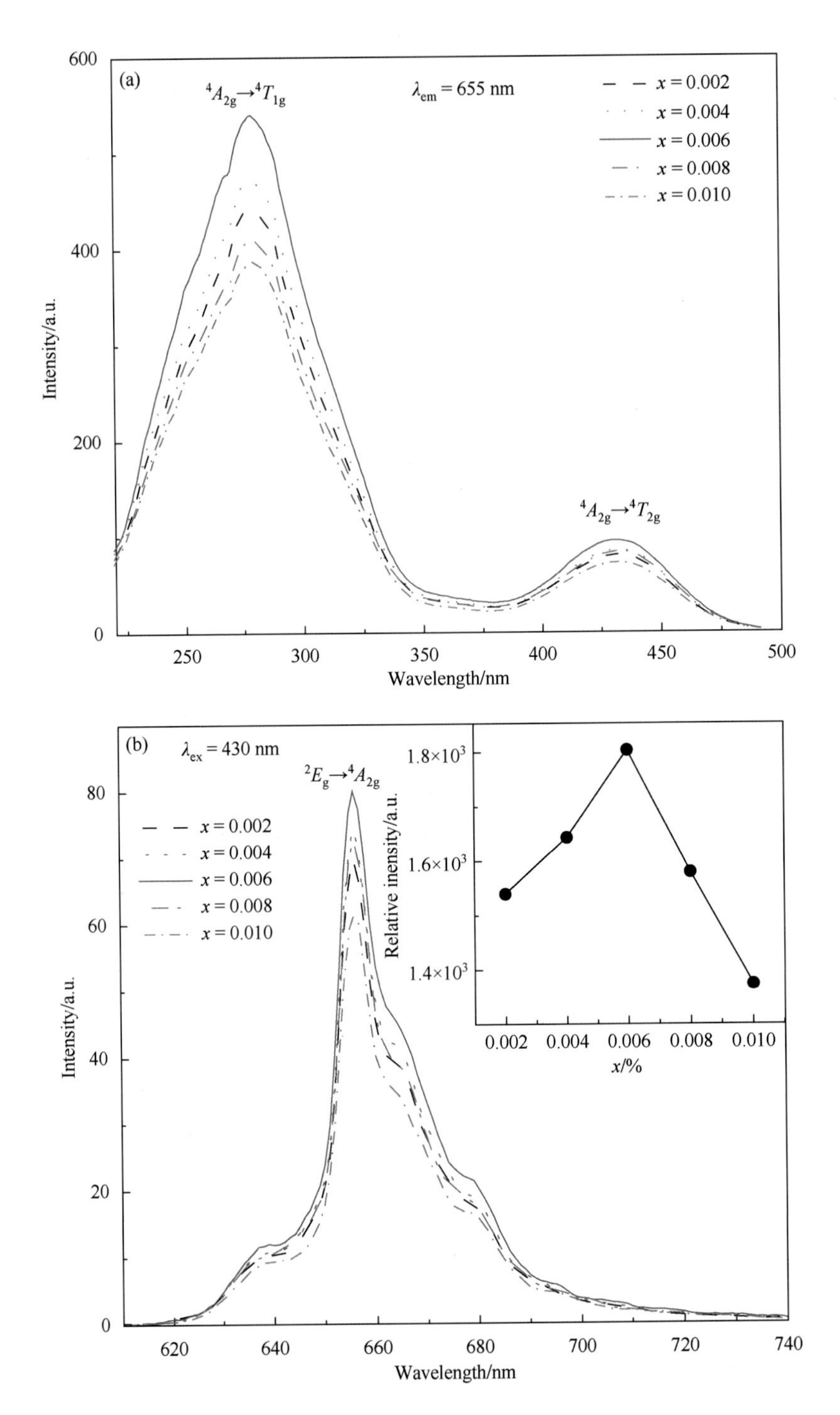


Fig.2-96 Excitation (a) and emission spectra (b) of the $SrGe_4O_9:4xMn^{4+}$ ($x$ = 0.002, 0.004, 0.006, 0.008, 0.010). The inserted figure is the concentration dependence of the relative emission intensity of $Mn^{4+}$ in $SrGe_4O_9:4xMn^{4+}$

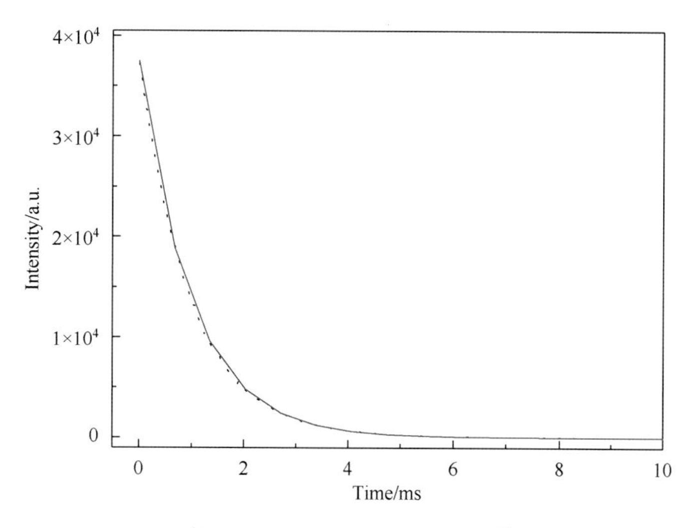


Fig.2-97 Decay curve of $Mn^{4+}$ emission of $SrGe_4O_9$:0.024$Mn^{4+}$ ($\lambda_{ex}$ = 430 nm; $\lambda_{em}$ = 655 nm)

$SrGe_4O_9$:0.024$Mn^{4+}$ exhibits intense red emission at 655 nm, which is due to the emission of $SrGe_4O_9$:0.024$Mn^{4+}$ excited by the emission of the blue LED chip. This result shows this red phosphor can be efficiently excited by the blue LED chip. Intensive emission of a LED chip can also be found in Fig.2-98. It is advantageous to obtain a w-LED by combining this phosphor with yellow phosphor YAG. Curve c is the EL spectrum of the w-LED with YAG. The white is obtained by combining the emission of the GaN chip with the greenish-yellow emission YAG. With the introduction of $SrGe_4O_9$:0.024$Mn^{4+}$, red emission peaks can be observed obviously at 655 nm from the spectrum d, which are due to the $^2E_g \rightarrow ^4A_{2g}$ transition of

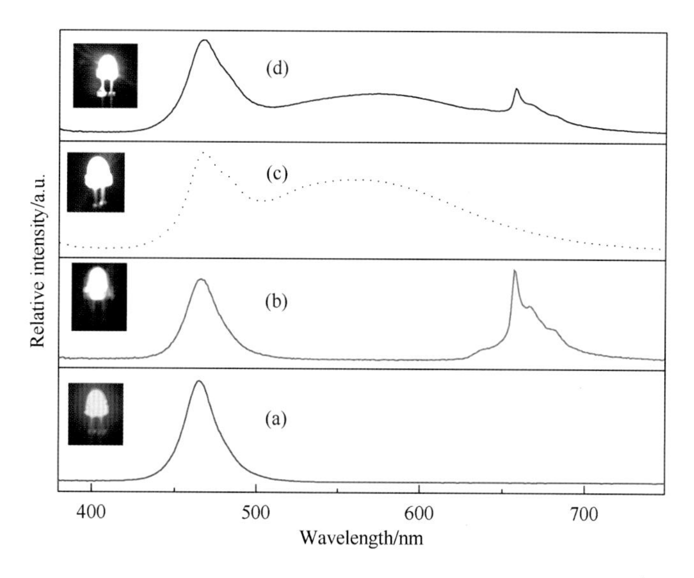


Fig.2-98 EL spectra of the GaN LED chip (a), red LED based on $SrGe_4O_9$:0.024$Mn^{4+}$ (b), w-LED based on YAG (c) and w-LED based on YAG and $SrGe_4O_9$:0.024$Mn^{4+}$ under 20 mA current excitation. The inserted figures are the images of these LEDs devices

$Mn^{4+}$. The bright white light can be observed by naked eyes from this w-LED when it is excited with 20 mA current. These results demonstrate that this red phosphor may find potential application in w-LEDs.

In summary, $SrGe_4O_9$:$Mn^{4+}$ red phosphors have been successfully synthesized by the solid-state method. The single-phase $SrGe_4O_9$:0.024$Mn^{4+}$ can be efficiently excited by the blue LED chip with intense red emission. The single red LED and w-LED based on this phosphor can emit intense red light and white light, respectively. So $SrGe_4O_9$:0.024$Mn^{4+}$ may be a promising red phosphor for the GaN-LED chip.

### 2.2.2 $BaGe_4O_9$:$Mn^{4+}$

$Mn^{4+}$ ions doped phosphors exhibit the efficient red emission of $Mn^{4+}$ with one broadband in the UV and blue light regions, which can be used in various applications, such as lighting, holographic recording, and thermoluminescence dosimetry. It is well known that $Mn^{4+}$ ions in different hosts with different crystal field share different luminescent properties since the d-d transition of $Mn^{4+}$ is highly influenced by the crystal field.

In recent years, the luminescent properties of $Mn^{4+}$ doped composite oxides have been investigated in detail. Such as, Dramićanin's group discussed the detailed spectroscopic and crystal-field analysis of $Mg_2TiO_4$:$Mn^{4+}$. $BaGe_4O_9$ has a similar crystal structure and the same space group *P*321 with $SrGe_4O_9$. $Ge^{4+}$ not only has the same valence state as $Mn^{4+}$ but also shares the identical ionic radius as $Mn^{4+}$ (53.0 pm, CN = 6). Hence, $Mn^{4+}$ ions will easily substitute the crystal sites of $Ge^{4+}$ ions, when $Mn^{4+}$ ions are doped into $BaGe_4O_9$. To the best of our knowledge, the luminescent properties of $BaGe_4O_9$ doped with $Mn^{4+}$ have not been reported. In this part, red phosphors $BaGe_4O_9$ doped with different contents of $Mn^{4+}$ have been prepared at high temperature, and their structure, luminescent properties, and application on w-LEDs were investigated.

Fig.2-99(a) exhibits the XRD pattern of the selected sample $BaGe_4O_9$:0.02$Mn^{4+}$, which is consistent with the JCPDS card of $BaGe_4O_9$ (No. 74-2156, space group *P*321). No other phase can be found from this pattern. This result indicates the red phosphor is of a single phase with a hexagonal structure. A small amount of $Mn^{4+}$ can easily be doped into the $BaGe_4O_9$ crystal lattice and does not change the crystal structure of this $BaGe_4O_9$ host. According to the crystal structure of $BaGe_4O_9$, as shown in Fig.2-99(b), there are two kinds of $Ge^{4+}$ sites observed in this crystal structure. One kind of $Ge^{4+}$ is coordinated with six $O^{2-}$ to form a regular octahedron. Another kind of $Ge^{4+}$ is in the center of one tetrahedron with four $O^{2-}$. Since $Mn^{4+}$ has the same charge and radius as $Ge^{4+}$, $Mn^{4+}$ maybe occupy the sites of $Ge^{4+}$ in the crystal structure of $BaGe_4O_9$:$Mn^{4+}$. The corresponding SEM image is shown in Fig.2-99(c). The as-prepared sample synthesized at high temperature is of irregular morphology with a particle size range from 10 μm to 30 μm.

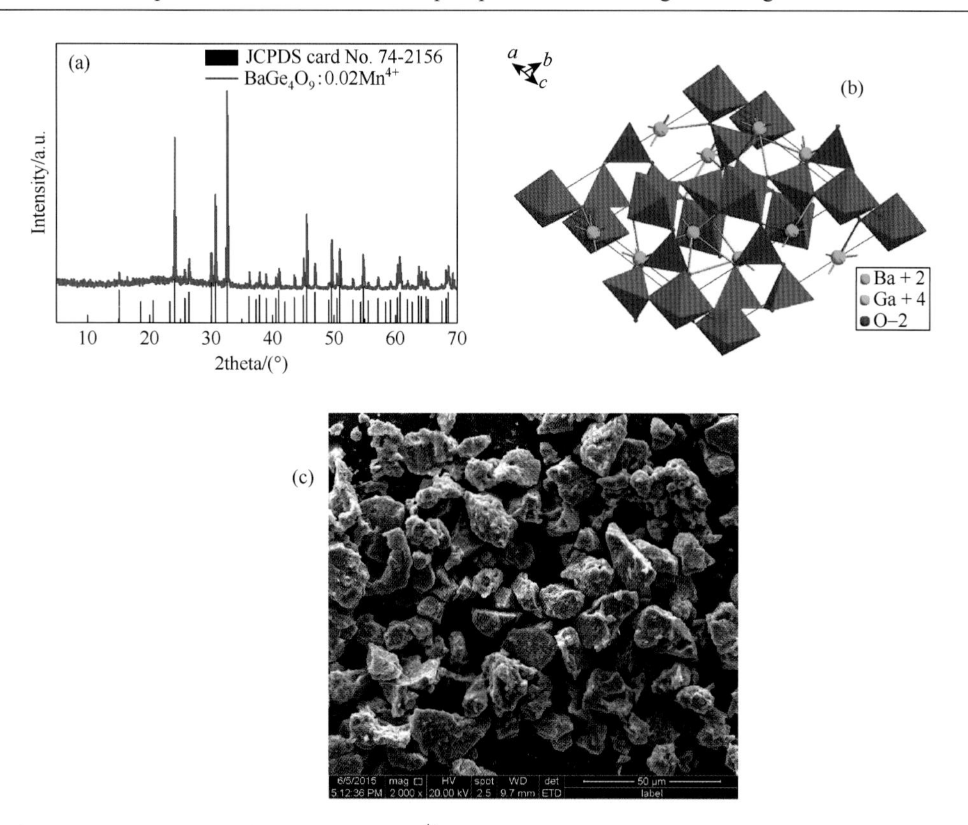


Fig.2-99 XRD pattern of $BaGe_4O_9$:$0.02Mn^{4+}$ (a), crystal structure of $BaGe_4O_9$ (b)and SEM image of $BaGe_4O_9$:$0.02Mn^{4+}$ (c)

Fig.2-100 is the excitation spectra of $BaGe_4O_9$:$4xMn^{4+}$ ($x$ = 0.0005, 0.001, 0.002, 0.003, 0.004) by monitoring emission at 667 nm. These samples exhibit similar excitation spectra with two broad excitation bands in the UV (290 nm) and blue light (435 nm) regions. The former is due to the spin-allowed between $^4A_{2g}$ and $^4T_{1g}$ transition of $Mn^{4+}$, and the latter is ascribed to the same transition of $^4A_{2g}$ to $^4T_{2g}$. When the content of $Mn^{4+}$ is 0.008, the excitation intensity reached the maximum. These excitation spectra show some red-shifts, compared with that of $SrGe_4O_9$:$Mn^{4+}$. According to the Tanabe-Sugano energy-level diagram, the excitation peak position of $Mn^{4+}$ in different hosts is largely affected by different levels of the crystal field strength. The radius of $Ba^{2+}$ ($r$ = 135 pm, CN = 6) is larger than that of $Sr^{2+}$ ($r$ = 118 pm, CN = 6), $BaGe_4O_9$:$Mn^{4+}$ possesses weaker crystal field strength, which leads to this red-shift in excitation spectra.

The emission spectra of these phosphors $BaGe_4O_9$:$4xMn^{4+}$ under 433 nm excitation are shown in Fig.2-101. The strongest emission peak is located at 667 nm, which is ascribed to the electric-dipole-forbidden transition $^2E_g \rightarrow ^4A_{2g}$ of $Mn^{4+}$. The inserted figure in Fig.2-101 is the concentration dependence of the relative emission intensity of $Mn^{4+}$ in $BaGe_4O_9$:$4xMn^{4+}$. When the concentration of $Mn^{4+}$ is 0.008, the emission intensity of $BaGe_4O_9$:$0.008Mn^{4+}$ is the strongest among these phosphors.

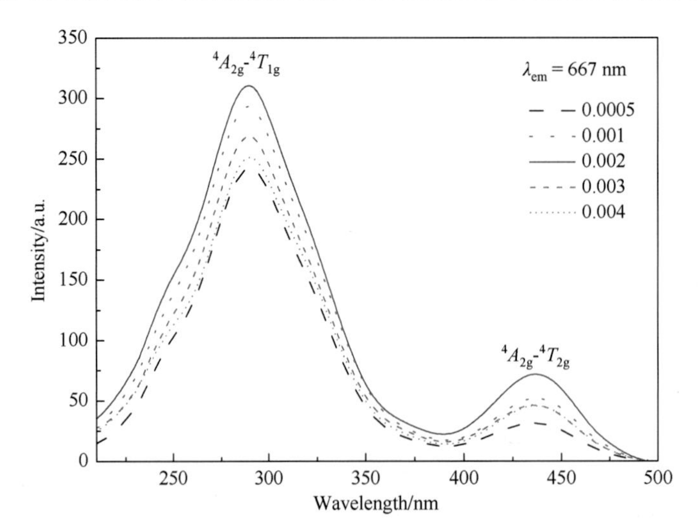


Fig.2-100 Excitation spectra of the $BaGe_4O_9$:4$x$$Mn^{4+}$ ($x$ = 0.0005, 0.001, 0.002, 0.003, 0.004) by monitoring emission at 667 nm

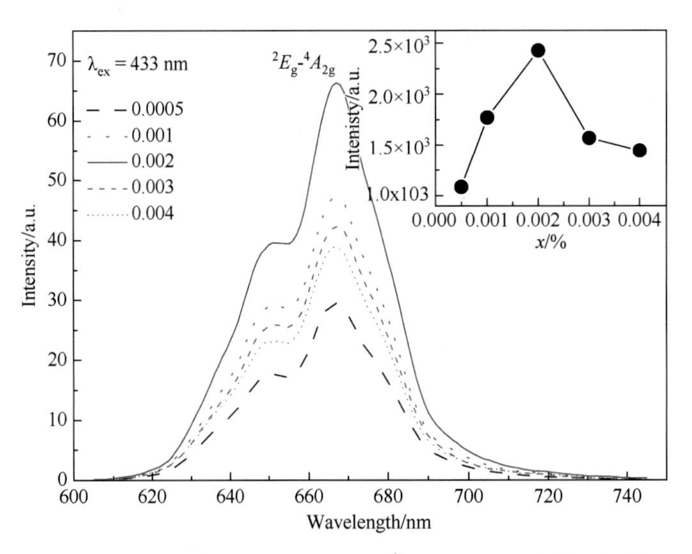


Fig.2-101 Emission Spectra of the $BaGe_4O_9$:4$x$$Mn^{4+}$ ($x$ = 0.0005, 0.001, 0.002, 0.003, 0.004) under 433 nm excitation. The inserted figure is concentration dependence of the relative emission intensity

We can calculate the $R_c$ of energy transfer among $Mn^{4+}$ ions using the equation (2-8). In this case, the quenching content ($x_c$), $N$, $V$ are 0.008, 3, 553.32 Å$^3$. The calculated $R_c$ value is about 35.31 Å for substitution of $Mn^{4+}$ at the $Ge^{4+}$ site. From this value, the $R_c$ of energy transfer is larger than the distance ($R$) between the $Mn^{4+}$ ions ($R$ is about 3.15 Å, seeing ICSD 28203). Since the value of the $R$ is smaller than that of the $R_c$, it is presumed that energy transfer between $Mn^{4+}$ ions dominates in $BaGe_4O_9$:$Mn^{4+}$.

Fig.2-102 shows the decay curve for the $^2E_g \rightarrow ^4A_{2g}$ transition (667 nm) of the $Mn^{4+}$ in $BaGe_4O_9$:0.008$Mn^{4+}$ red phosphor. This decay curve is well fitted into the single-exponential function, and the lifetime $\tau$ value of $BaGe_4O_9$:0.008$Mn^{4+}$ is about 1.69 ms.

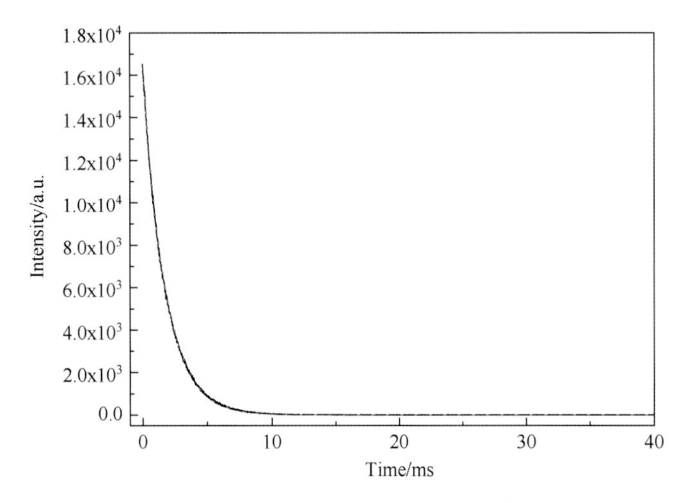


Fig.2-102 Decay curve of $Mn^{4+}$ emission of $BaGe_4O_9$:0.008$Mn^{4+}$($\lambda_{ex}$ = 433 nm; $\lambda_{em}$ = 667 nm)

To investigate the application of as-prepared phosphor on LEDs, a series of LED devices were fabricated by coating the phosphor $BaGe_4O_9$:0.008$Mn^{4+}$ with or without YAG on the blue GaN chips. Curve a in Fig.2-103 is the EL spectrum of the GaN chip with the strongest emission peak at 460 nm. The single red LED fabricated with $BaGe_4O_9$:0.008$Mn^{4+}$ exhibits red emission at 667 nm, which is due to the emission of $BaGe_4O_9$:0.008$Mn^{4+}$ excited by the emission of the blue LED chip. Intensive emission of LED the chip also can be found in the EL spectrum of red LED with this red phosphor. It is advantageous to obtain a w-LED by combining this phosphor with yellow phosphor YAG. The EL spectrum of w-LED with

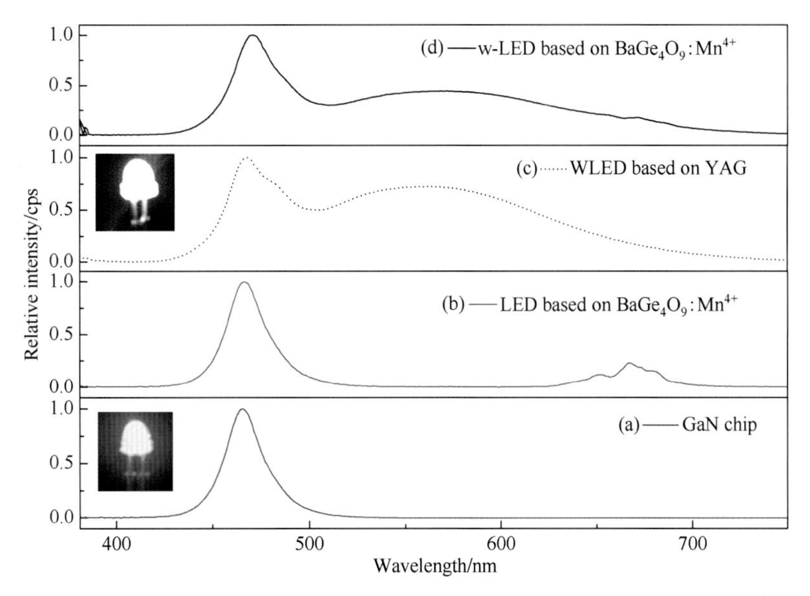


Fig.2-103 EL spectra of the InGaN LED chip (a), the red LED-based on $BaGe_4O_9$:0.008$Mn^{4+}$ (b), the w-LED based on YAG (c) and the w-LED based on YAG with $BaGe_4O_9$:0.008$Mn^{4+}$ (d) under 20 mA current excitation. The inserted figures are the images of these LEDs

YAG is shown in Fig.2-103(c). The white light is obtained by mixing the blue emission of the GaN chip with the greenish-yellow emission of YAG. With the introduction of $BaGe_4O_9$:0.008$Mn^{4+}$, red emission peaks of $Mn^{4+}$ can be observed at 667 nm, which are overlapped by the emission of YAG from curve d. The bright white light can be observed by naked eyes from this w-LED when it is excited with 20 mA current.

## References


Adachi S, Takahashi T. 2008. Direct synthesis and properties of $K_2SiF_6$: $Mn^{4+}$ phosphor by wet chemical etching of Si wafer. Journal of Applied Physics, 104 (2): 317.

Adachi S, Takahashi T. 2009. Photoluminescent properties of $K_2GeF_6$: $Mn^{4+}$ red phosphor synthesized from aqueous HF/$KMnO_4$ solution. Journal of Applied Physics, 106 (1): 339.

Arai Y, Adachi S. 2011. Optical properties of $Mn^{4+}$-activated $Na_2SnF_6$ and $Cs_2SnF_6$ red phosphors. Journal of Luminescence, 131 (12): 2652.

Arai Y, Adachi S. 2011. Optical transitions and internal vibronic frequencies of MnF[sub 6][sup 2-] ions in $Cs_2SiF_6$ and $Cs_2GeF_6$ red phosphors. Journal of The Electrochemical Society, 158 (6): 179.

Arai Y, Takahashi T, Adachi S. 2010. Photoluminescent properties of $K_2SnF_6$ ·$H_2O$: $Mn^{4+}$ red phosphor. Optical Materials, 32 (9): 1095.

Avram A M, Brik (Eds.) M G. 2013. Optical properties of 3d-ions in crystals: spectroscopy and crystal field analysis. Berlin: Springer and Tsinghua University Press.

Brik M G, Camardello S J, Srivastava A M, et al. 2016. Spin-forbidden transitions in the spectra of transition metal ions and nephelauxetic effect. ECS Journal of Solid State Science and Technology, 5 (1): 3067.

Brik M G, Srivastava A M, Avaram N M. 2011. Comparative analysis of crystal field effects and optical spectroscopy of six-coordinated $Mn^{4+}$ ion in the $Y_2Ti_2O_7$ and $Y_2Sn_2O_7$ pyrochlores. Optical Materials, 33 (11): 1671.

Brik M G, Srivastava A M. 2013. On the optical properties of the $Mn^{4+}$ ion in solids. Journal of Luminescence, 133 (1): 69.

Cao R P, Peng M Y, Song E H, et al. 2012. High efficiency $Mn^{4+}$ doped $Sr_2MgAl_{22}O_{36}$ red-emitting phosphor for white LED. ECS Journal of Solid State Science and Technology, 1 (4): 123.

Chan T S, Liu R S, Baginskiy I. 2008. Synthesis, crystal structure, and luminescence properties of a novel green-yellow emitting phosphor $LiZn_{1-x}PO_4$:$Mn_x$ for light-emitting diodes. Chemistry of Materials, 20 (4): 1215.

Chen W T, Shen H S, Liu R S, et al. 2012. Cation-size-mismatch tuning of photoluminescence in oxynitride phosphors. Journal of the American Chemical Society, 134 (19): 8022.

Chuang P H, Lin C C, Yang H, et al. 2013. Enhancing the color rendering index for phosphor-converted white LEDs using cadmium-free $cuInS_2$/ZnS QDs. Journal of the Chinese Chemical Society, 60 (7): 801.

Daicho H, Iwasaki T, Enomoto K, et al. 2012. A novel phosphor for glareless white light-emitting diodes. Nature Communications, 3: 1132.

Denault K A, George N C, Paden S R, et al. 2012. A green-yellow emitting oxyfluoride solid solution phosphor $Sr_2Ba$ $(AlO_4F)_{1-x}$ $(SiO_5)_x$: $Ce^{3+}$ for thermally stable, high color rendition solid-state white lighting. Journal of Materials Chemistry, 22 (35): 18204.

Dexter D L, Schulman J H. 1954. Theory of concentration quenching in inorganic phosphors. Journal of Chemical Physics, 22 (6): 1063.

Ding X, Wang Q, Wang Y H. 2016. Rare-earth-free red-emitting $K_2Ge_4O_9$:Mn (4+) phosphor excited by blue light for warm white LEDs. Physical Chemistry Chemical Physics, 18 (11): 8088.

Du M H. 2014. Chemical trends of Mn emission in solids. Journal of Materials Chemistry C, 2 (14): 2475.

Duan C J, Wang X J, Otten W M, et al. 2008.ChemInform abstract: preparation, electronic structure, and photoluminescence properties of $Eu^{2+}$-and $Ce^{3+}$/$Li^{+}$-Activated Alkaline Earth Silicon Nitride $MSiN_2$ (M: Sr, Ba). Chemistry of Materials, 39 (20):

1597.

Fang M H, Nguyen H D, Lin C C, et al. 2015. Preparation of a novel red $Rb_2SiF_6$: $Mn^{4+}$ phosphor with high thermal stability through a simple one-step approach. Journal of Materials Chemistry C, 3 (28): 7277.

Hoskins B F, Linden A, Mulvaney P C, et al. 1984. The structures of barium hexafluorosilicate and cesium hexafluororhenate (V). Inorganica Chimica Acta, 88 (2): 217

Jiang X Y, Chen Z, Huang S M, et al. 2014. A red phosphor $BaTiF_6$: $Mn^{4+}$: reaction mechanism, microstructures, optical properties, and applications for white LEDs. Dalton Transactions, 43 (25): 9414.

Jiang X Y, Pan Y X, Huang S M, et al. 2014. Hydrothermal synthesis and photoluminescence properties of red phosphor $BaSiF_6$: $Mn^{4+}$ for LED applications. Journal of Materials Chemistry C, 2 (13): 2301.

Kasa R, Adachi S. 2012. Red and deep red emissions from cubic $K_2SiF_6$: $Mn^{4+}$ and hexagonal $K_2MnF_6$ synthesized in $HF/KMnO_4/KHF_2/Si$ solutions. Journal of the Electrochemical Society, 159 (4): 89.

Kasa R, Arai Y, Takahashi T, et al. 2010. Photoluminescent properties of cubic $K_2MnF_6$ particles synthesized in metal immersed $HF/KMnO_4$ solutions. Journal of Applied Physics, 108: 113503.

Kubus M, Enseling D, Justel T, et al. 2013. Synthesis and luminescent properties of red-emitting phosphors: $ZnSiF_6·6H_2O$ and $ZnGeF_6·6H_2O$ doped with $Mn^{4+}$. Journal of Luminescence. 137: 88.

Li G G, Geng D L, Shang M M, et al. 2011. Tunable luminescence of $Ce^{3+}$ /$Mn^{2+}$-coactivated $Ca_2Gd_8$ $(SiO_4)_6O_2$ through energy transfer and modulation of excitation: potential single-phase white/yellow-emitting phosphors. Journal of Materials Chemistry, 21 (35): 13334.

Li X Q, Su X M, Liu P, et al. 2015. Shape-controlled synthesis of phosphor $K_2SiF_6$: $Mn^{4+}$ nanorods and their luminescence properties. CrystEngComm, 17 (4): 930.

Liao C X, Cao R P, Ma Z J, et al. 2013. Synthesis of $K_2SiF_6$: $Mn^{4+}$ phosphor from $SiO_2$ powders via redox reaction in $HF/KMnO_4$ solution and their application in warm-white LED. Journal of the American Ceramic Society, 96 (11): 3552.

Lin C C, Liu R S. 2011. Advances in phosphors for light-emitting diodes. The Journal of Physical Chemistry Letters, 2 (11): 1268.

Liu Y, Huang A Q, Yang S C, et al. 2022. Synthesis and optical properties of a new double-perovskite $Rb_2KInF_6$:$Mn^{4+}$ red phosphor used for blue LED pumped white lighting. Optical Materials, 127: 112307.

Loutts G B, Warren M, Taylor L, et al. 1998. Manganese-doped yttrium orthoaluminate: a potential material for holographic recording and data storage. Physical Review B, 57 (7): 3706.

Lv L F, Chen Z, Liu G K, et al. 2015. Optimized photoluminescence of red phosphor $K_2TiF_6$: $Mn^{4+}$ synthesized at room temperature and its formation mechanism. Journal of Materials Chemistry C, 3 (9): 1935.

Lv L F, Jiang X Y, Huang S M, et al. 2014. The formation mechanism, improved photoluminescence and LED applications of red phosphor $K_2SiF_6$: $Mn^{4+}$. Journal of Materials Chemistry C, 2 (20): 3879.

Medic M M, Brik M G, Drazic G, et al. 2015. Deep-red emitting $Mn^{4+}$ doped $Mg_2TiO_4$ nanoparticles. The Journal of Physical Chemistry C, 119 (1): 724.

Nguyen H D, Lin C C, Fang M H, et al. 2014. Synthesis of $Na_2SiF_6$: $Mn^{4+}$ red phosphors for white LED applications by co-precipitation. Journal of Materials Chemistry C, 2 (48): 10268.

Novita M, Ogasawara K. 2012. Comparative study of multiplet structures of $Mn^{4+}$ in $K_2SiF_6$, $K_2GeF_6$, and $K_2TiF_6$ based on first-principles configuration--interaction calculations. Japanese Journal of Applied Physics, 51 (2): 2604.

Oh J H, Kang H, Eo Y J, et al, 2014. Synthesis of narrow-band red-emitting $K_2SiF_6$: $Mn^{4+}$ phosphors for a deep red monochromatic LED and ultrahigh color quality warm-white LEDs. Journal of Materials Chemistry C, 3 (3): 607.

Palmer W G. 1954. Experimental inorganic chemistry. Cambridge: Cambridge University Press.

Park W B, Singh S P, Yoon C, et al. 2012. $Eu^{2+}$ luminescence from 5 different crystallographic sites in a novel red phosphor, $Ca_{15}Si_{20}O_{10}N_{30}$: $Eu^{2+}$. Journal of Materials Chemistry, 22 (28): 14068.

Paulusz A G. 1973. Efficient Mn (IV) emission in fluorine coordination. Journal of The Electrochemical Society, 120 (7): 942.

Perdew J P, Burke K, Ernzerhof M. 1996. Generalized gradient approximation made simple. Physical Review Letters, 77 (18): 3865.

Qiu Z X, Luo T T, Zhang J L, et al. 2015. Effectively enhancing blue excitation of red phosphor $Mg_2TiO_4:Mn^{4+}$ by $Bi^{3+}$ sensitization. Journal of Luminescence, 158: 130.

Reisfeld M J, Matwiyoff N A, Asprey L B. 1971. The electronic spectrum of cesium hexafluoromanganese (IV). Journal of Molecular Spectroscopy, 39 (1): 8.

Reisz A, Avram C N. 2007. Geometry of the $^4T_g$ excited state in $Cs_2SiF_6$: $Mn^{4+}$. Acta Physica Polonica A, 112 (5): 829.

Sakurai S, Nakamura T, Adachi S. 2016. Editors' Choice—$Rb_2SiF_6$: $Mn^{4+}$ and $Rb_2TiF_6$: $Mn^{4+}$ red-emitting phosphors. ECS Journal of Solid State Science and Technology, 5 (12): 206.

Schubert E F, Kim J K. 2005. Solid-state light sources getting smart. Science, 308 (5726): 1274.

Segall M D, Lindan P J D, Probert M J, et al. 2002. First-principles simulation: ideas, illustrations and the CASTEP code. Journal of Physics: Condensed Matter, 14 (11): 2717.

Seki K, Kamei S, Uematsu K, et al. 2013. Enhancement of the luminescence efficiency of $Li_2TiO_3:Mn^{4+}$ red emitting phosphor for white LEDs. Journal of Ceramic Processing Research, 14 (1): 67.

Sekiguchi D, Adachi S. 2014. Synthesis and optical properties of $BaTiF_6$: $Mn^{4+}$ red phosphor. ECS Journal of Solid State Science and Technology, 3 (4): 60.

Sekiguchi D, Nara J, Adachi S. 2013. Photoluminescence and raman scattering spectro scopies of $BaSiF_6$: $Mn^{4+}$ red phosphor. Journal of Applied Physics, 113 (18): 183516.

Setlur A A, Heward W J, Gao Y, et al. 2006. Crystal chemistry and luminescence of $Ce^{3+}$-doped $Lu_2CaMg_2(Si, Ge)_3O_{12}$ and its use in LED based lighting. Chemistry of Materials, 18 (14): 3314.

Song E H, Wang J Q, Ye S, et al.2016. Room-temperature synthesis and warm-white LED applications of $Mn^{4+}$ ion doped fluoroaluminate red phosphor $Na_3AlF_6$: $Mn^{4+}$. Journal of Materials Chemistry C, 4 (13): 2480.

Tan H Y, Rong M Z, Zhou Y Y, et al. 2016. Luminescence behavior of $Mn^{4+}$ ions in seven coordination environments of $K_3ZrF_7$. Dalton Transactions, 45 (23): 9654.

Van Ipenburg M E, Dirksen G J, Blasse G. 1995. Charge-transfer excitation of transition-metal-ion luminescence. Materials Chemistry and Physics, 39 (3): 236.

Wang H, He P, Liu S, et al. 2009. A europium (III) organic ternary complex applied in fabrication of near UV-based white light-emitting diodes. Applied Physics B, 97 (2): 481.

Wang L, Wang X J, Kohsei T, et al. 2015. Highly efficient narrow-band green and red phosphors enabling wider color-gamut LED backlight for more brilliant displays. Optics Express, 23 (22): 28707.

Wang L, Xu Y, Wang D, et al. 2013. Deep red phosphors $SrAl1_2O_{19}:Mn^{4+}$, M (M = $Li^+$, $Na^+$, $K^+$, $Mg^{2+}$) for high colour rendering white LEDs. Physics Status Solidi A, 210 (7): 1433.

Wei L L, Lin C C, Fang M H, et al. 2015. A low-temperature co-precipitation approach to synthesize fluoride phosphors $K_2MF_6$: $Mn^{4+}$ (M = Ge, Si) for white LED applications. Journal of Materials Chemistry C, 3 (8): 1655.

Wei L L, Lin C C, Wang Y Y, et al. 2015. Photoluminescent evolution induced by structural transformation through thermal treating in the red narrow-band phosphor $K_2GeF_6$: $Mn^{4+}$. ACS Applied Materials & Interfaces, 7 (20): 10656.

Wen D W, Feng J J, Li J H, et al. 2015. $K_2Ln$ ($PO_4$) ($WO_4$): $Tb^{3+}$, $Eu^{3+}$ (Ln = Y, Gd and Lu) phosphors: highly efficient pure red and tuneable emission for white light-emitting diodes. Journal of Materials Chemistry C, 3 (9): 2107.

Xu H P, Hong F, Liu G X, et al. 2020. Green route synthesis and optimized luminescence of $K_2SiF_6:Mn^{4+}$ red phosphor for warm WLEDs. Optical Materials, 99: 109500.

Xu H P, Hong F, Pang G, et al. 2020. Co-precipitation synthesis, luminescent properties and application in warm WLEDs of $Na_3GaF_6:Mn^{4+}$ red phosphor. Journal of Luminescence, 219: 116960.

Xu Y K, Adachi S. 2011. Properties of $Mn^{4+}$-activated hexafluorotitanate phosphors. Journal of The Electrochemical Society, 158 (3): J58.

Yamada S, Emoto H, Ibukiyama M, et al. 2012. Properties of SiAlON powder phosphors for white LEDs. Journal of the European Ceramic Society, 32 (7): 1355.

Yan X, Li W, Wang X, et al. 2012. Facile synthesis of $Ce^{3+}$, $Eu^{3+}$ co-doped YAG nanophosphor for white light-emitting diodes. Journal of The Electrochemical Society, 159 (2): 195.

Ye T, Li S, Wu X, et al. 2013. Sol-gel preparation of efficient red phosphor $Mg_2TiO_4$:$Mn^{4+}$ and XAFS investigation on the substitution of $Mn^{4+}$ for $Ti^{4+}$. Journal of Materials Chemistry C, 1 (28): 4327.

Yen William M, Shigeo S, Hajime Y. 2006. Phosphor handbook (second ed). Boca Raton: CRC Press.

Zeng Q H, He P, Pang M, et al. 2009. $Sr_9R_{2-x}Eu_xW_4O_{24}$ (R = Gd and Y) red phosphor for near-UV and blue InGaN-based white LEDs. Solid State Communications, 149 (21-22): 880.

Zhou Q, Zhou Y Y, Liu Y, et al. 2015. A new and efficient red phosphor for solid-state lighting: $Cs_2TiF_6$: $Mn^{4+}$. Journal of Materials Chemistry C, 3 (37): 9615.

Zhou Q, Zhou Y Y, Liu Y, et al. 2015. A new red phosphor $BaGeF_6$: $Mn^{4+}$: hydrothermal synthesis, photo-luminescence properties, and its application in warm white LED devices. Journal of Materials Chemistry C, 3 (13): 3055.

Zhou Q, Zhou Y Y, Wang Z L, et al. 2015. Fabrication and application of non-rare earth red phosphors for warm white-light-emitting diodes. RSC Advances, 5 (103): 84821.

Zhu H M, Lin C C, Luo W Q, et al. 2014. Highly efficient non-rare-earth red-emitting phosphor for warm white light-emitting diodes. Nature Communications, 5: 4312.

# Chapter III Green phosphors for white light-emitting diodes

## 3.1 $Tb^{3+}$-activated double molybdates and tungstates

### 3.1.1 Introduction

w-LEDs with the advantages of long-lifetime, low energy consumption and environmental-friendly characteristics are thought to be the most important solid-state light sources for substitution the widely used incandescent lamps and fluorescent lamps. Presently, the emission bands of LED chips are shifted to the near-UV range since this wavelength can offer a higher efficiency solid-state lighting. However, the main tricolor phosphors for the near-UV InGaN-based LEDs are still some classic phosphors, such as $BaMgAl_{10}O_{17}:Eu^{2+}$ for blue, $ZnS:(Cu^{+}, Al^{3+})$ for green and $Y_2O_2S:Eu^{3+}$ for red. Conventional phosphors used in fluorescent lighting are not suitable for the near-UV LED chips because of their poor absorption in the near-UV region. Hence, it is interesting to research new tricolor phosphors.

Alkaline rare-earth double molybdates and tungstates, $AB(MO_4)_2$ ($A = Li^{+}$, $Na^{+}$; B = trivalent rare-earth ions; M = Mo, W), share a tetragonal scheelite-like ($CaWO_4$) isostructure. The alkali-metal ions and rare-earth ions are disordered in the same site. Mo/W (VI) ion is coordinated by four oxygen atoms in a tetrahedral site and the alkali-metal ion/rare-earth ion is coordinated by eight $O^{2-}$ of near four $MoO_4^{2-}$/$WO_4^{2-}$. They can find many applications, such as phosphors for lighting, display, scintillators, laser and amplifiers for fiber-optic communication. Recently, the powder samples of these compounds were investigated for they might be promising candidates as phosphors for the visual display and solid-state lighting. $Tb^{3+}$ ion is an excellent luminescent center in some hosts, and it also shows excellent green emission in this tetragonal scheelite structure.

As we know, excessive activator ions usually decrease the emission intensity markedly. This phenomenon is called “concentration quenching”, which is caused by the migration of excitation energy between luminescent centers and closely related to the crystal structure. The absence of concentration quenching has been reported in some hosts. However, this phenomenon for $Tb^{3+}$ ions in some hosts was not reported before. The phosphors, $NaLa(MoO_4)_2{:}xTb^{3+}$ and $NaLa(WO_4)_2{:}xTb^{3+}$, were prepared by the solid-state reaction technique in this book. Their PL properties and the concentration quenching phenomenon were investigated.

### 3.1.2 NaLa(MoO$_4$)$_2$:$x$Tb$^{3+}$

The XRD patterns of NaLa(MoO$_4$)$_2$:$x$Tb$^{3+}$ ($x$ = 0, 0. 05, 0.11) are shown in Fig.3-1. Curve a is the XRD pattern of NaLa(MoO$_4$)$_2$, which is consistent with the JCPDS card No. 24-1103 [NaLa(MoO$_4$)$_2$]. This shows that the compound is of the single-phase with the tetragonal structure. NaLa(MoO$_4$)$_2$:0.05Tb$^{3+}$ and NaLa(MoO$_4$)$_2$:0.11Tb$^{3+}$ are of similar patterns with that of NaLa(MoO$_4$)$_2$. It indicates that these phosphors doped with Tb$^{3+}$ are also of the single phase.

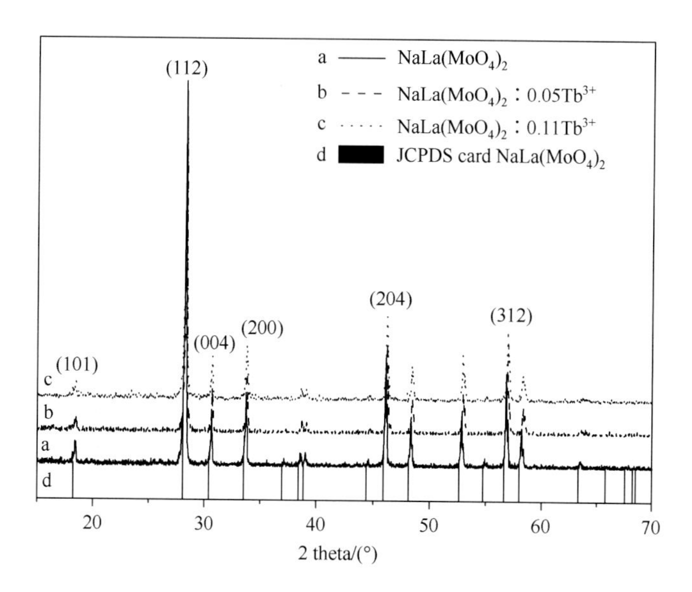


Fig.3-1 The XRD patterns of NaLa(MoO$_4$)$_2$:$x$Tb$^{3+}$ ($x$ = 0, 0. 05, 0.11)

Fig.3-2 shows the excitation and emission spectra of the phosphor NaLa(MoO$_4$)$_2$:0.11Tb$^{3+}$. The strong band from 250 nm to 350 nm is mainly ascribed to the O→Mo CT band. Some 4f-4f transitions of Tb$^{3+}$, such as the transitions of $^7F_6 \rightarrow ^5K_9$, $^7F_6 \rightarrow ^5H_7$ are overlapped by the Mo-O charge transfer band. The sharp peaks from 360 nm to 450 nm are ascribed to the 4f-4f transitions of Tb$^{3+}$ in the host. The emission spectra of NaLa(MoO$_4$)$_2$:0.11Tb$^{3+}$ are obtained under 378 nm and 270 nm excitation. The two strong emission peaks at 545 nm and 490 nm are due to $^5D_4 \rightarrow ^7F_5$ and $^5D_4 \rightarrow ^7F_6$ transitions of Tb$^{3+}$, respectively. Other $^5D_4 \rightarrow ^7F_J$ transitions are weaker. The concentration dependence of the relative emission intensity of Tb$^{3+}$ $^5D_4 \rightarrow ^7F_5$ transition in NaLa(MoO$_4$)$_2$:$x$Tb$^{3+}$ is shown in the insert of Fig.3-2. It can be observed that NaLa(MoO$_4$)$_2$:0.11Tb$^{3+}$ shows the strongest emission under 270 nm excitation.

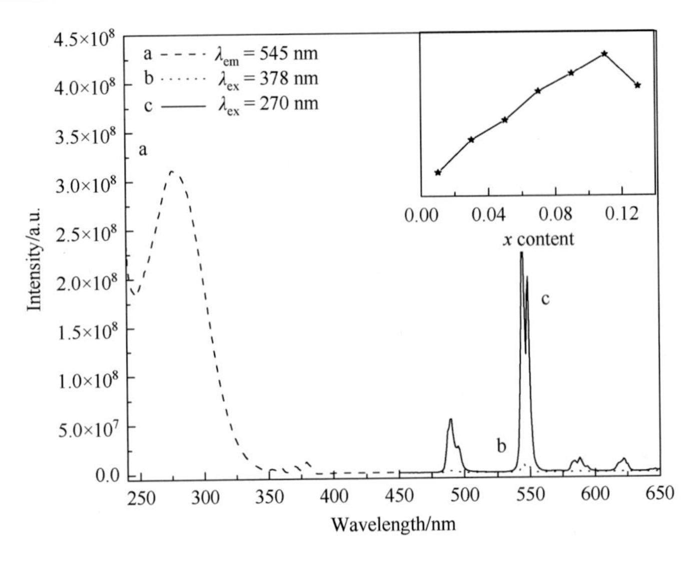


Fig.3-2　The excitation (a, $\lambda_{em}$ = 545 nm) and emission (b, $\lambda_{ex}$ = 378 nm; c, $\lambda_{ex}$ = 270 nm) spectra of $NaLa(MoO_4)_2$:0.11$Tb^{3+}$. The inserted figure is the concentration dependence of the relative emission intensity of $Tb^{3+}$ $^5D_4 \rightarrow {}^7F_5$ transition in $NaLa(MoO_4)_2$:$xTb^{3+}$

### 3.1.3　$NaLa(WO_4)_2$:$xTb^{3+}$

The XRD patterns of $NaLa(WO_4)_2$:$xTb^{3+}$ ($x$ = 0, 0.10, 0.20, 0.30, 0.40, 0.50, 0.60, 0.70, 0.80, 0.90, 1.00) were measured. As examples, the patterns of $NaLa(WO_4)_2$, $NaLa(WO_4)_2$:0.10$Tb^{3+}$ and $NaTb(WO_4)_2$ are shown in Fig.3-3. The patterns are very consistent with the JCPDS card No. 79-1118 [$NaLa(WO_4)_2$]. This result indicates that the samples have a single phase with the scheelite structure. In the structure, $Tb^{3+}$ ions occupy the same sites as $La^{3+}$

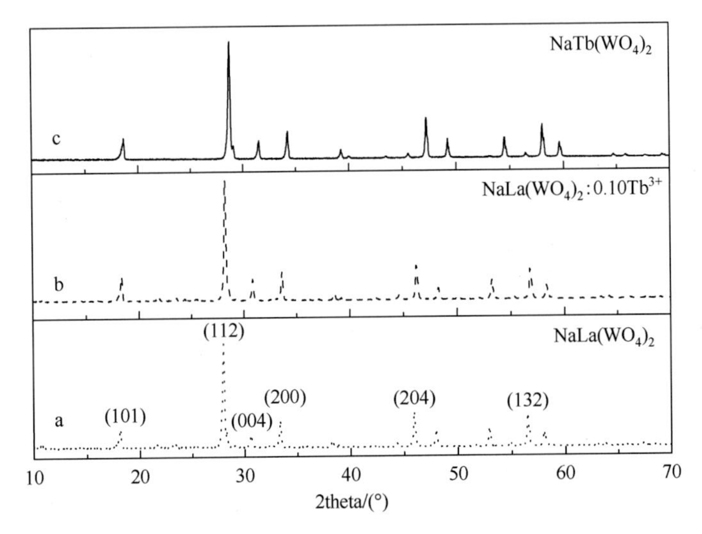


Fig.3-3　The XRD patterns of $NaLa(WO_4)_2$, $NaLa(WO_4)_2$:0.10$Tb^{3+}$ and $NaTb(WO_4)_2$

ions in the samples. Their *d* values are slightly different because of the distinct ionic radii between $Tb^{3+}$ and $La^{3+}$.

The excitation spectra of $NaLa(WO_4)_2$:$x$$Tb^{3+}$ ($x$ = 0.10, 0.40, 0.70, 1.00) for emission at 545 nm are shown in Fig.3-4. The broadband from 250 nm to 300 nm is mainly ascribed to the O→W CT band, which overlaps some 4f-4f transitions of $Tb^{3+}$. The sharp peaks from 300 nm to 450 nm are ascribed to the intra-configurational 4f-4f transitions of $Tb^{3+}$ in the host. The excitation intensities of the O→W CT band and the 4f-4f transitions of $Tb^{3+}$ are increasing with the increase of the $Tb^{3+}$ content. When $Tb^{3+}$ ions take the place of $La^{3+}$ ions in the host completely, the excitation intensity is the strongest.

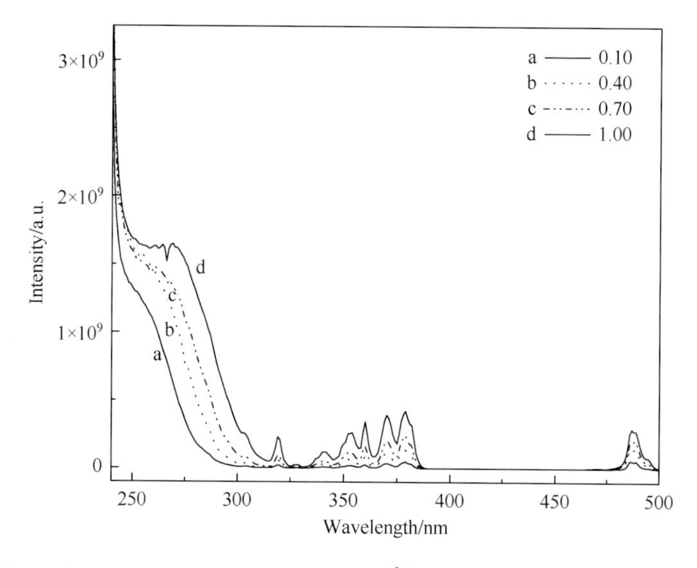


Fig.3-4 The excitation spectra of $NaLa(WO_4)_2$:$x$$Tb^{3+}$ ($x$ = 0.10, 0.40, 0.70, 1.00; $\lambda_{em}$ = 545 nm)

Fig.3-5 shows the emission spectra of $NaLa(WO_4)_2$:$x$$Tb^{3+}$ ($x$ = 0.10, 0.40, 0.70, 1.00) under the 378 nm excitation. The strong green emission peak at 545 nm is due to the $^5D_4 \rightarrow {}^7F_5$ transition of $Tb^{3+}$. Other $^5D_4 \rightarrow {}^7F_J$ transitions are weaker. With the increase of the $Tb^{3+}$ content, the emission intensities are increasing. The green-emitting phosphor $NaTb(WO_4)$ shares the strongest emission.

Fig.3-6 is the concentration dependence of the relative emission intensity of $Tb^{3+}$ $^5D_4 \rightarrow {}^7F_5$ transition in $NaLa(WO_4)_2$:$x$$Tb^{3+}$ ($\lambda_{ex}$ = 378 nm). The absence of concentration quenching of $Tb^{3+}$ is observed in $NaLa(WO_4)_2$:$x$$Tb^{3+}$. This phenomenon is not observed in other hosts as we know. The concentration quenching is caused by the interaction among luminescent center ions. $NaLa(MoO_4)_2$:$x$$Tb^{3+}$ and $NaLa(WO_4)_2$:$x$$Tb^{3+}$ share the similar scheelite structure. However, the ionic radius of the tetrahedral $Mo^{6+}$ (0.41 Å) is smaller than that of $W^{6+}$ (0.42 Å). Then the bond length of Tb—O—W—O—Tb is longer than that of Tb—O—Mo—O—Tb. So the $Tb^{3+}$-$Tb^{3+}$ nearest- neighbor distance in $NaLa(WO_4)_2$:$x$$Tb^{3+}$ is longer than that in $NaLa(MoO_4)_2$:$x$$Tb^{3+}$.

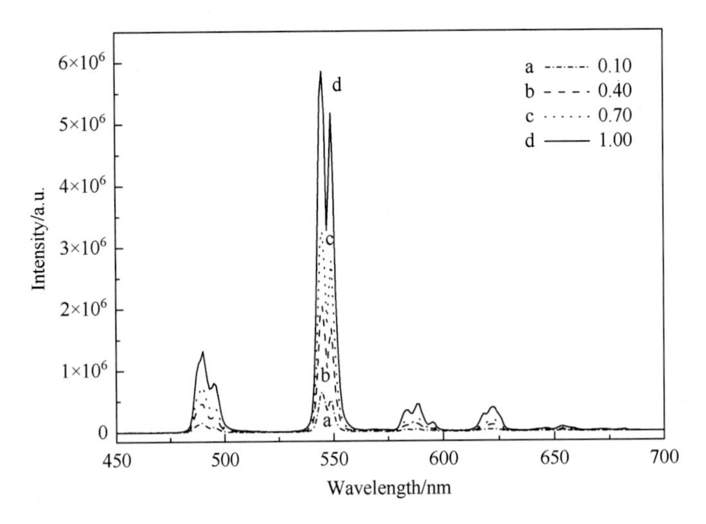


Fig.3-5　The emission spectra of $NaLa(WO_4)_2$:$x$$Tb^{3+}$ ($x$ = 0.10, 0.40, 0.70, 1.00; $\lambda_{ex}$ = 378 nm)

This suggests that the energy migration between $Tb^{3+}$ ions in $NaLa(WO_4)_2$:$x$$Tb^{3+}$ is inefficient, compared with $NaLa(MoO_4)_2$:$x$$Tb^{3+}$.

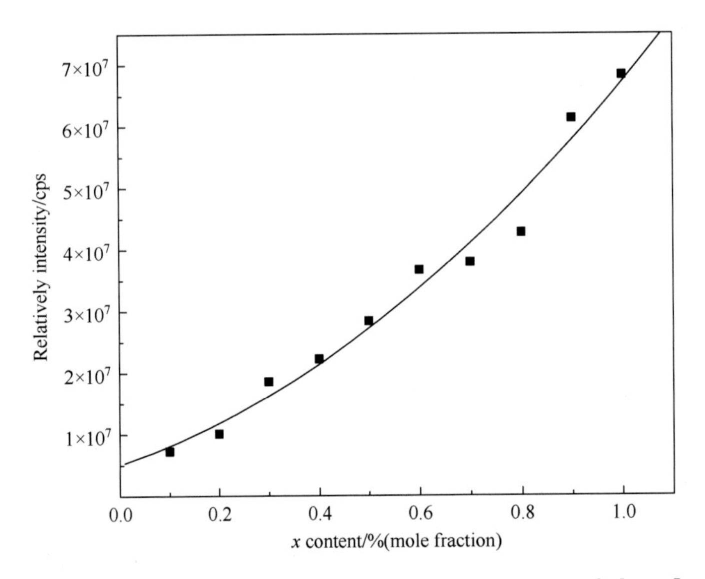


Fig.3-6　Concentration dependence of the relative emission intensity of $Tb^{3+}$ $^5D_4\rightarrow{}^7F_5$ transition in $NaLa(WO_4)_2$:$x$$Tb^{3+}$ ($\lambda_{ex}$ = 378 nm)

The decay curves for the transition of $^5D_4\rightarrow{}^7F_5$ (545 nm) of $Tb^{3+}$ ion in $NaTb(WO_4)_2$ under 270 nm and 378 nm excitation are shown in Fig.3-7. The decay curves can be well fitted by a single-exponential function as $I$ = A exp ($-t/\tau$), and the values of the decay time of $NaTb(WO_4)_2$ are 0.583 ms and 0.575 ms, respectively.

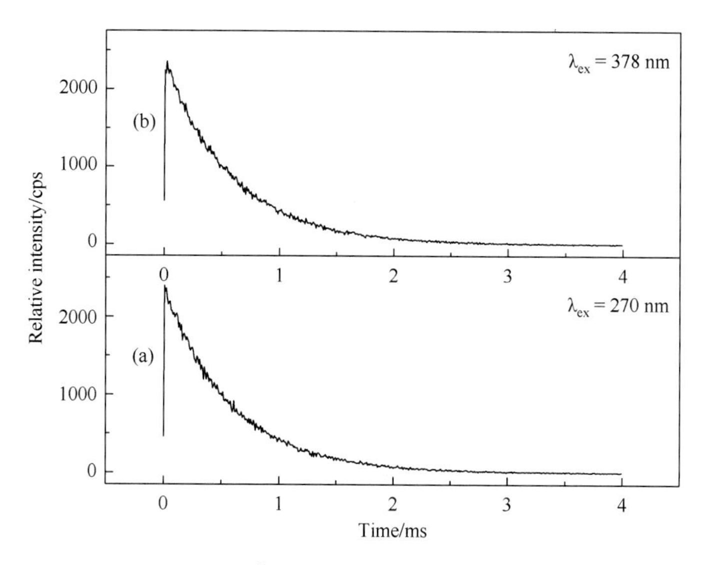


Fig.3-7 The decay curves of the $Tb^{3+}$ emission of $NaTb(WO_4)_2$: $\lambda_{ex}$ = 270 nm (a), 378 nm (b)

### 3.1.4 $NaLa(MoO_4)_2$:$0.11Tb^{3+}$ and $NaTb(WO_4)_2$

Since both $NaLa(MoO_4)_2$:$xTb^{3+}$ and $NaLa(WO_4)_2$:$xTb^{3+}$ share the scheelite-like ($CaWO_4$) isostructure and exhibit intense green emission, their PL spectra are compared in Fig.3-8. Curves a and c are the excitation spectra of $NaLa(MoO_4)_2$:$0.11Tb^{3+}$ and $NaTb(WO_4)_2$, respectively. The O→Mo CT band shows an obvious red-shift, comparing with the O→W CT band. The 4f-4f transitions of $Tb^{3+}$ in $NaTb(WO_4)_2$ are higher than that in $NaLa(MoO_4)_2$:$0.11Tb^{3+}$.

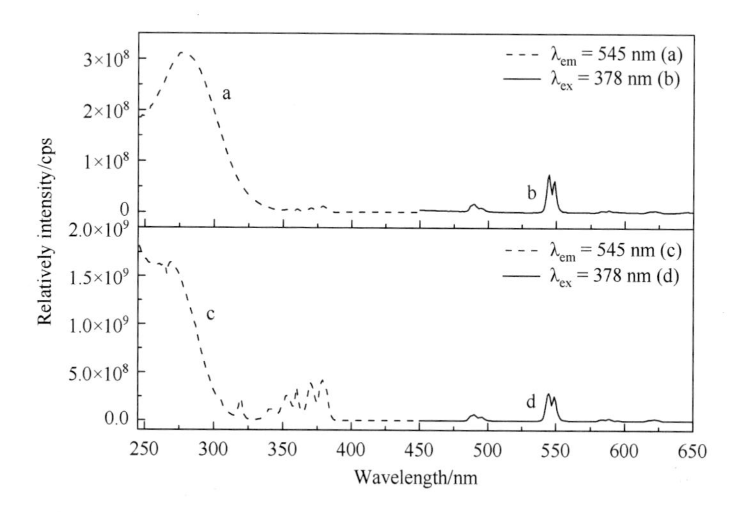


Fig.3-8 The excitation ($\lambda_{em}$ = 545 nm) and emission ($\lambda_{ex}$ = 378 nm) spectra of $NaLa(MoO_4)_2$:$0.11Tb^{3+}$ (a, b) and $NaTb(WO_4)_2$ (c, d)

In our previous work, we discussed that the charge transfer transition from the coordination ions (L) to center ions (M) is parity-allowed, whose energy is sensitive to the degree of the M—L bond covalency. The CT energy decreases with the increase of the M—L bond covalency. The M—L bond covalency is influenced by the electronegativity difference between M and L, the ligand number and the M—L bond length. The bond covalency increases with decreasing electronegative difference of M and L and decreasing the M—L bond length. The electronegativity of Mo(1.8) is larger than that of W (1.7), and the ionic radius of the tetrahedral $Mo^{6+}$(0.41 Å) is smaller than that of $W^{6+}$ (0.42 Å). Then the overlapping magnitude of the electronic cloud for the Mo—O bond is higher than that of the W—O bond. The Mo—O bond covalency is higher than that of the W—O bond. So the O→M CT band shows a red-shift, compared with the O→M CT band.

Curves b and d in Fig.3-8 are the emission spectra of $NaLa(MoO_4)_2$:$0.11Tb^{3+}$ and $NaTb(WO_4)_2$ under 378 nm excitation. These two curves are of a similar shape, the strongest emission is at 545 nm. However, the intensity of $NaTb(WO_4)_2$ is about 3.37 times higher than that of $NaLa(MoO_4)_2$:$0.11Tb^{3+}$. The CIE values for the phosphor $NaTb(WO_4)_2$ are calculated to be $x = 0.31$ and $y = 0.62$ in terms of its emission spectrum.

In conclusion, a series of $Tb^{3+}$-doped double molybdates and tungstates green phosphors, $NaLa(MoO_4)_2$:$xTb^{3+}$ and $NaLa(WO_4)_2$:$xTb^{3+}$ were prepared by the solid-state reaction. By investigating PL properties of these phosphors, the phenomenon of no concentration quenching of $Tb^{3+}$ was observed in $NaLa(WO_4)_2$:$xTb^{3+}$. However, the quenching concentration of $Tb^{3+}$ can be found in $NaLa(MoO_4)_2$:$xTb^{3+}$. Under 378 nm excitation, the intensity of emission in $NaTb(WO_4)_2$ is about 3.37 times higher than that in $NaLa(MoO_4)_2$:$0.11Tb^{3+}$.

## 3.2 $Tb^{3+}$-activated $LaBSiO_5$

### 3.2.1 Introduction

The rare-earth ion $Tb^{3+}$ has been widely used as a green-emitting activator for luminescent materials that have been applied in extensive fields, such as fluorescent lamps, plasma display panels, field emission displays, and w-LEDs. However, $Tb^{3+}$ shows weak absorption peaks in the near-UV range, because 4f-4f transitions are parity forbidden. To improve the absorption of $Tb^{3+}$ in phosphors, the sensitizer was co-doped. $Ce^{3+}$ not only exhibits good luminescent properties with one broadband but also can act as an efficient sensitizer. It can efficiently absorb the UV light and transfer the energy to the luminescent center, and the emission intensity of which is then strengthened. Many phosphors co-doped with $Tb^{3+}$ and $Ce^{3+}$ have been developed to generate efficient green light via energy transfer from $Ce^{3+}$ to $Tb^{3+}$ ions. For example, in

$Ba_2Y(BO_3)_2Cl:Ce^{3+},Tb^{3+}$ phosphors, the energy-transfer efficiency from the $Ce^{3+}$ ion to the $Tb^{3+}$ ion can reach up to 95%.

The compound $LaBSiO_5$ (marked as LBS) shows excellent thermal and hydrolytic stability and is considered to be an efficient luminescent host. LBS shares a trigonal crystal structure with a space group of *P*31 (the corresponding lattice parameters are $a = 6.874$ Å, $b = 6.874$ Å, $c = 6.717$ Å, $Z = 3$, $V = 274.87$ Å$^3$). In this crystal structure, the $La^{3+}$ ion is in six coordinations with $O^{2-}$ ions. $BO_4$ and $SiO_4$ are interlinked by corner-sharing to form a six-membered ring system. LBS doped with $Eu^{3+}$ and $Ce^{3+}$ has been investigated by Xue et al. and Nehru et al. To the best of our knowledge, further work on $LaBSiO_5$ doped with rare-earth ions has not been reported.

In this study, LBS doped with $Tb^{3+}$ and $Ce^{3+}$ was synthesized by using a solid-state reaction, and the structure and PL properties of the compounds were investigated. Finally, the energy transfer process from $Ce^{3+}$ to $Tb^{3+}$ in LBS was investigated.

### 3.2.2 XRD

The XRD patterns of $LaBSiO_5$ phosphors doped with rare-earth ions were been measured. Fig.3-9 presents the XRD patterns of the products $LaBSiO_5$:$0.15Tb^{3+}$, $LaBSiO_5$:$0.15Tb^{3+}$,$0.25Ce^{3+}$ and $LaBSiO_5$:$0.15Tb^{3+}$,$0.35Ce^{3+}$. The XRD pattern of $LaBSiO_5$:$0.15Tb^{3+}$ fired at 1100 ℃ corresponds to the standard JCPDS card of LBS (No. 87-2172). This result shows that

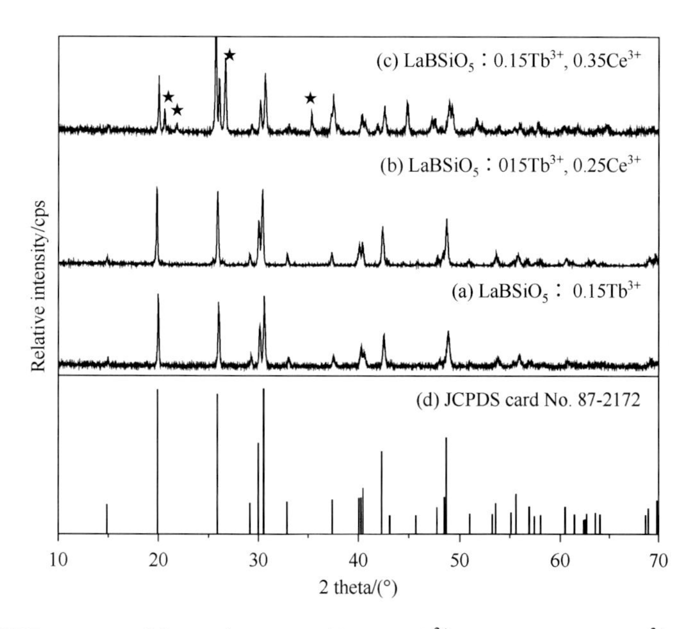


Fig.3-9 XRD patterns of the products $LaBSiO_5$:$0.15Tb^{3+}$ (a), $LaBSiO_5$:$0.15Tb^{3+}$,$0.25Ce^{3+}$ (b), $LaBSiO_5$:$0.15Tb^{3+}$,$0.35Ce^{3+}$ (c) and JCPDS card of $LaBSiO_5$ (d)

the phosphor has the same phase as LBS with a trigonal crystal structure. The doping of $Tb^{3+}$ does not cause significant changes in the structure of LBS, and $Tb^{3+}$ ions occupy the sites of $La^{3+}$ ions. Curve b is the XRD pattern of $LaBSiO_5$:0.15$Tb^{3+}$,0.25$Ce^{3+}$. No other obvious phases can be found. At a $Ce^{3+}$ content of up to 0.35, other phases appear in the XRD pattern of $LaBSiO_5$:0.15$Tb^{3+}$,0.35$Ce^{3+}$.

### 3.2.3 PL properties

Fig.3-10 shows the excitation and emission spectra of $LaBSiO_5$:$x$$Tb^{3+}$ ($x$ = 0.05, 0.10, 0.15, 0.20, 0.25). The broadband from 270 nm to 300 nm is mainly ascribed to the 4f-5d transition of $Tb^{3+}$. The sharp peaks from 300 nm to 390 nm are ascribed to the intra-configurational 4f-4f transitions of $Tb^{3+}$ in the host. The excitation intensities of the 4f-5d and 4f-4f transitions of $Tb^{3+}$ increase as the $Tb^{3+}$ content increases, and they are the strongest when the $Tb^{3+}$ content is 0.15. Fig.3-10(b) shows the emission spectra of $LaBSiO_5$:$x$$Tb^{3+}$ ($x$ = 0.05, 0.10, 0.15, 0.20, 0.25) under the 378 nm excitation. These phosphors can emit green light with main peaks at 492 nm, 544 nm, 587 nm and 621 nm, which are due to the $^5D_4 \rightarrow ^7F_J$ transitions of $Tb^{3+}$ ($J$ = 6, 5, 4, 3, respectively). Among these, the strongest green emission peak is at 544 nm (the $^5D_4 \rightarrow ^7F_5$ transition), because it is a magnetic dipole allowed transition with $J = \pm 1$. With the increase of the $Tb^{3+}$ content, the emission intensities also increase. The sample $LaBSiO_5$:0.15$Tb^{3+}$ shows the strongest green emission among these five phosphors. The CIE coordinates for the phosphor $LaBSiO_5$:0.15$Tb^{3+}$ are $x$ = 0.304, $y$ = 0.594, in terms of the emission spectrum.

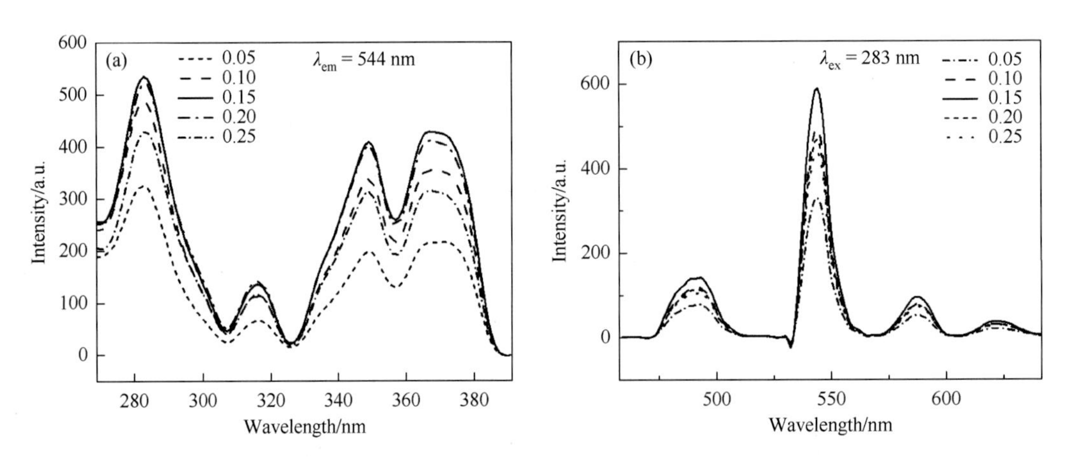


Fig.3-10 Excitation (a) and emission (b) spectra of $LaBSiO_5$: $x$$Tb^{3+}$ ($x$ = 0.05, 0.10, 0.15, 0.20, 0.25)

According to Dexter and Schulman's equation (2-8), the $R_c$ value is calculated to be 10.53 Å for the substitution of $Tb^{3+}$ at the $La^{3+}$ site. It is presumed that energy transfer between $Tb^{3+}$ ions dominates in the case of $LaBSiO_5$:$Tb^{3+}$ phosphor.

The excitation and emission spectra of the phosphors $LaBSiO_5$:$y$$Ce^{3+}$ ($y$ = 0.01, 0.02, 0.03, 0.04, 0.05) are shown in Fig.3-11. Fig.3-11(a) is the excitation spectra of these phosphors on monitoring emission at 376 nm. There are three broad excitation bands at 240 nm, 270 nm and 330 nm, which are due to the 4f-5d transitions of $Ce^{3+}$. When the content of $Ce^{3+}$ is 0.03, the excitation intensity is the strongest. Fig.3-11(b) shows the emission spectra of $LaBSiO_5$:$y$$Ce^{3+}$ ($y$ = 0.01, 0.02, 0.03, 0.04, 0.05) under the 330 nm excitation. These phosphors show two broad overlapping peaks at 355 nm and 376 nm, respectively. Both are of almost equal intensity. The two broad emission peaks are due to the 5d-$^2F_{5/2}$ and 5d-$^2F_{7/2}$ transitions of $Ce^{3+}$. When the content of $Ce^{3+}$ is 0.03, the emission intensity is the strongest. According to the equation (2-8), the $R_c$ value for $Ce^{3+}$ in LBS is 18.00 Å.

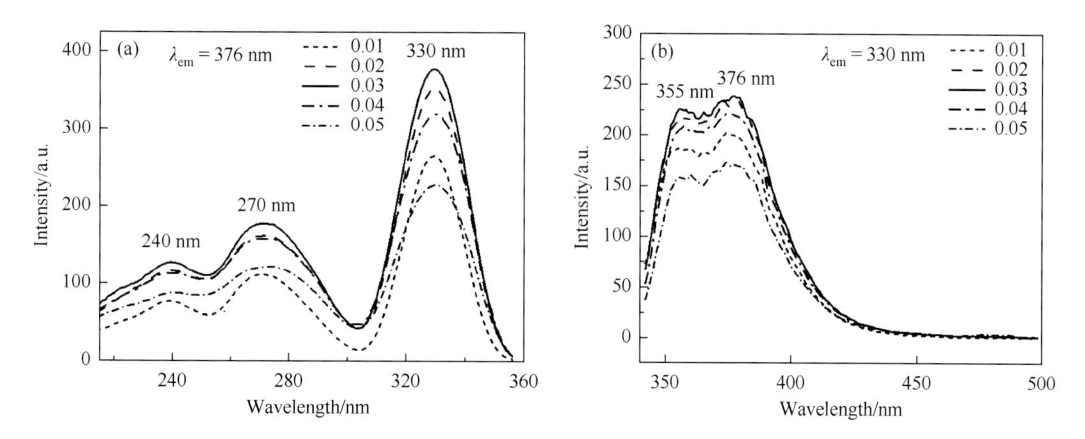


Fig.3-11 Excitation (a) and emission (b) spectra of the phosphors $LaBSiO_5$:$y$$Ce^{3+}$ ($y$ = 0.01, 0.02, 0.03, 0.04, 0.05)

Fig.3-12 is the excitation and emission spectra of $LaBSiO_5$:0.15$Tb^{3+}$, $LaBSiO_5$:0.03$Ce^{3+}$ and $LaBSiO_5$:0.15$Tb^{3+}$,0.25$Ce^{3+}$. Comparing Fig.3-12(a) with Fig.3-12(b), the spectral overlap between the excitation band of $Tb^{3+}$ and the emission band of $Ce^{3+}$ can be observed. This result supports the occurrence of energy transfer from $Ce^{3+}$ to $Tb^{3+}$. Therefore, the energy transfer phenomenon between $Ce^{3+}$ and $Tb^{3+}$ is investigated. The excitation and emission spectra of $LaBSiO_5$:0.15$Tb^{3+}$,0.25$Ce^{3+}$ are shown in Fig.3-12(c). The excitation spectrum of $LaBSiO_5$:0.15$Tb^{3+}$,0.25$Ce^{3+}$ is similar with that of $LaBSiO_5$:0.03$Ce^{3+}$. The 4f-5d transition of $Tb^{3+}$ from 250 nm to 300 nm is overlapped by the excitation band of $Ce^{3+}$, and the 4f-4f transitions from 300 nm to 390 nm can not be found. This result is in agreement with the work reported by Sun et al. From the emission spectra of $LaBSiO_5$:0.15$Tb^{3+}$,0.25$Ce^{3+}$ under the 330 nm excitation, the emission of $Ce^{3+}$ and $Tb^{3+}$ can be observed. The broadband from 360 nm to 420 nm is due to the 5d-$^2F_{7/2}$ transitions of $Ce^{3+}$, the sharp peaks 492 nm, 544 nm, 587 nm and 621 nm are due to the $^5D_4 \rightarrow {}^7F_J$ transitions of $Tb^{3+}$ ($J$ = 6, 5, 4, 3, respectively). These results suggest that energy transfer from $Ce^{3+}$ to $Tb^{3+}$ takes place.

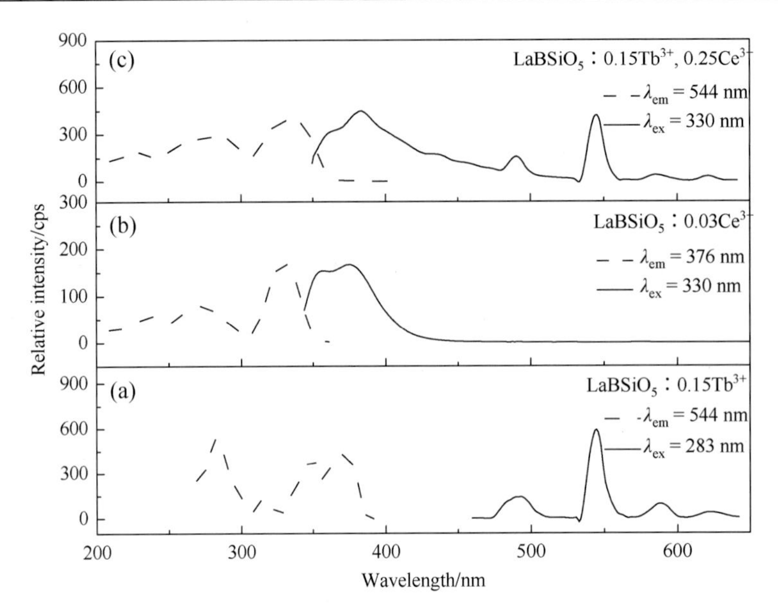


Fig.3-12 Excitation and emission spectra of $LaBSiO_5$:0.15$Tb^{3+}$ (a), $LaBSiO_5$:0.03$Ce^{3+}$ (b) and $LaBSiO_5$:0.15$Tb^{3+}$,0.25$Ce^{3+}$ (c)

The energy transfer process from $Ce^{3+}$ to $Tb^{3+}$ is shown schematically in Fig.3-13. When $Ce^{3+}$ is excited by the UV light, the electrons on $Ce^{3+}$ are excited from the ground state 4f to the excited state 5d. Because the 5d level of $Ce^{3+}$ is close to the $^5D_J$ level, the energy is transferred from the 5d level of $Ce^{3+}$ to the $^5D_{3,4}$ levels of $Tb^{3+}$. Finally, the $^5D_{3,4}$ levels can give strong emission of $Tb^{3+}$ (the $^5D_4$-$^7F_J$ transitions).

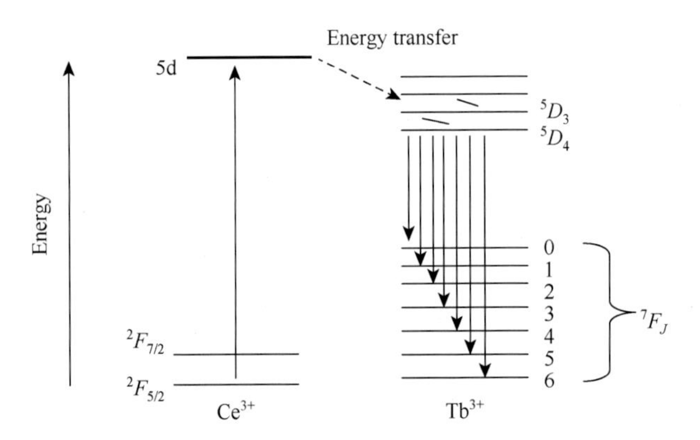


Fig.3-13 Energy levels for energy transfer from $Ce^{3+}$ to $Tb^{3+}$ in LBS

Fig.3-14 shows the emission spectra of $LaBSiO_5$:0.15$Tb^{3+}$,$z$$Ce^{3+}$ ($z$ = 0.03, 0.05, 0.10, 0.15, 0.25) under the 330 nm excitation. Emissions characteristic of $Ce^{3+}$ and $Tb^{3+}$ are observed. From these five curves, on doping with $Ce^{3+}$, the emission intensity of $Tb^{3+}$ increases, and the emission intensity of the 5d-$^2F_{5/2}$ transition ($Ce^{3+}$) decreases rapidly with the increase of the $Ce^{3+}$

content. When the $Ce^{3+}$ content reaches an excess of 10%, the emissions of the 5d-$^2F_{5/2}$ and 5d-$^2F_{7/2}$ ($Ce^{3+}$) are also enhanced. This result shows that $Ce^{3+}$ can efficiently transfer energy to $Tb^{3+}$ when $Ce^{3+}$ is at lower concentrations. With the increasing content of $Ce^{3+}$, the efficiency of the energy transfer from $Ce^{3+}$ to $Tb^{3+}$ is reduced. The dependence of the emission intensity of $Ce^{3+}$ and $Tb^{3+}$ on the concentration of $Ce^{3+}$ is shown in Fig.3-15. When the $Ce^{3+}$ content is 25%, the blue-violet and green emissions of $LaBSiO_5$:0.15$Tb^{3+}$,0.25$Ce^{3+}$ are the strongest.

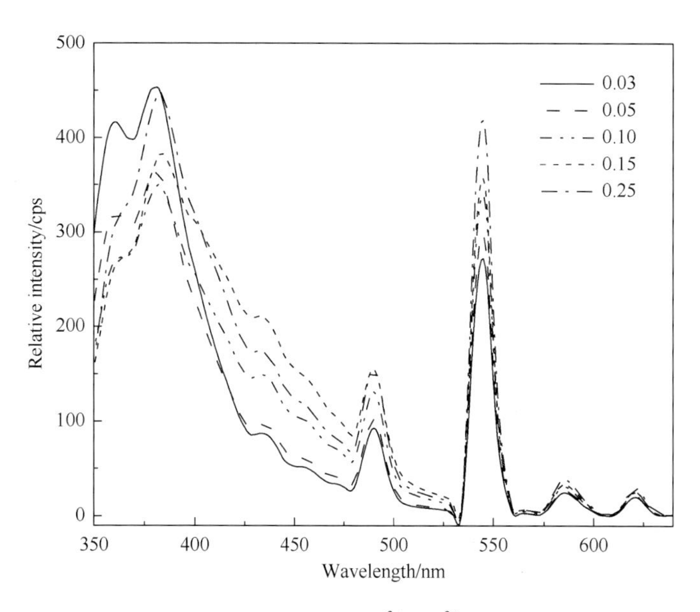


Fig.3-14 Emission spectra of $LaBSiO_5$:0.15$Tb^{3+}$,$z$$Ce^{3+}$ ($z$ = 0.03, 0.05, 0.10, 0.15, 0.25) under the 330 nm excitation

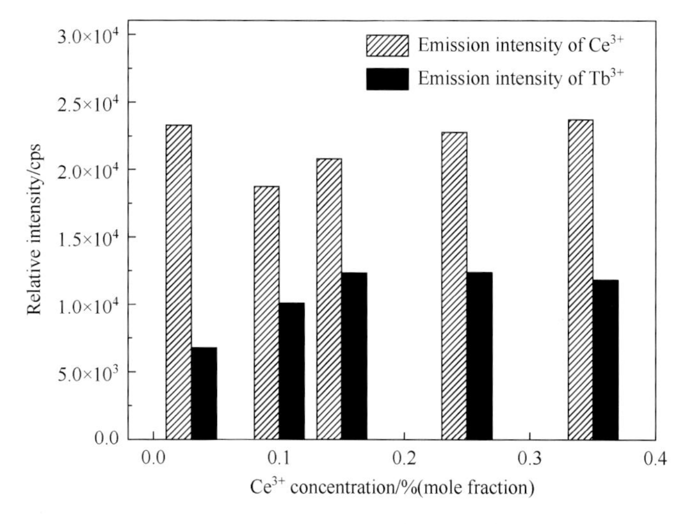


Fig.3-15 Dependence of the emission intensity of $Ce^{3+}$ and $Tb^{3+}$ on the concentration of $Ce^{3+}$ in $LaBSiO_5$:0.15$Tb^{3+}$,$z$$Ce^{3+}$

In conclusion, LBS doped with $Ce^{3+}$ and $Tb^{3+}$ was prepared by using the solid-state method at 1100 ℃. The phosphor LBS:$Tb^{3+}$ shows intense green emission, and LBS:$Ce^{3+}$ shows the blue-violet emission under the UV light excitation. With the doping of $Ce^{3+}$, the green emission in LBS:$Tb^{3+}$,$Ce^{3+}$ was enhanced. $Ce^{3+}$ can efficiently absorb the UV light and transfer the energy to $Tb^{3+}$. Finally, the energy transfer process from $Ce^{3+}$ to $Tb^{3+}$ is discussed.

## 3.3 $Eu^{2+}$-activated $Ca_3SiO_4Cl_2$

### 3.3.1 Introduction

Phosphor-converted light-emitting diodes (LEDs) are an important kind of solid-state illumination. Commercial w-LEDs are mainly obtained by combining a yellow-emitting phosphor YAG pumped with a blue-emitting GaN LED chip. At present, the emission bands of LED chips are shifted to the near-ultraviolet (UV) range and this wavelength can offer a higher efficiency of the solid-state lighting. Conventional phosphors used in fluorescence lighting are not suitable for the near-UV LED chips, because of their poor absorption in the near-UV light region. Hence, there is a need to investigate novel phosphors for the near-UV LED chips.

Alkaline earth halo-silicates are well-known hosts for inorganic luminescent materials due to their low synthesis temperature and high luminescence efficiency. The compound $Ca_3SiO_4Cl_2$ was reported first by Chaterlier. Many researchers have investigated its phase structure transformation, and the luminescent properties in different crystal structures doped with $Eu^{2+}$. Little information on this host doped with other rare-earth ions has been reported, according to our investigation.

$Ce^{3+}$ ions are very good luminescent centers as well as sensitizers of luminescence in many hosts. They can efficiently absorb the UV light and transfer the energy to the luminescent centers. Thus the emission intensity of the luminescent center would be strengthened. Due to the overlap of spectra between the $Ce^{3+}$ emission band and the $Eu^{2+}$ excitation band, the 4f-5d transitions of $Ce^{3+}$ to $Eu^{2+}$ are very sensitive to the nature of the host lattice, and the absorption and emission of $Ce^{3+}$ and $Eu^{2+}$ are efficient in many hosts.

In this paper, $Ca_3SiO_4Cl_2$ co-doped with $Ce^{3+}$, $Eu^{2+}$ was prepared by the high-temperature reaction. The luminescent properties and the energy transfer process of $Ca_3SiO_4Cl_2$:$Ce^{3+}$,$Eu^{2+}$ were investigated. Finally, a single green LED was made by combining the green phosphor with a near-UV InGaN chip.

### 3.3.2 XRD

The XRD patterns of $Ca_3SiO_4Cl_2$:$0.003Ce^{3+}$ and $Ca_3SiO_4Cl_2$:$0.03Eu^{2+}$ are shown in

Fig.3-16. They are consistent with the JCPDS card No. 70-2447 [$Ca_3SiO_4Cl_2$]. This indicates that the as-prepared phosphors share the single-phase and the doping of $Ce^{3+}$ and $Eu^{2+}$ does not change the crystal structure of $Ca_3SiO_4Cl_2$. The slight blue-shift of diffraction peaks of $Ca_3SiO_4Cl_2$:0.003$Ce^{3+}$ and $Ca_3SiO_4Cl_2$:0.03$Eu^{2+}$ is due to the distinct ionic radii between $Ca^{2+}$ and $Eu^{2+}$, $Ce^{3+}$. This host lattice $Ca_3SiO_4Cl_2$ has a monoclinic structure with the space group of *P*21/*c*, and its lattice parameters are $a = 9.782$, $b = 6.738$, $c = 10.799$. The doped $Eu^{2+}$ and $Ce^{3+}$ will occupy the sites of $Ca^{2+}$ in the octahedral sites.

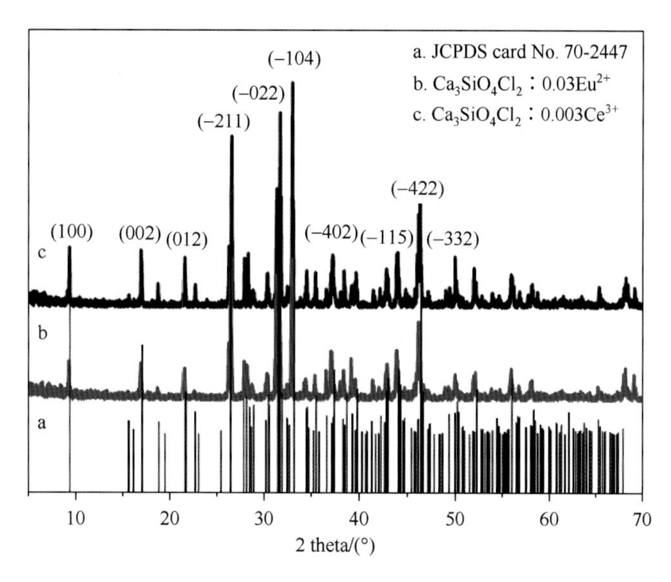


Fig.3-16 The XRD pattern of $Ca_3SiO_4Cl_2$:0.003$Ce^{3+}$ and $Ca_3SiO_4Cl_2$:0.03$Eu^{2+}$

### 3.3.3 PL properties of $Ca_3SiO_4Cl_2$:$Ce^{3+}$

Fig.3-17 shows the excitation spectra of the phosphors $Ca_3SiO_4Cl_2$:3*x*$Ce^{3+}$ ($x = 0.0010$, 0.0015, 0.0020, 0.0025) by monitoring emission at 390 nm. These four curves are of similar shapes. There are three excitation bands at 260 nm, 290 nm and 330 nm, which are due to the 4f→5d transitions of $Ce^{3+}$. When the content of $Ce^{3+}$ is 0.15%, the excitation intensity is the strongest.

Fig.3-18 shows the emission spectra of $Ca_3SiO_4Cl_2$:3*x*$Ce^{3+}$ ($x = 0.0010$, 0.0015, 0.0020, 0.0025) under the 330 nm excitation.The broad blue emission band is due to the 5d→4f transition of $Ce^{3+}$. With the increase of the $Ce^{3+}$ content, the intensity of the blue emission becomes higher. When the content of $Ce^{3+}$ is 0.15%, the emission intensity is the strongest.

We also calculated the $R_c$ value which is about 60 Å for substitution of $Ce^{3+}$ at the $Ca^{2+}$ site. From this value, the $R_c$ of energy transfer is larger than the distance ($R$) between the $Ce^{3+}$ ions ($R$ is about 5.41 Å) (seeing ICSD 38359). As it can be seen that $R$ is far less than $R_c$, it is presumed that energy transfer between $Ce^{3+}$ ions dominates in the case of $Ca_3SiO_4Cl_2$:$Ce^{3+}$ phosphor.

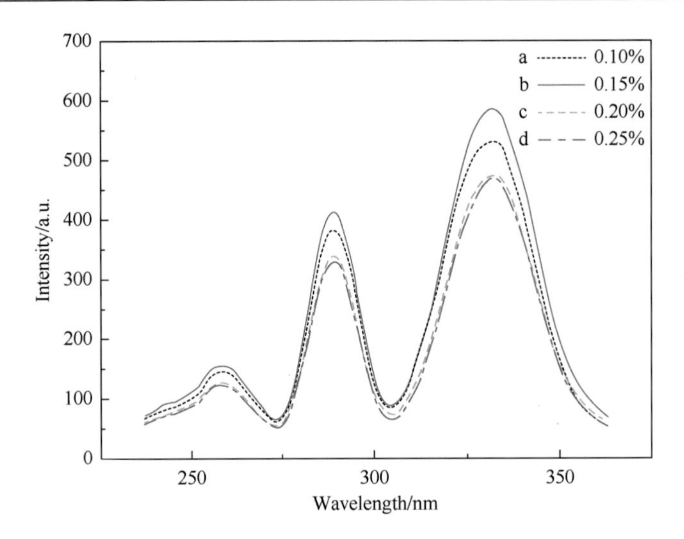


Fig.3-17 The excitation spectra of $Ca_3SiO_4Cl_2{:}3xCe^{3+}$ ($x$ = 0.0010, 0.0015, 0.0020, 0.0025) by monitoring emission at 390 nm

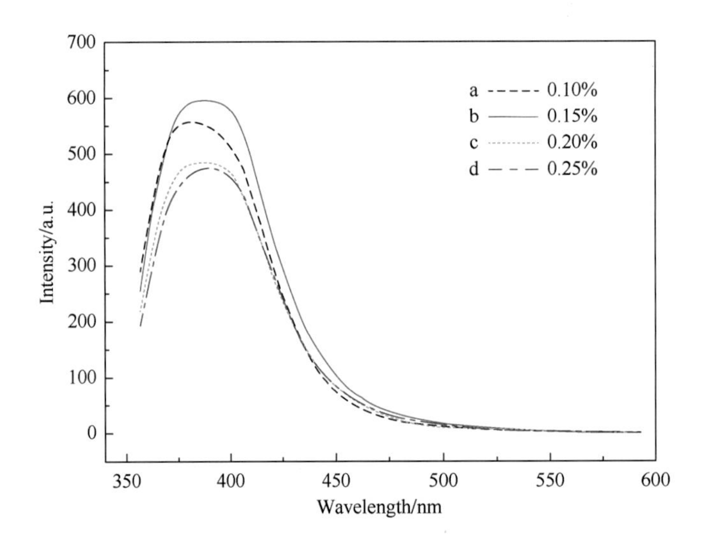


Fig.3-18 The emission spectra of $Ca_3SiO_4Cl_2{:}3xCe^{3+}$ ($x$ = 0.0010, 0.0015, 0.0020, 0.0025) under 330 nm excitation

### 3.3.4 PL properties of $Ca_3SiO_4Cl_2{:}Ce^{3+},Eu^{2+}$

The excitation and emission spectra of the phosphor $Ca_3SiO_4Cl_2{:}0.03Eu^{2+}$ are shown in Fig.3-19. There are three excitation bands at 250 nm, 340 nm and 440 nm, which are due to the 4f→5d transitions of $Eu^{2+}$. These samples show an intense green emission, which is located at 505 nm under the 340 nm excitation. This result is in agreement with that reported by Liu et al.

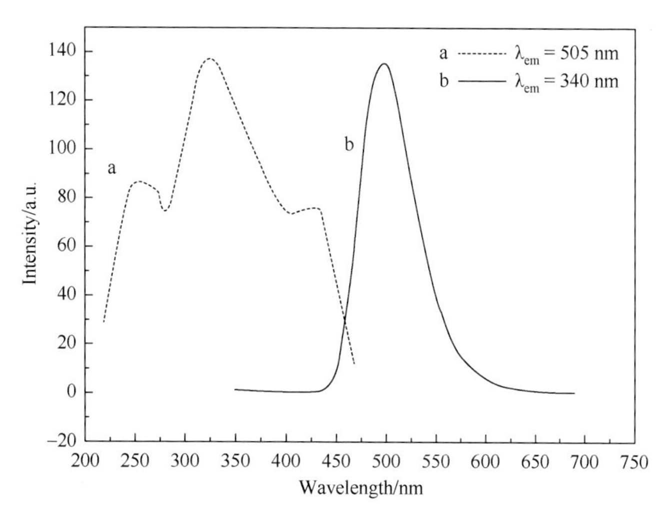


Fig.3-19 Excitation and emission spectra of the phosphor $Ca_3SiO_4Cl_2$:$0.03Eu^{2+}$

One of the conditions usually required for energy transfer is the spectrum overlapping of the donor emission to the acceptor excitation. The excitation spectra of the phosphors $Ca_3SiO_4Cl_2$:$0.0045Ce^{3+}$,$3yEu^{2+}$($y$ = 0, 0.002, 0.004, 0.006, 0.008, 0.010) by monitoring emission at 505 nm are shown in Fig.3-20. With the introduction of $Eu^{2+}$, the excitation spectra of the phosphors are similar to that of $Ca_3SiO_4Cl_2$:$0.03Eu^{2+}$. The excitation bands of $Ce^{3+}$ are overlapped by $Eu^{2+}$. This result shows that $Ce^{3+}$ can efficiently transfer energy to $Eu^{2+}$. When the $Eu^{2+}$ content is 0.6%, the excitation intensity is the strongest.

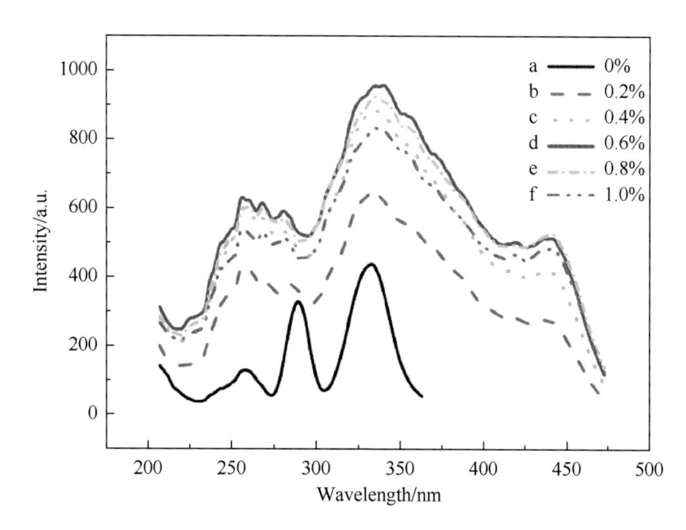


Fig.3-20 The excitation spectra of $Ca_3SiO_4Cl_2$:$0.0045Ce^{3+}$, $3yEu^{2+}$ ($y$ = 0, 0.002, 0.004, 0.006, 0.008, 0.010) by monitoring emission at 390 nm and 505 nm

There are two emission peaks in the spectra of $Ca_3SiO_4Cl_2$:$0.0045Ce^{3+}$,$3yEu^{2+}$ (Fig.3-21). The emission bands at 400 nm and 504 nm are due to the $4f^05d^1 \rightarrow 4f^15d^0$ transition of $Ce^{3+}$ and the $4f^65d \rightarrow 4f^7$ transitions of $Eu^{2+}$, respectively. $Eu^{2+}$ can give enhanced green emission.

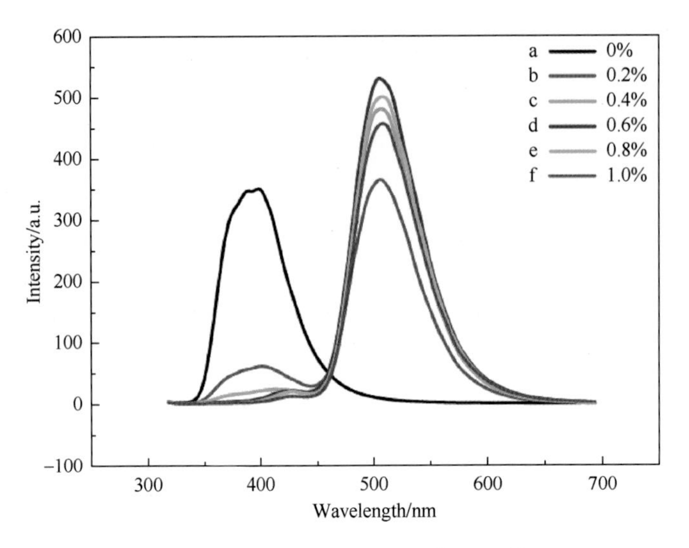


Fig.3-21　The emission spectra of $Ca_3SiO_4Cl_2$:0.0045$Ce^{3+}$, 3$y$$Eu^{2+}$ ($y$ = 0, 0.002, 0.004, 0.006, 0.008, 0.010) under the 340 nm excitation

through $Ce^{3+} \rightarrow Eu^{2+}$ energy transfer. With the introduction of $Eu^{2+}$, the emission of $Ce^{3+}$ decreases, and the $Eu^{2+}$ emission increases. When the content of $Eu^{2+}$ is 0.6%, the emission intensity of $Eu^{2+}$ is the strongest.

Our target is to search novel phosphors for near-UV LED chips. Therefore one single LED has been made with the as-obtained phosphor. Fig.3-22 shows the EL spectrum of the

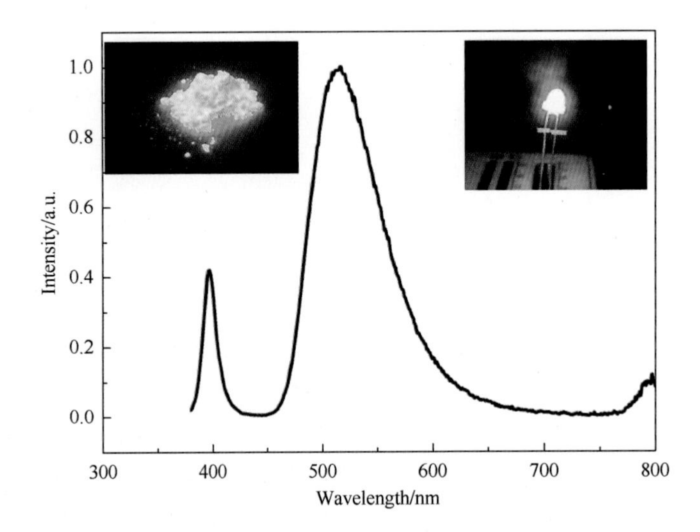


Fig.3-22　The EL spectrum of the LED made with $Ca_3SiO_4Cl_2$:0.0045$Ce^{3+}$,0.018$Eu^{2+}$ under 20 mA current excitation. The inserted figures are the photograph of the phosphor under 365 nm excitation and the green LED

intense green LED made with $Ca_3SiO_4Cl_2$:0.0045$Ce^{3+}$,0.018$Eu^{2+}$ under 20 mA current excitation. There are two emission bands in the EL spectrum. The emission at 395 nm is due to the emission of the near-UV GaN chip. The green emission at 505 nm is ascribed to the emission of the green phosphor. The bright green light from the LED can be observed by naked eyes (Fig.3-22). Its CIE chromaticity coordinates are $x = 0.261$, $y = 0.582$.

In summary, a series of phosphors, $Ca_3SiO_4Cl_2:3xCe^{3+}$, $Ca_3SiO_4Cl_2$:0.0045$Ce^{3+}$,3$y$$Eu^{2+}$ are prepared by the solid-state reaction at high temperature, and their structures and PL properties are investigated. The green phosphor $Ca_3SiO_4Cl_2$:0.0045$Ce^{3+}$,0.018$Eu^{2+}$ shows intense green emission with one broader excitation in the near-UV range. The bright green light from the LED based on this phosphor can be observed by naked eyes. Hence, it is considered to be a good candidate for the green component of a three-band wLED.

## References


Aznar A, Solé R, Aguiló M, et al. 2004. Growth, optical characterization, and laser operation of epitaxial Yb: KY $(WO_4)_2$/KY $(WO_4)_2$Yb: KY $(WO_4)_2$/KY $(WO_4)_2$ composites with monoclinic structure. Applied Physics Letters, 85 (19): 4313.

Baginskiy I, Liu R S. 2009. Significant improved luminescence intensity of $Eu^{2+}$-doped $Ca_3SiO_4Cl_2$green phosphor for white LEDs synthesized through two-stage method.Journal of the Electrochemical Society, 156 (5): G29.

Barber D B, Pollock C R, Beecroft L L, et al. 1997. Amplification by optical composites. Optics Letters, 22 (16): 1247.

Blasse G. 1994. Scintillator materials. Chemistry of Materials, 6 (9): 1465.

Cavalli E, Boutinaud P, Cucchietti T, Bettinelli M. 2009. Spectroscopy and excited states dynamics of $Tb^{3+}$-doped KLa $(MoO_4)_2$ crystals. Optical Materials, 31 (3): 470.

Chi L, Chen H, Zhuang H, et al. Crystal structure of $LaBSiO_5$. Journal of Alloys Compounds, 252 (1): L12.

Chiu C H, Liu C H, Huang S B, et al. 2007. White-light-emitting diodes using red-emitting LiEu $(WO_4)_{2-x}$ $(MoO_4)_x$ phosphors. Journal of the Electrochemical Society, 154 (7): J181.

Dexter D L, Schulman J A. 1954. Theory of concentration quenching in inorganic phosphors. Journal of Chemical Physics, 22 (6): 1063.

Do Y R, Huh Y D, 2000. Optical properties of potassium europium tungstate phosphors. Journal of the Electrochemical Society, 147 (11): 4385.

Fang M H, Mariano C O M, Chen K C, et al. 2021. High-performance $NaK_2Li[Li_3SiO_4]_4$:Eu green phosphor for backlighting light-emitting diodes. Chemistry of Materials, 33 (5): 1893.

Feldmann C, Jüstel T, Ronda C R, et al. 2003. Inorganic luminescent materials: 100 years of research and application. Advanced Functional Materials, 13 (7): 511.

Hao Z, Zhang X, Luo Y, et al. 2013. Preparation and photoluminescence properties of single-phase $Ca_2SiO_3Cl_2$: $Eu^{2+}$ bluish-green emitting phosphor. Materials Letters, 93 (93): 272.

Honma T, Toda K, Ye Z G, et al. 1998.Concentration quenching of the $Eu^{3+}$-activated luminescence in some layered perovskites with two-dimensional arrangement. Journal of Physics Chemistry of Solids, 59 (8): 1187.

Kato A, Oishi S, Shishido T, et al. 2005. Evaluation of stoichiometric rare-earth molybdate and tungstate compounds as laser materials. Journal of Physics & Chemistry of Solids, 66 (11): 2079-2081.

Khizar M, Fan Z Y, Kim K H, et al. 2005. Nitride deep-ultraviolet light-emitting diodes with microlens array. Applied Physics Letters, 86 (17): 531.

Kim J S, Jeon P E, Park Y H, et al. 2004. White-light generation through ultraviolet-emitting diode and white-emitting phosphor. Applied Physics Letters, 85 (17): 3696.

Lee S, Seo S Y. 2002. Optimization of yttrium aluminum garnet: $Ce^{3+}$ phosphors for white light-emitting diodes by combinatorial chemistry method. Journal of the Electrochemical Society, 149 (11): J85.

Lei Y, Zou Q, Du D. 2009. Energy transfer between $Ce^{3+}$ and $Eu^{2+}$ in doped $KSrPO_4$. Applied Physics A, 97 (3): 635.

Leonyuk N I, Belokoneva E L, Bocelli G, et al. 1999 Crystal growth and structural refinements of the $Y_2SiO_5$, $Y_2Si_2O_7$, and $LaBSiO_5$ single crystals. Crystal Research Technology, 34 (9): 1175.

Leonyuk N I, Belokoneva E L, Bocelli G, et al. 1999. High-temperature crystallization and X-ray characterization of $Y_2SiO_5$, $Y_2Si_2O_7$, and $LaBSiO_5$. Journal of Crystal Growth, 205 (3): 361.

Li G, Geng D L, Shang M M, et al. 2011. Tunable luminescence of $Ce^{3+}$ /$Mn^{2+}$ /$Tb^{3+}$-tri activated $Mg_2Gd_8$ $(SiO_4)_6O_2$ via energy transfer: potential single-phase white-light-emitting phosphors. Journal of Physical Chemistry C, 115: 21882.

Liao H X, Zhao M, Zhou Y Y, et al. 2019. Polyhedron transformation toward stable narrow-band green phosphors for wide-color-gamut liquid crystal display. Advanced Functional Materials, 29 (30): 1901988.

Lin H, Liu X R, Pun E Y B. 2002. Sensitized luminescence and energy transfer in $Ce^{3+}$ and $Eu^{2+}$ codoped calcium magnesium chlorosilicate. Optical Materials, 18 (4): 397.

Liu J, Lian H, Shi C, et al. 2005. $Eu^{3+}$-doped high-temperature phase $Ca_3SiO_4Cl_2$ a yellowish-orange phosphor for white light-emitting diodes. Journal of the Electrochemical Society, 152 (11): G880.

Liu J, Lian H, Sun J, et al. 2005. Characterization and properties of green-emitting $Ca_3SiO_4Cl_2$: $Eu^{2+}$ powder phosphor for white light-emitting diodes. Chemistry Letter, 34 (10): 1340.

Liu X, Lin J. 2008. $LaGaO_3$: A (A = $Sm^{3+}$ and/or $Tb^{3+}$) as promising phosphors for field emission displays. Journal of Materials Chemistry, 18 (2): 221.

Lü W, Guo N, Jia Y, et al. 2013. Tunable color of $Ce^{3+}$ /$Tb^{3+}$ /$Mn^{2+}$-coactivated $CaScAlSiO_6$ via energy transfer: a single-component red/white-emitting phosphor. Inorganic Chemistry, 44 (21): 3007-12.

Mueller-Mach R, Mueller G O, Krames M R, et al. 2002. High-power phosphor-converted light-emitting diodes based on III-nitrides. IEEE Journal of Selected Topics in Quantum Electronics, 8 (2): 339.

Nazarov M V, Jeon D Y, Kang J H, et al. 2004. Luminescence properties of europium-terbium double activated calcium tungstate phosphor. Solid State Communications, 131 (5): 307.

Neeraj S, Kijima N, Cheetham A K. 2004. Novel red phosphors for solid-state lighting: the system $NaM(WO_4)_{2-x}(MoO_4)_x$:$Eu^{3+}$ (M = Gd, Y, Bi). Chemical Physics Letters. 387 (1-3): 2.

Nehru L C, Marimuthu K, Jayachandran M, et al. 2001. $Ce^{3+}$-doped stillwellites: a new luminescent system with strong ion lattice coupling. Journal of Physics D Applied Physics, 34 (17): 2599.

Ronda C R. 1997. Recent achievements in research on phosphors for lamps and displays. Journal of Luminescence, 72-74 (6): 49.

Setlur A A, Heward W J, Hannah M E, et al. 2008. Incorporation of $Si_4N_3$into $Ce^{3+}$-doped garnets for warm white LED phosphors. Chemistry of Materials, 20 (19): 6277.

Shi F, Meng J, Ren Y, et al. 1998. Structure, luminescence, and magnetic properties of $AgLnW_2O_8$ (Ln = Eu, Gd, Tb, and Dy) compounds. Journal of Physics Chemistry of Solids, 59 (1): 105.

Sivakumar V, Varadaraju U V. 2005. Intense red-emitting phosphor for white light-emitting diodes. Journal of the Electrochemical Society, 152 (10): H168.

Sivakumar V, Varaduaraju U V. 2006. A promising orange-red phosphor under near UV excitation. Electrochem and Solid-State Letters, 9 (6): H35.

Steigerwald D A, Bhat J C, Collins D, et al. 2002. Illumination with solid-state lighting technology. IEEE Journal of Selected Topics in Quantum Electronics, 8 (2): 310.

Sun J, Lai J, Xia Z, et al. 2012. Luminescence properties and energy transfer in $Ba_2Y(BO_3)_2Cl$:$Ce^{3+}$, $Tb^{3+}$ phosphors. Applied Physics B, 107 (3): 827-831.

Van Vliet J P M, Blasse G, Brixner L H, 1988. Luminescence properties of alkali europium double tungstates and molybdates $AEuM_2O_8$. Journal of Solid State Chemistry, 76 (1): 160.

Wang Z, Liang H, Gong M, et al. 2005. A potential red-emitting phosphor for LED solid-state lighting. Electrochemical and Solid-State Letters, 8 (4): H33.

Wang Z, Liang H, Gong M, et al. 2007. Luminescence investigation of $Eu^{3+}$ activated double molybdates red phosphors with scheelite structure. Journal of Alloys Compounds, 38 (25): 308.

Wen D, Shi J. 2013. A novel narrow-line red-emitting $Na_2Y_2B_2O_7$:$Ce^{3+}$, $Tb^{3+}$, $Eu^{3+}$ phosphor with high efficiency activated by terbium chain for near-UV white LEDs. Dalton Transactions, 42 (47): 16621.

Wu C, Wang Y, Wang D. 2008. Optical properties of $Tb^{3+}$ ion in $LnP_3O_9$ (Ln = Y, Gd) host matrix. Electrochemical and Solid-State Letters, 11 (2): J9.

Xia Z, Liu R S. 2012. Tunable blue-green color emission and energy transfer of $Ca_2Al_3O_6F$:$Ce^{3+}$, $Tb^{3+}$ phosphors for near-UV white LEDs. Journal of Physical Chemistry C, 116 (29): 15604.

Xia Z, Wu W. 2013. Preparation and luminescence properties of $Ce^{3+}$ and $Ce^{3+}$ /$Tb^{3+}$-activated $Y_4Si_2O_7N_2$ phosphors. Dalton Transactions, 42 (36): 12989.

Xue Y N, Xiao F, Zhang Q Y. 2011. Enhanced red light emission from $LaBSiO_5$:$Eu^{3+}$, $R^{3+}$ (R = Bi or Sm) phosphors. Spectrochimica Acta Part A: Molecular and Biomolecular Spectroscopy, 78 (2): 607-611.

Yang F, Ma H, Liu Y, et al. 2013. A new green luminescent material $Ba_3Bi$ $(PO_4)_3$: $Tb^{3+}$. Ceramics International, 39 (2): 2127.

Yang W J, Luo L, Chen T M, et al. 2005. Luminescence and energy transfer of Eu-and Mn-coactivated $CaAl_2Si_2O_8$ as a potential phosphor for white-light UVLED. Chemistry of Materials, 17 (15): 3883.

Ye S, Wang X M, Jing X P. 2008. Energy transfer among $Ce^{3+}$, $Eu^{2+}$, and $Mn^{2+}$ in $CaSiO_3$. Journal of the Electrochemical Society, 155 (6): J143.

Z Xia, J Zhuang, L Liao. 2012. Novel red-emitting $Ba_2Tb(BO_3)_2Cl$:Eu phosphor with efficient energy transfer for potential application in white light-emitting diodes. Inorganic Chemistry, 51 (13): 7202.

Zhang X, Gong M. 2014. Single-phased white-light-emitting $NaCaBO_3$:$Ce^{3+}$, $Tb^{3+}$, $Mn^{2+}$ phosphors for LED applications. Dalton Transactions, 43 (6): 2465.

Zhang X, Gong X. 2011. Photoluminescence properties and energy transfer of thermal-stable $Ce^{3+}$, $Mn^{2+}$-codoped barium strontium lithium silicate red phosphors. Journal of Alloys and Compounds, 509 (6): 2850.

Zhang X, Park B, Choi N, et al. 2009. A novel blue-emitting $Sr_3Al_2O_5Cl_2$:$Ce^{3+}$, $Li^{+}$ phosphor for near UV-excited white light-emitting diodes. Materials Letters, 63 (8): 700.

Zhang X, Zhou L, Pang Q, et al. 2014. Novel broadband excited and linear red-emitting $Ba_2Y$ $(BO_3)_2Cl$: $Ce^{3+}$, $Tb^{3+}$, $Eu^{3+}$ phosphor: luminescence and energy transfer. Journal of the American Ceramic Society, 97 (7): 2124.

Zhang X, Zhou L, Pang Q, et al. 2014. Tunable Luminescence and $Ce^{3+}$→$Tb^{3+}$→$Eu^{3+}$ energy transfer of broadband-excited and narrow line red-emitting $Y_2SiO_5$: $Ce^{3+}$, $Tb^{3+}$, $Eu^{3+}$ phosphor. Journal of Physical Chemistry C, 118 (14): 7591.

Zhao W, An S, Fan B, et al. 2013. Cathodoluminescence properties of $Tb^{3+}$-doped $Na_3YSi_2O_7$ phosphors. Applied Physics A, 111 (2): 601.

Zhuo Y, Hariyani S, Zhong J, et al. 2021. Creating a Green-emitting phosphor through selective rare-earth site preference in $NaBaB_9O_{15}$:$Eu^{2+}$. Chemistry of Materials, 33 (9): 3304.

# Chapter IV Blue phosphors for white light-emitting diodes

## 4.1 $Eu^{2+}$-activated $BaMgAl_{10}O_{17}$ for solid-state lighting

### 4.1.1 Introduction

Increasingly strong interest was focused on the solid-state lighting of light-emitting diodes (LEDs) since Nakamura et al. fabricated a blue-emitting LED in 1993. Presently, the emission bands of LEDs are close to the near-UV range around 400 nm, and this wavelength can offer a higher efficiency solid-state lighting. The w-LEDs can be generated by several different methods. The most commonly used method is to combine the red/green/blue tricolor phosphors with a GaN/InGaN chip. Because the red and green phosphors can share the absorption in the blue range, the blue phosphor should exhibit higher efficiency. Presently used blue phosphor for near-UV InGaN-based LEDs is mainly $BaMgAl_{10}O_{17}:Eu^{2+}$ (BAM). BAM is used as a commercial blue phosphor in fluorescent lamps, because of its efficient blue emission. But its intensity of excitation around the near-UV range is lower, so it is important to broaden its excitation band around 400 nm and strengthen its emission intensity.

The BAM particle has a $\beta$-alumina structure consisting of a spinel block ($MgAl_{10}O_{16}$) and a mirror plane (BaO). The $Eu^{2+}$ ions are substituted for the barium site of the mirror plane. The broad-band emission of the $Eu^{2+}$ is due to the electronic transition between the $4f^7$ ground state and the $4f^65d^1$ excited state. Since the 5d orbital is exposed to the surrounding ions, this transition is highly influenced by the crystal field. Hence, the PL properties of the phosphors doped with $Eu^{2+}$ may be different, by adjusting their sub-lattice structure around the luminescent center ions.

$B^{3+}$ ($1s^2$) ion shares the similar electronic configurations of $Al^{3+}$ ($2s^22p^6$). But their ionic radii and electronegativities are different. The introduction of boric acid maybe change the sub-lattice structure and influence the crystal field around $Eu^{2+}$ ions, then the PL properties of the phosphors are expected to be optimized. Hence, the BAM phosphors doped with different contents of $H_3BO_3$ were prepared in the present work, and their structure and PL properties were investigated.

### 4.1.2 Structure and PL properties

The phosphors BAM doped with different contents of boric acid have been prepared. The XRD pattern of $BAM_6$ in Fig.4-1 is consistent with the JCPDS card No. 26-0163 [$BaMgAl_{10}O_{17}$]. The result shows the phosphor is of a single-phase with a hexagonal structure.

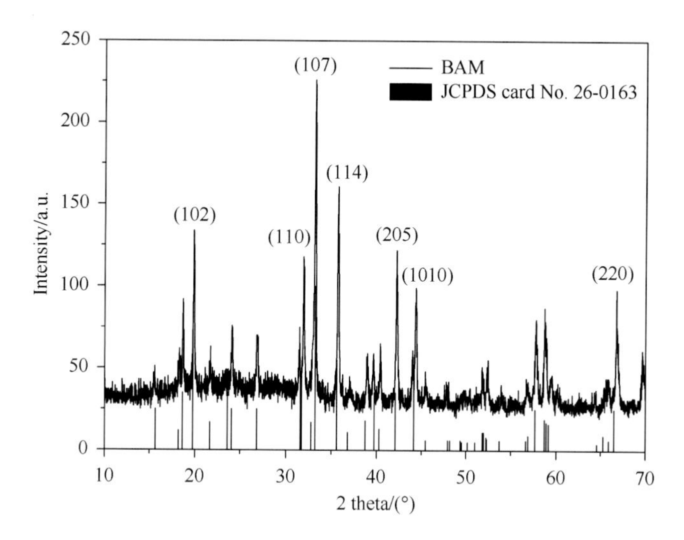


Fig.4-1 The XRD pattern of $BAM_6$

The PL spectra of these phosphors are investigated at room temperature. The excitation spectra of these phosphors are shown in Fig. 4-2. The broad excitation band from 300 nm to 420 nm is due to the 4f-5d transitions of $Eu^{2+}$ in the host lattices. Compared with the curve a,

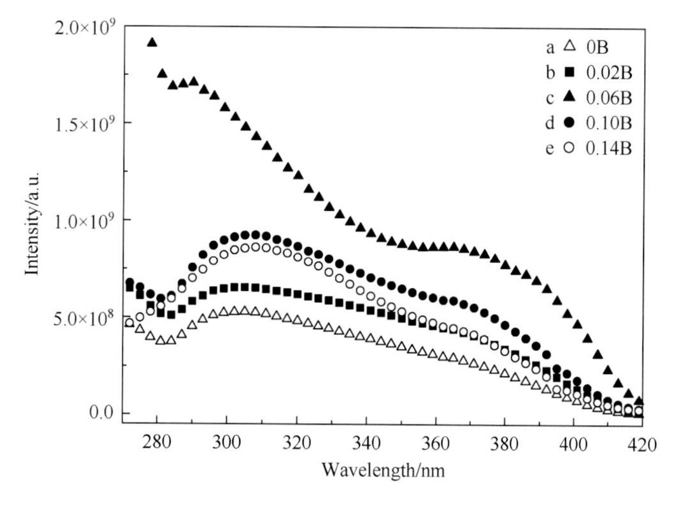


Fig.4-2 The Excitation spectra of BAM:0.1Eu doped with different contents $H_3BO_3$ by monitoring 440 nm emission at room temperature

a broad excitation band overlaps on the long-wavelength direction of the $Eu^{2+}$ 4f-5d transitions from 360 nm to 420 nm in curves b, c, d and e. This excitation band is due to the introduction of $H_3BO_3$. The $B^{3+}$ ion radius (41 pm) is smaller than that of $Al^{3+}$ ion (67 pm), then the doping of $H_3BO_3$ will decrease the size of a spinel block $[Mg(Al_{1-x}B_x)_{10}O_{16}]$. The distance from the spinel block ($MgAl_{10}O_{16}$) to the mirror plane (BaO) decreases, so the energy transfer from the host to $Eu^{2+}$ is decreased. This would induce the red-shift in the excitation band, and this red-shift is beneficial for the application on near-UV LEDs. Optimum excitation intensity is obtained when the content of $H_3BO_3$ is at 6 %.

The phosphors for near-UV LED chips should show intense emission and excitation band in the near-UV range. Our target is to find blue phosphors for near-UV LEDs, so the emission spectra of these phosphors under the 400 nm light excitation are studied (Fig.4-3). The main emission of these phosphors is blue light, which is due to the $4f^65d^1 \rightarrow 4f^7$ transition of $Eu^{2+}$ ion. However the emission peak shows a little red-shift with the increasing content of $H_3BO_3$, and the emission intensities of these phosphors are different. When the content of $H_3BO_3$ is at 6 %, the sample emission intensity is the strongest, and its intensity is about 6 times higher than that of undoped BAM. Its CIE chromaticity coordinates are calculated to be $x = 0.151$, $y = 0.058$.

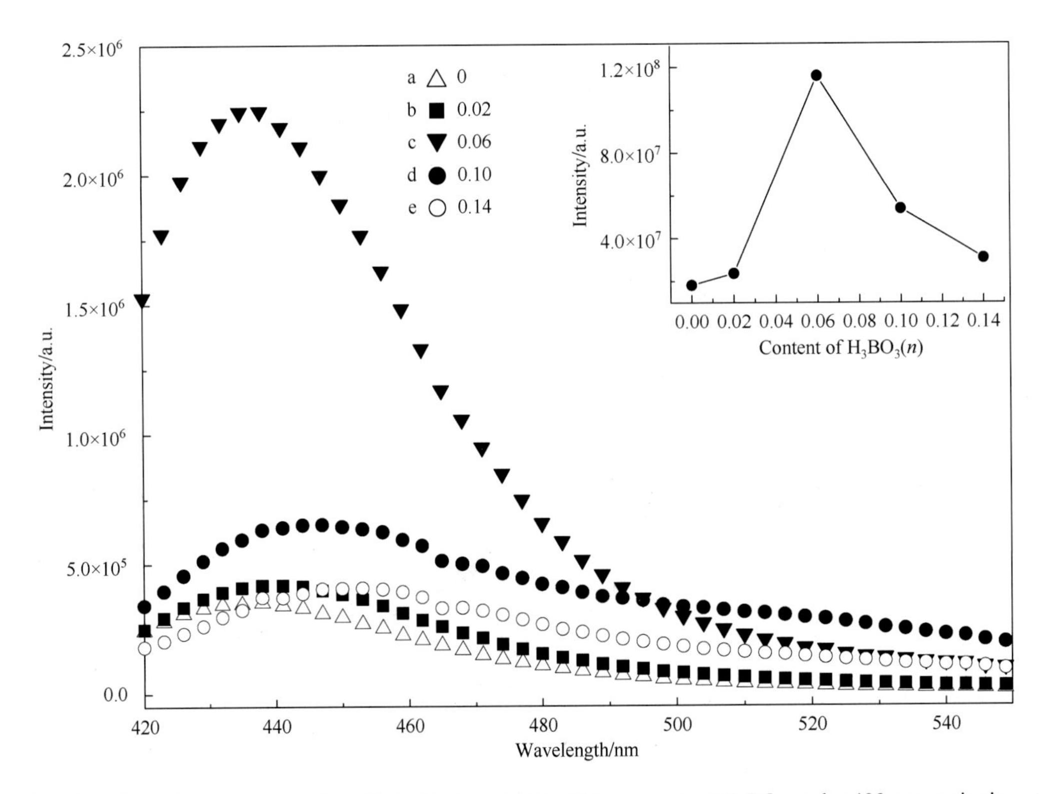


Fig.4-3 The emission spectra of BAM:0.1Eu doped with different content $H_3BO_3$ under 400 nm excitation at room temperature. The inserted figure is the concentration quenching curve for $BaMgAl_{10}O_{17}:Eu^{2+}$ doped different contents $H_3BO_3$ under the 400 nm light excitation

The emission intensity for phosphors $BaMgAl_{10}O_{17}:Eu^{2+}$ doped different contents of $H_3BO_3$ is investigated as a function of the dopant concentrations $x$ as exhibited in the insert of Fig.4-3. It can be observed that the phosphor $BAM_6$ shows the strongest emission under the 400 nm light excitation.

The low-temperature emission spectrum (77 K) of the sample $BAM_6$ is shown in Fig.4-4. The full width at half maximum (FWHM) of the emission peak at 77 K is 47 nm, which is narrower than that of the peak at 297 K (53 nm). This may be due to the thermal quenching at the configurational coordinate diagram. The thermally activated luminescent center is strongly interacted with thermally active phonon, contributing to FWHM of the emission spectrum. At higher temperature, the population density of phonon is increased, and the electron-phonon interaction is dominant, and the FWHM of the emission spectrum is broadened. This result is consistent with that of some other phosphors in Ref.

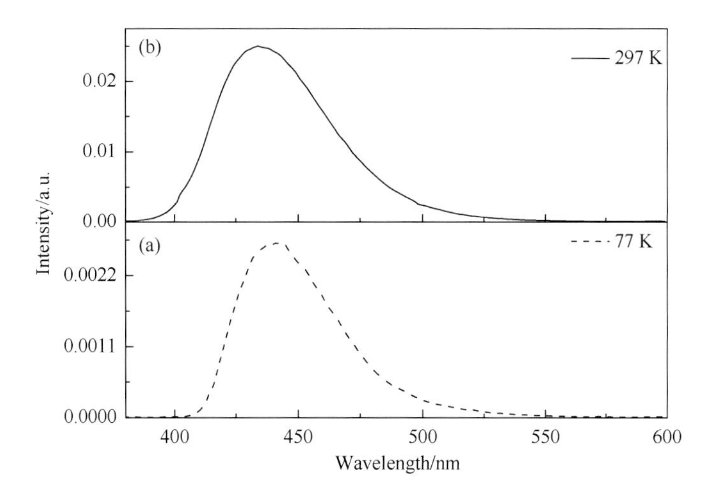


Fig.4-4 The emission spectra of $BAM_6$ at 77 K (a) and 297 K (b) excited by a 10 Mw He-Cd laser with a 325 nm line

The decay curves of the $Eu^{2+}$ emission of the phosphors are shown in Fig.4-5. They are all well fitted into the single-exponential function [$I = I_0\exp(-t/\tau)$ ]. The lifetime values are derived to be 649 ns, 609 ns, 577 ns, 631 ns and 909 ns, with the $H_3BO_3$ content $n = 0$, 0.02, 0.06, 0.10, 0.14, respectively. From the dependence of the lifetime of the phosphors on $H_3BO_3$ content $n$ in the insert of Fig.4-5, the lifetime of the phosphor $BAM_6$ is the shortest among these phosphors. As discussed above, the $B^{3+}$ ion radius is smaller than that of $Al^{3+}$ ion, and the distance from the spinel block ($MgAl_{10}O_{16}$) to the mirror plane (BaO) will decrease with the introduction of $H_3BO_3$. So the energy transfer from the host to $Eu^{2+}$ will be easier, and the lifetime of the phosphor decreases. With the increase of the $H_3BO_3$ content, the crystal volume decreases, and the distance of $Eu^{2+}$-$Eu^{2+}$ gets shorter. Then the concentration quenching will occur. This result is in agreement with the result of the concentration quenching curve for $BaMgAl_{10}O_{17}:Eu^{2+}$ doped different contents of $H_3BO_3$ (Fig.4-3).

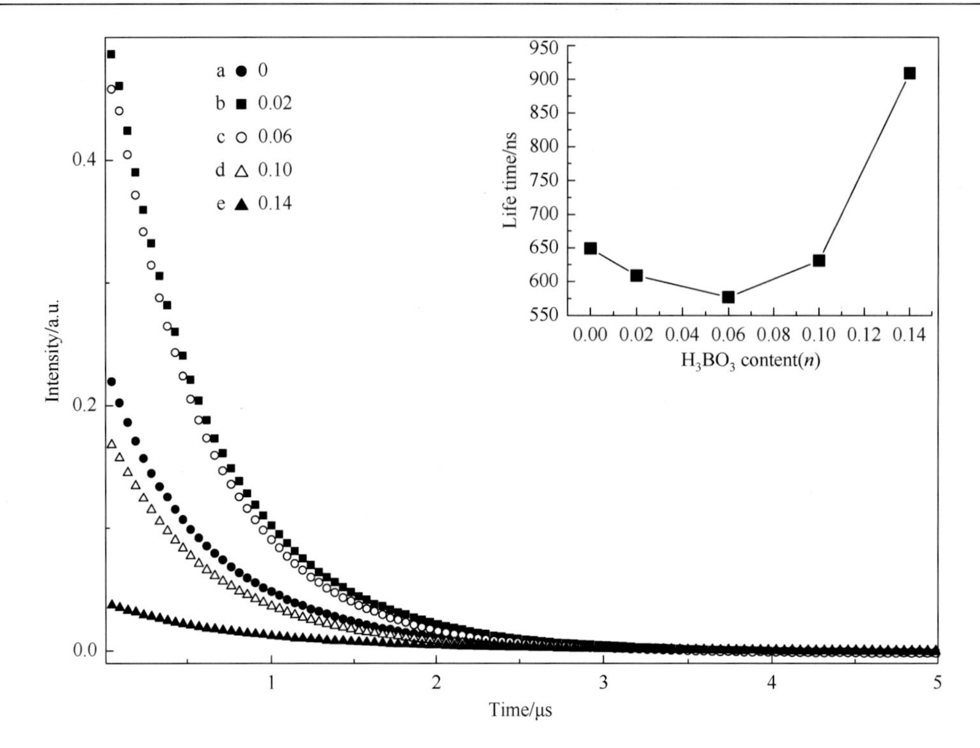


Fig.4-5 Decay curves of the $Eu^{2+}$ emission of the phosphors [$\lambda_{ex}$ = 337 nm (laser); $\lambda_{em}$ = 440 nm (Xe lamp) ]. The inserted figure is the dependence of the lifetime of the phosphors on $H_3BO_3$ content $n$

In summary, the phosphors BAM doped with different contents $H_3BO_3$ were prepared by the solid-state reaction. And their PL properties at room temperature and low temperature were studied. From the standpoint of application, each proper mono-color LED phosphor must meet the following necessary conditions. ①The phosphor must show high thermal stability. ②The phosphor must efficiently absorb the 400 nm excitation energy that the InGaN chip emitted, and higher luminescent intensity under 400 nm excitation. ③The chromaticity coordinates of the phosphor are close to the NTSC standard values. The phosphor $BAM_6$ exhibits intense blue emission under the 400 nm light excitation, and its excitation band around 400 nm is broadened by the introduction of $H_3BO_3$. So it can be a new deep blue phosphor for the near-UV LED.

## 4.2 $Eu^{2+}$-activated $Sr_4Si_3O_8Cl_4$ for solid-state lighting

### 4.2.1 Introduction

At present, the phosphors for near-UV LED chips should exhibit intense broad emission with one broad excitation band in the near UV range. Compared with the phosphors activated by some trivalent rare-earth ions (such as $Eu^{3+}$), $Eu^{2+}$ ions in many phosphors have broader emission with one broad excitation band in the near-UV range, which is assigned to the

$4f^6 5d^1 \rightarrow 4f^7$ transitions. To strengthen and broaden the absorption of the phosphors in the near-UV range, the one important approach is to introduce a sensitizer to strengthen the excitation band in the phosphor. We consider that $Bi^{3+}$ is probably an eligible co-activator. On the one hand, the introduction of $Bi^{3+}$ will influence their sub-lattice structure around the luminescent center ion. On the other hand, $Bi^{3+}$ is a very good sensitizer of luminescence in many hosts. It can efficiently absorb the UV-light and transfer the energy to the luminescent center. Then the excitation band would be broadened and the emission intensity of the luminescent center would be strengthened.

Alkaline earth halo-silicates are well-known good hosts for inorganic luminescent materials, due to their low synthesis temperature and high luminescence efficiency. $Sr_4Si_3O_8Cl_4$ can be served as the host of phosphors, which has an orthorhombic crystal structure. The luminescent properties of $Sr_4Si_3O_8Cl_4$ doped with $Eu^{2+}$ were reported by Burrus firstly. The luminescent properties of $Sr_4Si_3O_8Cl_4$ doped with $Eu^{2+}$ were optimized by co-doping with some divalent metal ions, such as $Ca^{2+}$, $Mg^{2+}$, $Zn^{2+}$.

In this part, several $Sr_4Si_3O_8Cl_4$ co-doped with $Eu^{2+}$, $Bi^{3+}$ are prepared by the high-temperature reaction in the reduction atmosphere. The luminescent properties of $Sr_4Si_3O_8Cl_4$:$Eu^{2+}$,$Bi^{3+}$ are investigated. At last, a single blue-green LED is fabricated by combining the blue-green phosphor with a 395 nm emitting LED chips.

### 4.2.2 Structure, PL properties and application on LEDs

The XRD patterns of $Sr_{3.90}Si_3O_8Cl_4$:0.10$Eu^{2+}$ and $Sr_{3.50}Si_3O_8Cl_4$:0.10$Eu^{2+}$,0.40$Bi^{3+}$ are shown in Fig.4-6. They are almost consistent with the JCPDS card No. 40-0074 [$Sr_4Si_3O_8Cl_4$]. This

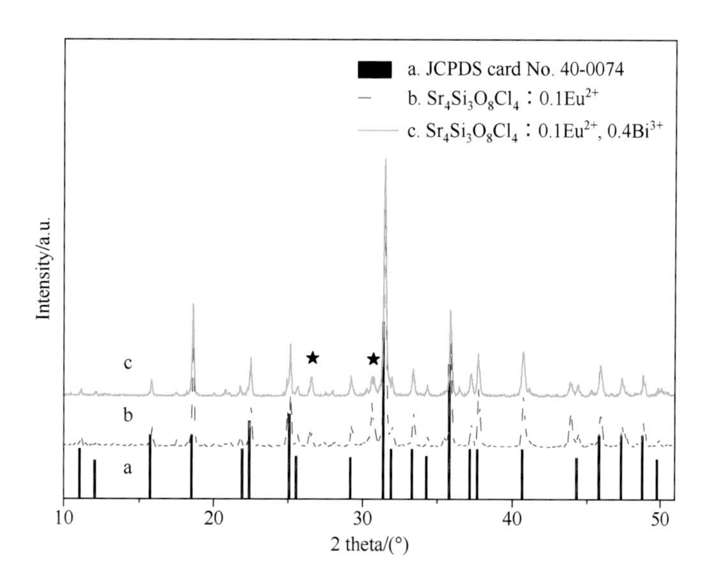


Fig.4-6 XRD patterns of $Sr_4Si_3O_8Cl_4$:0.10$Eu^{2+}$ and $Sr_4Si_3O_8Cl_4$:0.10$Eu^{2+}$,0.40$Bi^{3+}$

indicates that the phosphor $Sr_4Si_3O_8Cl_4$:$Eu^{2+}$ almost shares the same phase as $Sr_4Si_3O_8Cl_4$, except for a few peaks of $SrSiO_3$ (marked by star shape). This result is in agreement with the references. The impurity phase may be due to loss of chlorine content during the firing process. $Eu^{2+}$ and $Bi^{3+}$ maybe occupy the site of $Sr^{2+}$ in an octahedral site, because the ion radii of $Bi^{3+}$ (103 pm) and $Eu^{2+}$ (117 pm) are close to that of $Sr^{2+}$ (118 pm). Meanwhile, with the doping of $Bi^{3+}$/$Eu^{2+}$, a slight shift in the diffraction peaks toward higher angles can be found in Fig.4-6. This shift is due to the difference in ionic radius between $Bi^{3+}$ /$Eu^{2+}$ and $Sr^{2+}$.

Fig.4-7(a) is the excitation spectra of the phosphors $Sr_4Si_3O_8Cl_4$:$4xEu^{2+}$ ($x = 0.015$, 0.020, 0.025, 0.030) by monitoring emission at 490 nm. These four curves are of similar shapes. The broad excitation band from 230 nm to 430 nm is due to the 4f-5d transitions of $Eu^{2+}$. These broad and intense excitation bands in the near-UV range match well with the near-UV LED chip. When the content of $Eu^{2+}$ is 0.10, the excitation intensity is the strongest.

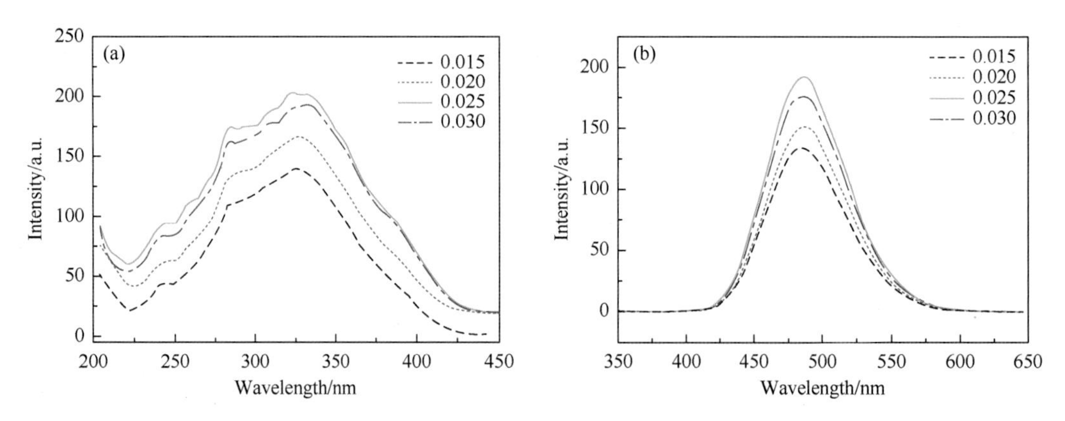


Fig.4-7 (a) Excitation spectra and (b) emission spectra of $Sr_4Si_3O_8Cl_4$:$4xEu^{2+}$ ($x = 0.015$, 0.020, 0.025, 0.030)

The emission spectra of $Sr_4Si_3O_8Cl_4$:$4xEu^{2+}$ ($x = 0.015$, 0.020, 0.025, 0.030) under the 320 nm excitation are shown in Fig.4-6(b). The blue-green emission band is due to the $4f^6 5d$ - $4f^7$ transition of $Eu^{2+}$. With the increase of the $Eu^{2+}$ content, the intensity of the blue-green emission becomes higher. When the content of $Eu^{2+}$ is 0.10, the emission intensity is the strongest.

Since the luminescent mechanism of $Eu^{2+}$ is due to the 4f-5d allowed electric-dipole transition, the energy transfer process of $Eu^{2+}$ in the $Sr_4Si_3O_8Cl_4$ phosphors would be attributed to an electric multipole-multipole interaction. When the energy transfer occurs between the same sites of activators, the intensity of multipole interaction can be determined by the change of the emission intensity from the emitting level. According to the report of Van Uitert, the emission intensity ($I$) of per activator ion follows the equation.

$$\frac{I}{x} = K[1 + \beta x^{\frac{Q}{3}}]^{-1} \tag{4-1}$$

where $I$ is the emission intensity, $x$ is the $Eu^{2+}$ concentration, $K$ and $\beta$ are constants for the same excitation condition for a given host crystal. $Q$ is a constant of multipole interaction equals (3, 6, 8 or 10) for energy transfer among the nearest-neighbor ions, dipole-dipole (d-d), dipole-quadruple (d-q), or quadruple-quadruple (q-q) interaction, respectively. To get a $Q$ value for the emission center, the plot of log $(I/x)$ as a function of $\log x$ for the $Sr_4Si_3O_8Cl_4{:}4xEu^{2+}$ phosphors is plotted, as shown in Fig.4-8. It can be seen that a linear relation between log $(I/x)$ and $\log x$ is found and the slope is −0.5299. The $Q$ value can be obtained as 1.5897. That means the quenching is directly proportional to the activator ion concentration, which indicates that the concentration quenching is caused by the energy transfer among the nearest-neighbor $Eu^{2+}$ ions in the $Sr_4Si_3O_8Cl_4{:}4xEu^{2+}$ phosphors.

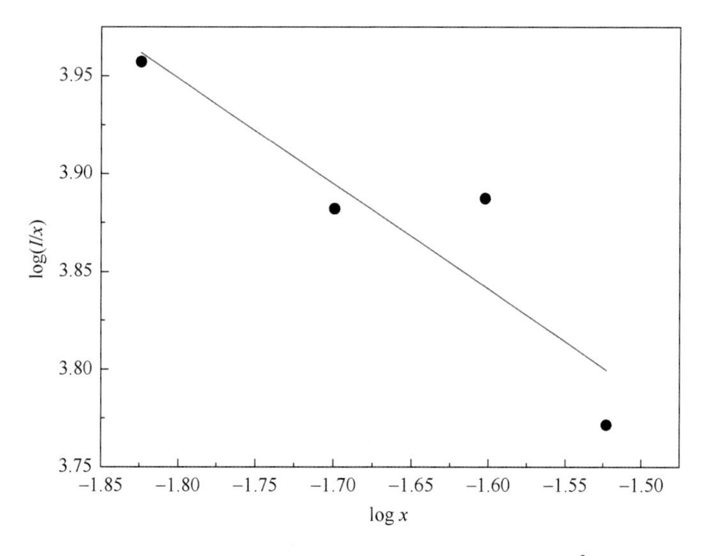


Fig.4-8 The plot of log $(I/x)$ as a function of $\log x$ for the $Sr_4Si_3O_8Cl_4{:}4xEu^{2+}$ phosphors ($\lambda_{ex}$ = 320 nm)

The excitation spectra of the phosphors $Sr_4Si_3O_8Cl_4{:}0.10Eu^{2+}$, $4yBi^{3+}$ ($y$ = 0.00, 0.05, 0.10, 0.15, 0.20) by monitoring emission 490 nm are shown in Fig.4-9. With the doping of $Bi^{3+}$, the shoulder peak around 400 nm is broadened with a little red-shift, and the excitation intensity is enhanced. When the content of $Bi^{3+}$ is 0.10, the excitation intensity is the strongest.

Fig.4-10 is the emission spectra of $Sr_4Si_3O_8Cl_4{:}0.10Eu^{2+},4yBi^{3+}$ ($y$ = 0.00, 0.05, 0.10, 0.15, 0.20) under the 320 nm excitation. The broad emission from 400 nm to 600 nm is due to the $4f^65d^1 \rightarrow 4f^7$ transition of $Eu^{2+}$ ions. The strongest peak is located at 490 nm. The concentration dependence of the relative integral emission intensity of $Sr_4Si_3O_8Cl_4{:}0.10Eu^{2+},4yBi^{3+}$ is shown in the inserted figure of Fig.4-10. When the content of $Bi^{3+}$ is at 0.10, its emission intensity is the strongest, and its intensity is about 1.3 times higher than that of $Sr_4Si_3O_8Cl_4{:}0.10Eu^{2+}$ without $Bi^{3+}$. This result is consistent with their excitation spectra. The bright blue-green light can be observed from $Sr_4Si_3O_8Cl_4{:}0.10Eu^{2+},0.40Bi^{3+}$ under the 365 nm light

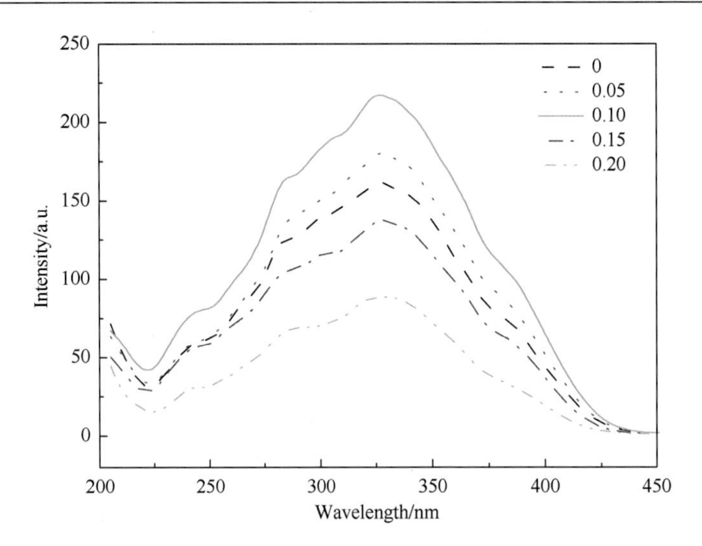


Fig.4-9 The excitation spectra of $Sr_4Si_3O_8Cl_4$:0.10$Eu^{2+}$,4*y*$Bi^{3+}$ ($y$ = 0.00, 0.05, 0.10, 0.15, 0.20) by monitoring emission at 490 nm

excitation (Fig.4-10). The compositions $Sr_4Si_3O_8Cl_4$:0.10$Eu^{2+}$,4*y*$Bi^{3+}$ are not charge-balanced and slight excess of positive charge because of $Bi^{3+}$ substituting $Sr^{2+}$. The excess charge could be compensated by several mechanisms, for example, cation vacancies, varying slightly O contents, etc. In this series of phosphors, this kind of substitution did not influence the emission spectra shapes of $Sr_4Si_3O_8Cl_4$:0.10$Eu^{2+}$,4*y*$Bi^{3+}$ (Fig.4-10).

Fig.4-11 shows the EL spectra of the intense blue-green LED fabricated with $Sr_4Si_3O_8Cl_4$:0.10$Eu^{2+}$ and $Sr_4Si_3O_8Cl_4$:0.10$Eu^{2+}$,0.40$Bi^{3+}$ under 20 mA current excitation. The emission band at 400 nm is attributed to the emission of the LED chip and the broadband from 430 nm to 600 nm is due to the emission of the phosphor. Comparing with curve a, the ratio of LED chip emission to phosphor emission in curve b is smaller. The result indicates phosphor $Sr_{3.50}Si_3O_8Cl_4$:0.10$Eu^{2+}$, 0.40$Bi^{3+}$ can more efficiently absorb the emission of LED chip, and exhibit stronger blue-green emission compared with $Sr_4Si_3O_8Cl_4$:0.10$Eu^{2+}$. The bright blue-green light is observed from these two LEDs by naked eyes. This result is in accord with their emission spectra. Their CIE chromaticity coordinates are calculated to be $x$ = 0.169, $y$ = 0.356 and $x$ = 0.170, $y$ = 0.362, respectively.

A series of phosphors, $Sr_4Si_3O_8Cl_4$:4*x*$Eu^{2+}$, $Sr_4Si_3O_8Cl_4$:0.10$Eu^{2+}$,4*y*$Bi^{3+}$ were prepared by the solid-state reaction at high temperature. Their structure and PL properties were investigated. The doping of $Bi^{3+}$ enhances the excitation and emission intensity of these phosphors. The blue-green phosphor $Sr_4Si_3O_8Cl_4$:0.10$Eu^{2+}$,0.40$Bi^{3+}$ exhibits the intense blue-green emission with broader excitation in the near-UV range. The LED based on this phosphor can emit intense blue-green light, which can be observed by naked eyes. Hence it is considered to be a good candidate for the blue-green component of a w-LED.

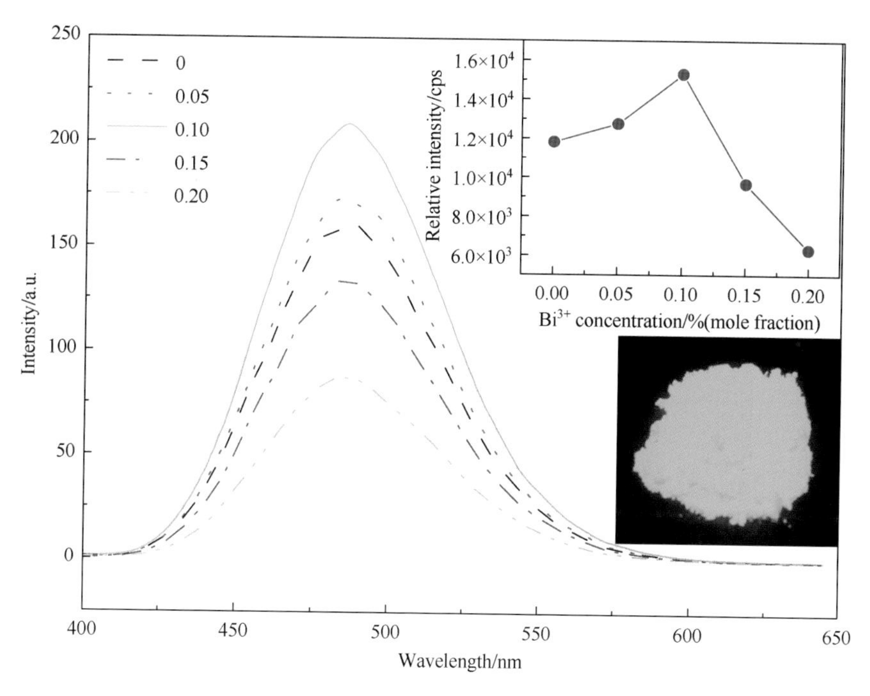


Fig.4-10 The emission spectra of $Sr_4Si_3O_8Cl_4$:$0.10Eu^{2+}$,$4yBi^{3+}$ ($y$ = 0.00, 0.05, 0.10, 0.15, 0.20) under 320 nm excitation. The inserted figures are the concentration dependence of the relative integral emission intensity of $Sr_4Si_3O_8Cl_4$:$0.10Eu^{2+}$,$4yBi^{3+}$ and the image of $Sr_4Si_3O_8Cl_4$:$0.10Eu^{2+}$,$0.40Bi^{3+}$ under the 365 nm excitation

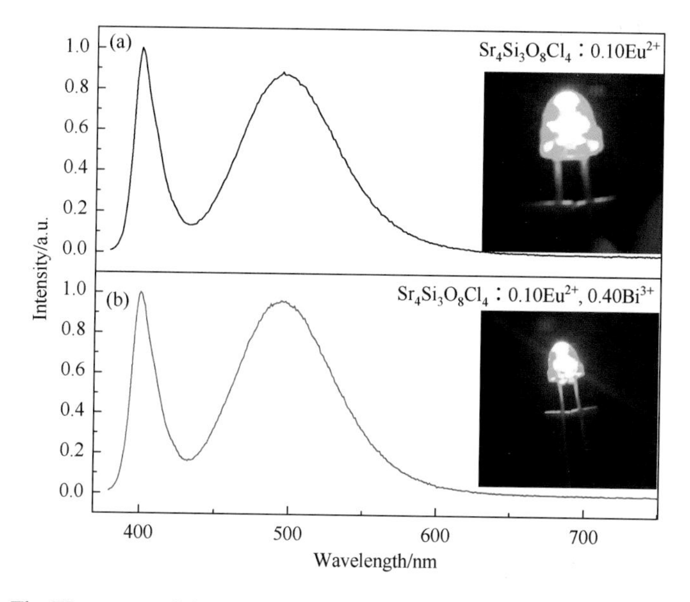


Fig.4-11 The EL spectra of the blue-green LEDs-based on (a) $Sr_4Si_3O_8Cl_4$:$0.10Eu^{2+}$ and (b) $Sr_4Si_3O_8Cl_4$:$0.10Eu^{2+}$,$0.40Bi^{3+}$ under 20 mA current excitation. The inserted figures are the images of these two LEDs

## 4.3 $Eu^{2+}$-activated $AAl_2Si_2O_8$ ($A = Ca^{2+}, Mg^{2+}, Ba^{2+}$)

### 4.3.1 Introduction

The excitation band of $Eu^{2+}$ in phosphors is a broadband, which is due to the d-f transition. It can exhibit intense emission from the violet to red light region in the different hosts with different crystal filed. So the excitation and emission bands of $Eu^{2+}$ can be controlled by changing the crystal field. The different cations with different radii in different host compounds would induce some change in the sub-lattice structure around the luminescent center ions, even change the host structure. Hence the crystal filed will be changed, which leads to different PL properties.

Recently, $CaAl_2Si_2O_8$ doped with the luminescent center ion is investigated, and it shows excellent blue emission with one broadband in the near-UV range, which can match the excitation of the near-UV GaInN chip. Anorthite ($CaAl_2Si_2O_8$) was reported to be crystallized in a triclinic crystal system under ambient pressure by Angel in 1988. In the crystal lattice, there are six crystallographically independent cation sites, namely four $Ca^{2+}$ sites, one $Al^{3+}$ site and one $Si^{4+}$ site. One type of $Ca^{2+}$ ion occupies an octahedral site with six oxygen atoms, and other $Ca^{2+}$ ions occupy three kinds of polyhedral sites with seven coordinated oxygen atoms. Al and Si atoms both occupy tetrahedral sites with four coordinated oxygen atoms.

In this section, the phosphors $CaAl_2Si_2O_8:Eu^{2+}$ doped with different contents $Sr^{2+}$ and $Mg^{2+}$ were prepared by the solid-state reaction, and their luminescent properties were investigated.

### 4.3.2 $CaAl_2Si_2O_8:xEu^{2+}$

The XRD pattern of $CaAl_2Si_2O_8:0.04Eu^{2+}$ is shown in Fig.4-12. It is consistent with the JCPDS card No. 41-1486 [$CaAl_2Si_2O_8$]. This indicates that the phosphor $CaAl_2Si_2O_8:0.04Eu^{2+}$ shares a single-phase with the triclinic structure, and the doping of $Eu^{2+}$ does not change the crystal structure.

Fig.4-13 is the excitation spectra of the phosphors $CaAl_2Si_2O_8:xEu^{2+}$ ($x = 0.02$, 0.0.04, 0.06, 0.08) by monitoring emission at 435 nm. The broad excitation band from 300 nm to 415 nm is due to the 4f-5d transitions of $Eu^{2+}$. The excitation bands from 360 nm to 400 nm show a flatform. This flat absorption is useful to find an application in near-UV LED chips. When the content of $Eu^{2+}$ is 4 %, the excitation intensity of $CaAl_2Si_2O_8:0.04Eu^{2+}$ is the strongest among these phosphors.

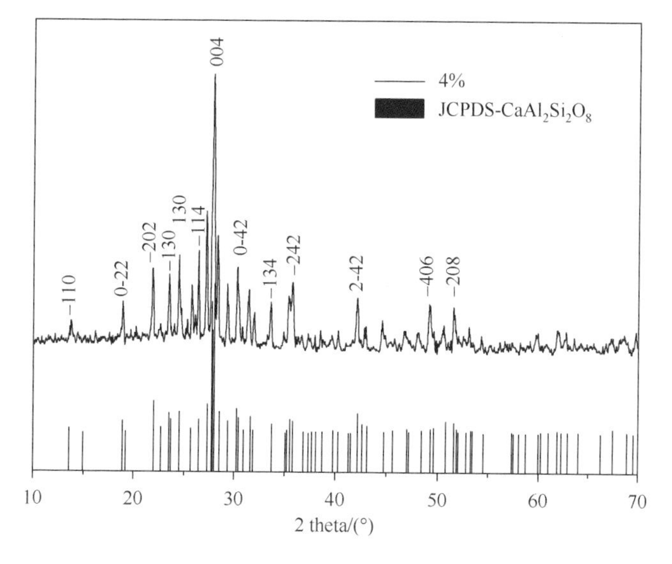


Fig.4-12 The XRD pattern of $CaAl_2Si_2O_8$:0.04$Eu^{2+}$

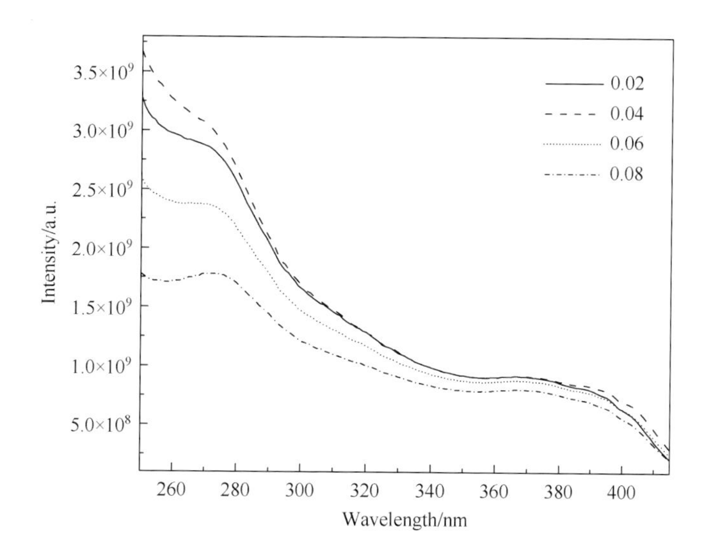


Fig.4-13 The excitation spectra of $CaAl_2Si_2O_8$: $x Eu^{2+}$ ($x$ = 0.02, 0.0.04, 0.06, 0.08) by monitoring emission at 435 nm

Our target is to investigate the blue phosphor for near-UV LED chips, then the emission spectra of $CaAl_2Si_2O_8$:$x Eu^{2+}$ ($x$ = 0.02, 0.0.04, 0.06, 0.08) under 400 nm excitation are shown in Fig.4-14. The broad blue emission is due to the 5d-4f transitions of $Eu^{2+}$. It is observed that the emission bands of the phosphors show some red-shifts with an increase of $Eu^{2+}$ concentration. Qiu et al. had reported that the probability of an energy transfer among $Eu^{2+}$ ions increases when the $Eu^{2+}$ concentration increases. With the increase of the $Eu^{2+}$ content, the distance between $Eu^{2+}$ ions becomes less, and the probability of energy transfer among $Eu^{2+}$ ions increases. In other words, the energy transfer probability of $Eu^{2+}$ ions from the higher 5d levels to the lower 5d levels of $Eu^{2+}$

ions increases. Hence the emission band of the sample shows the red-shift with an increase of $Eu^{2+}$ concentration. The sample $CaAl_2Si_2O_8$:$0.04Eu^{2+}$ shows the strongest emission with appropriate CIE chromaticity coordinates ($x$ = 0.163, $y$ = 0.112).

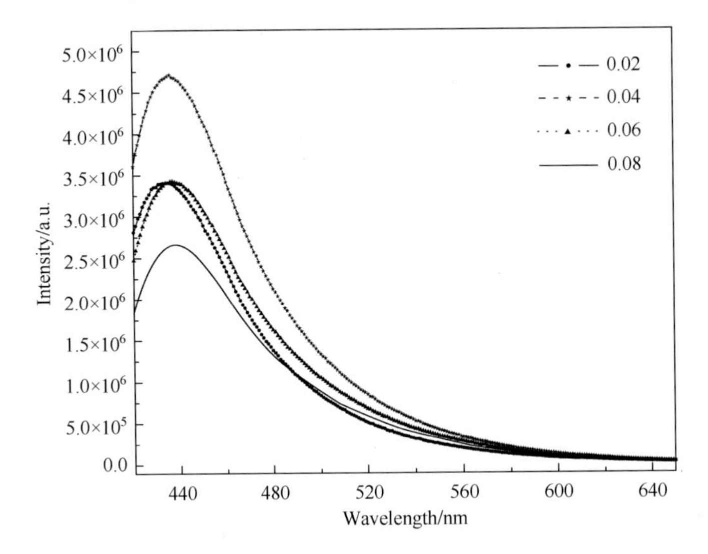


Fig.4-14 The emission spectra of $CaAl_2Si_2O_8$:$xEu^{2+}$ ($x$ = 0.02, 0.0.04, 0.06, 0.08) under 400 nm excitation

### 4.3.3 $CaAl_2Si_2O_8$:$0.04Eu^{2+}$, $ySr^{2+}$

The serial of samples $CaAl_2Si_2O_8$:$0.04Eu^{2+}$, $ySr^{2+}$ ($y$ = 0.10, 0.20, 0.30, 0.40, 0.50, 0.60, 0.70, 0.80, 0.90) were prepared. As examples, the XRD patterns of $CaAl_2Si_2O_8$:$0.04Eu^{2+}$,$ySr^{2+}$ ($y$ = 0.10, 0.30, 0.50, 0.70, 0.90) are shown in Fig.4-15. In this system, for $y \leqslant 0.03$, all the compositions crystallize

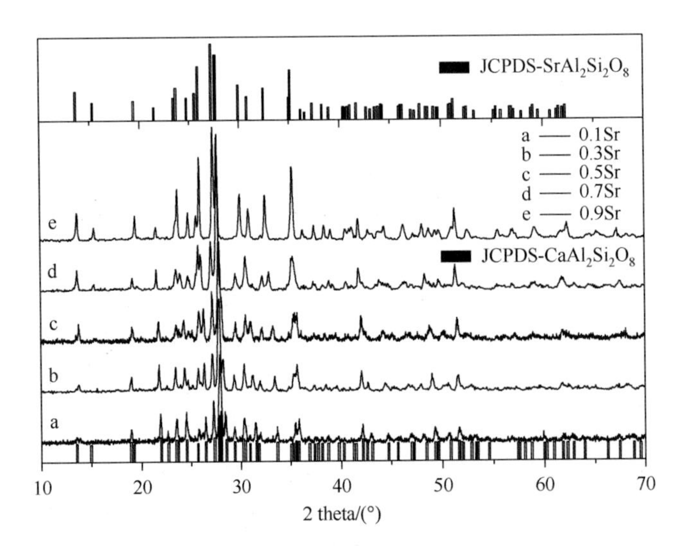


Fig.4-15 The XRD patterns of $CaAl_2Si_2O_8$:$0.04Eu^{2+}$, $ySr^{2+}$ ($y$ = 0.10, 0.30, 0.50, 0.70, 0.90)

in the triclinic structure of $CaAl_2Si_2O_8$ and on a further increase of $y$, the system undergoes a compositionally induced phase transition from the triclinic to the monoclinic structure. The XRD pattern of $CaAl_2Si_2O_8$:0.04$Eu^{2+}$,0.90$Sr^{2+}$ is very consistent with the JCPDS card of $SrAl_2Si_2O_8$.

Fig.4-16 is the excitation spectra of $CaAl_2Si_2O_8$:0.04$Eu^{2+}$,$y$$Sr^{2+}$ ($y$ = 0.10, 0.30, 0.50, 0.90) by monitoring emission at 435 nm. The broad excitation band from 300 nm to 415 nm is due to the 4f-5d transitions of $Eu^{2+}$. The phosphor $CaAl_2Si_2O_8$:0.04$Eu^{2+}$,0.30$Sr^{2+}$ shows the strongest excitation intensity in the near-UV range (Fig.4-16).

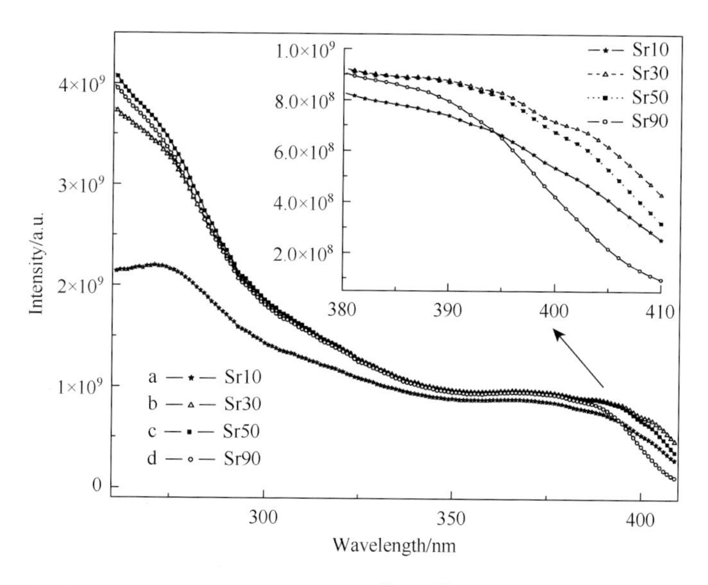


Fig.4-16 The excitation spectra of $CaAl_2Si_2O_8$:0.04$Eu^{2+}$,$y$$Sr^{2+}$ ($y$ = 0.10, 0.30, 0.50, 0.90) by monitoring emission at 435 nm. The inserted figure is the enlarged spectra from 380 nm to 410 nm

The emission spectra of $CaAl_2Si_2O_8$:0.04$Eu^{2+}$,$y$$Sr^{2+}$ ($y$ = 0.10, 0.30, 0.50, 0.90) are shown in Fig.4-17. The emission band shifts to a shorter wavelength with an increase of the $Sr^{2+}$ concentration. It is well known that the ionic radius of $Sr^{2+}$ is bigger than that of $Ca^{2+}$. In the triclinic structure of $CaAl_2Si_2O_8$, the length of the $c$ axis increases by replacing part of the $Ca^{2+}$ by $Sr^{2+}$ ions, the effect of preferential orientation of a $d$-orbital in the chain direction decreases, so that the $Eu^{2+}$ emission shifts to the shorter wavelength. The sample $CaAl_2Si_2O_8$:0.04$Eu^{2+}$,$y$$Sr^{2+}$ shares the intense emission, and its CIE chromaticity coordinates are calculated to be $x$ = 0.165, $y$ = 0.123.

### 4.3.4 $CaAl_2Si_2O_8$:0.04$Eu^{2+}$, $z$$Mg^{2+}$

The XRD patterns of $CaAl_2Si_2O_8$:0.04$Eu^{2+}$ doped with different contents of $Mg^{2+}$ are shown in Fig.4-18. In this serial of samples, for $z \leqslant 0.03$, all the compositions crystallize in the triclinic structure of $CaAl_2Si_2O_8$, which is similar to the JCPDS card No. 41-1486 [$CaAl_2Si_2O_8$]. With the further increase of $z$, another phase appears.

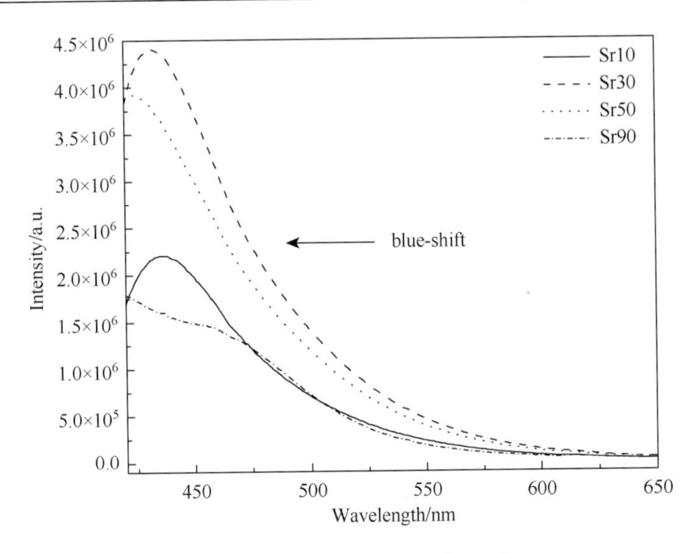


Fig.4-17　The emission spectra of $CaAl_2Si_2O_8$:$0.04Eu^{2+}$,$ySr^{2+}$ ($y$ = 0.10, 0.30, 0.50, 0.90) under the 400 nm excitation

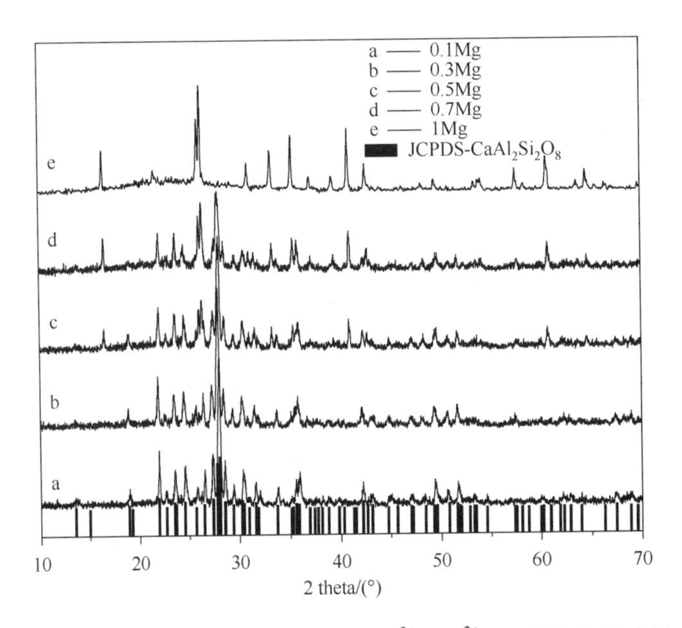


Fig.4-18　The XRD patterns of $CaAl_2Si_2O_8$:$0.04Eu^{2+}$,$zMg^{2+}$ ($z$ = 0.10, 0.30, 0.50, 0.70, 1.0)

Fig.4-19 is the excitation spectra of $CaAl_2Si_2O_8$:$0.04Eu^{2+}$,$zMg^{2+}$ ($z$ = 0.10, 0.30, 0.50) by monitoring emission at 435 nm. These three curves share a similar shape. When the $z$ is 0.30, the sample shows the strongest excitation. The emission spectra are exhibited in Fig.4-20. The emission band shows a red-shift with an increase of the $Mg^{2+}$ concentration. The reason may be the difference between the ionic radius of $Mg^{2+}$ and the ionic radius of $Ca^{2+}$. In the triclinic structure of $CaAl_2Si_2O_8$, the length of the $c$ axis increases by replacing part of the $Ca^{2+}$ by $Mg^{2+}$ ions, the effect of preferential orientation of a $d$-orbital in the chain direction increases, so that the $Eu^{2+}$ emission shifts to the longer wavelength. The phosphor exhibits the strongest emission with appropriate CIE chromaticity coordinates ($x$ = 0.166, $y$ = 0.125).

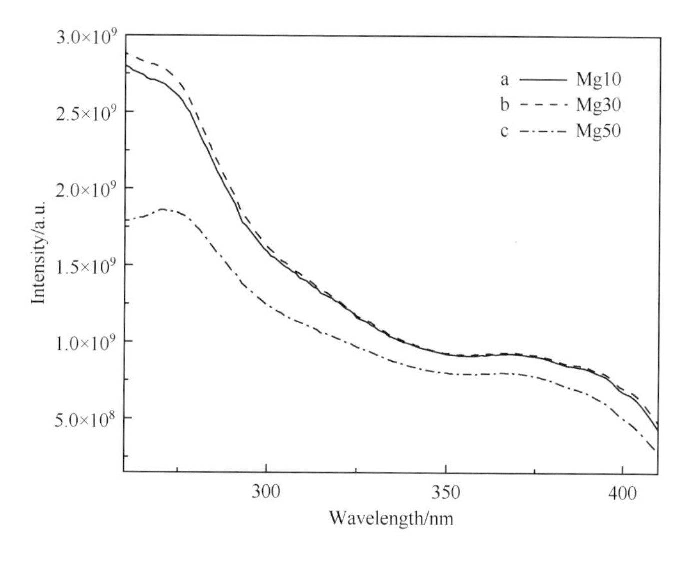


Fig.4-19 The excitation spectra of $CaAl_2Si_2O_8$:0.04$Eu^{2+}$,$z$$Mg^{2+}$ ($z$ = 0.10, 0.30, 0.50) by monitoring emission at 435 nm

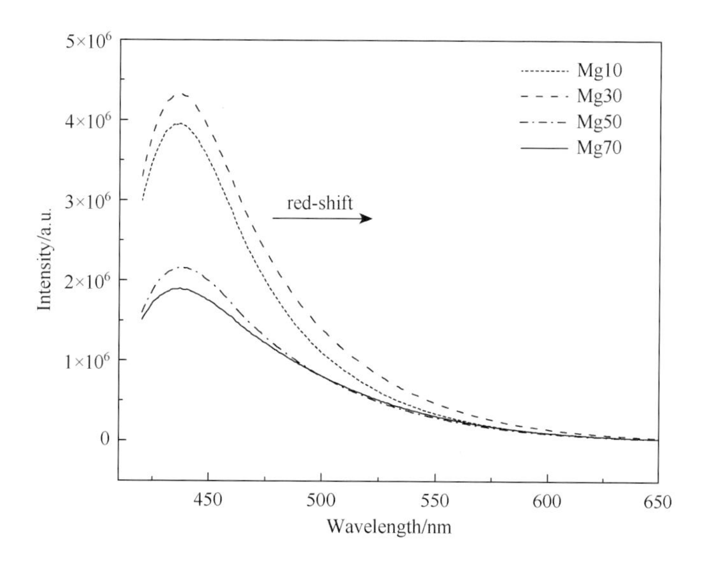


Fig.4-20 The emission spectra of $CaAl_2Si_2O_8$:0.04$Eu^{2+}$,$z$$Mg^{2+}$ ($z$ = 0.10, 0.30, 0.50, 0.70) under the 400 nm excitation

We have prepared blue phosphors, $CaAl_2Si_2O_8:Eu^{2+}$ doped with different content $Sr^{2+}$ and $Mg^{2+}$, and investigated their PL properties. The emission spectra show a blue-shift with the introduction of $Sr^{2+}$, and a red-shift with $Mg^{2+}$. This is due to the ionic radius difference of $Ca^{2+}$,

$Sr^{2+}$ and $Mg^{2+}$. These phosphors show intense blue emission with broad excitation in the near-UV range, which could find application in the near-UV InGaN-based LED.

# 4.4 $Eu^{2+}$-activated $SrAl_2Si_2O_8$

## 4.4.1 Introduction

Since the excitation band of $Eu^{2+}$ in phosphors is a broadband, which is due to the d-f transition. It can exhibit intense emission from the violet to red light region in the different hosts with different crystal filed. The broad excitation and emission band are ideal for application in the phosphor-conversion LEDs. $CaAl_2Si_2O_8$ and $BaAl_2Si_2O_8$ have been investigated as good hosts for the rare-earth ions by Park and Im et al. But the luminescent properties of $SrAl_2Si_2O_8$ doped with the rare-earth ions were seldom reported.

In this part, the phosphors $SrAl_2Si_2O_8$ doped with different contents of $Eu^{2+}$ were prepared by the solid-state reaction, and their luminescent properties were investigated. Their PL spectra were compared with that of $BaMgAl_{10}O_{17}:Eu^{2+}$, which was prepared by us.

## 4.4.2 Structure and PL properties

The XRD patterns of the phosphors $SrAl_2Si_2O_8:xEu^{2+}$ ($x = 0.02$, 0.10, 0.30, 0.40) are shown in Fig.4-21. These four curves a, b, c and d are of similar shapes, and they are consistent with the JCPDS card No. 38-1454 ($SrAl_2Si_2O_8$). This result shows that these phosphors doped different contents of $Eu^{2+}$ are still of the same monoclinic structure as $SrAl_2Si_2O_8$, and no other phase is observed in these patterns.

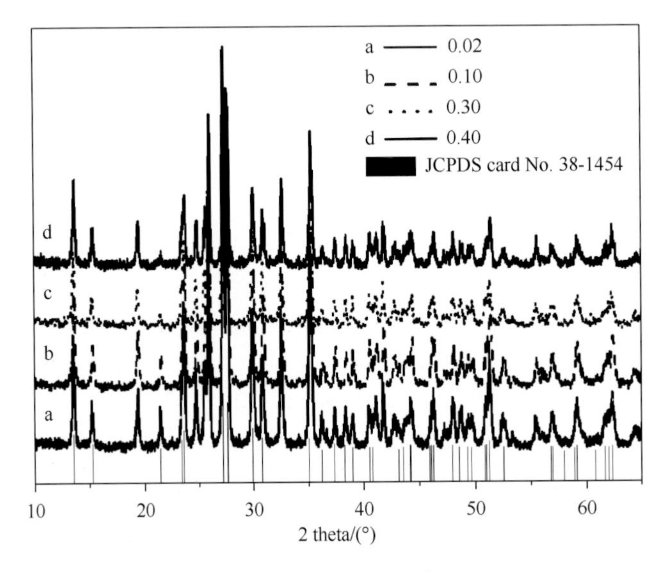


Fig.4-21 The XRD patterns of $SrAl_2Si_2O_8:xEu^{2+}$ ($x = 0.02$, 0.10, 0.30, 0.40)

The excitation spectra of the phosphors $SrAl_2Si_2O_8$:$xEu^{2+}$ (for $x = 0.02, 0.16, 0.40$) are shown in Fig.4-22. The broad excitation band from 300 nm to 390 nm is due to the 4f-5d transitions of $Eu^{2+}$. With the increase of the $Eu^{2+}$ content, the excitation intensity between 300 nm and 380 nm of the phosphor is decreasing. Whereas the intensity between 380 nm and 390 nm of the phosphor is increasing. This is useful to strengthen the absorption of the near-UV light. Our target is to find a novel blue phosphor for near-UV LED chips, and $BaMgAl_{10}O_{17}$:$Eu^{2+}$ (BAM) is the class blue phosphor for the near-UV LED, so the BAM was prepared by the same method as $SrAl_2Si_2O_8$:$Eu^{2+}$. Its excitation spectrum is shown in Fig.4-22(d). Compared with the excitation spectrum of BAM, the excitation band from 350 nm to 390 nm of $SrAl_2Si_2O_8$:$xEu^{2+}$ shows a flatform. The excitation intensities in this range are stronger than that of BAM. These characteristics of $SrAl_2Si_2O_8$:$xEu^{2+}$ are useful to find an application in w-LEDs.

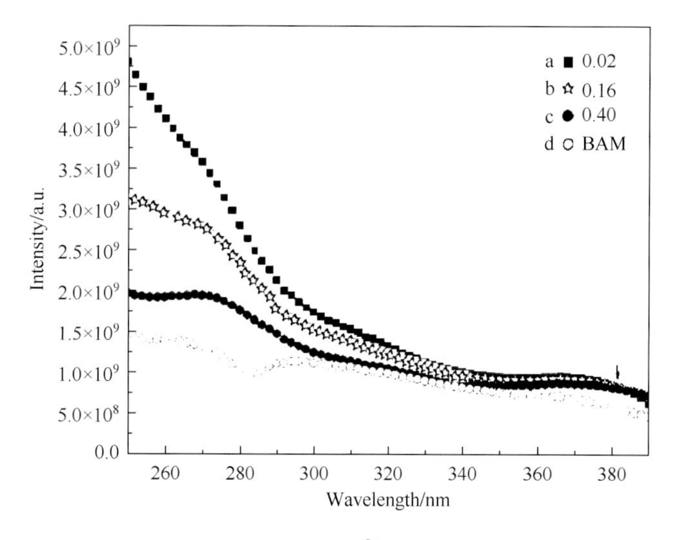


Fig.4-22 The excitation spectra of $SrAl_2Si_2O_8$:$xEu^{2+}$ (a, 0.02; b, 0.16; c, 0.40; $\lambda_{em} = 410$ nm), and $BaMgAl_{10}O_{17}$:$Eu^{2+}$ (d), $\lambda_{em} = 430$ nm

The emission spectra of the phosphors $SrAl_2Si_2O_8$:$xEu^{2+}$ (for $x = 0.02, 0.16, 0.40$) under the 270 nm light excitation are shown in Fig.4-23. The broad blue emission is due to the 5d-4f transition of $Eu^{2+}$. It is observed that the emission bands of the phosphors show some red-shifts with an increase of $Eu^{2+}$ concentration. Qiu et al. had reported that the probability of an energy transfer among $Eu^{2+}$ ions increased when the $Eu^{2+}$ concentration increased. With the increase of $Eu^{2+}$ content, the distance between $Eu^{2+}$ ions becomes less, and the probability of energy transfer among $Eu^{2+}$ ions increases. In other words, the energy transfer probability of $Eu^{2+}$ ions from the higher 5d levels to the lower 5d levels of $Eu^{2+}$ ions increases. Hence the emission band of the sample shows the red-shift with an increase of the $Eu^{2+}$ concentration.

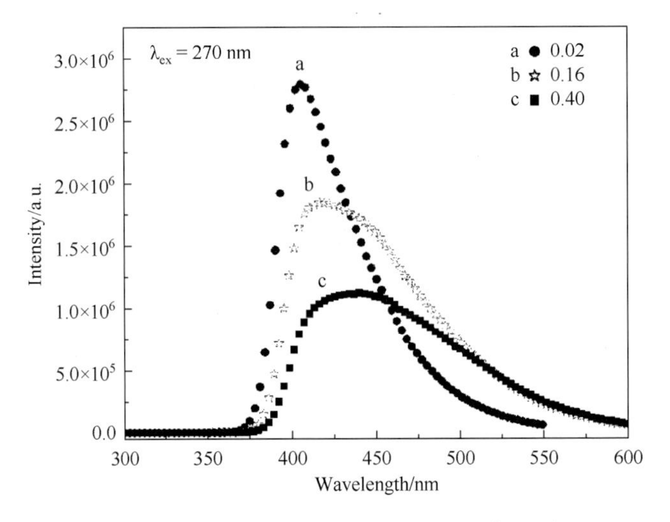


Fig.4-23　The emission spectra of $SrAl_2Si_2O_8$:$x$$Eu^{2+}$ ($\lambda_{ex}$ = 270 nm)

Fig.4-24 presents the emission spectra of the phosphors $SrAl_2Si_2O_8$ doped different contents of $Eu^{2+}$. Their emission intensities are increasing with the increase of the $Eu^{2+}$ content. This result is consistent with that of their excitation spectra (Fig.4-22). The emission band shows some red-shifts as same as the curves in Fig.4-23. The reason is same as the result of emission spectra under the 270 nm light excitation discussed above. The spectrum d in Fig.4-24 is the emission spectrum of the BAM under the 400 nm light excitation. The broad blue emission is also due to the 5d-4f transition of $Eu^{2+}$. The emission intensity of the phosphor $SrAl_2Si_2O_8$:0.40$Eu^{2+}$ is about 3.9 times higher than that of BAM.

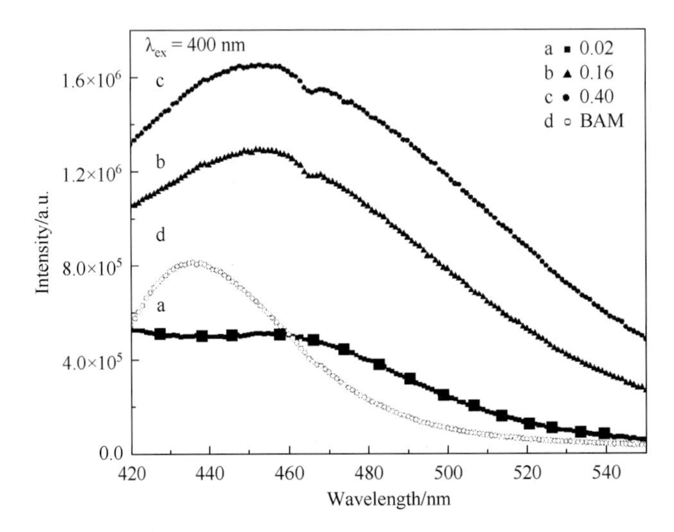


Fig.4-24　The emission spectra of $SrAl_2Si_2O_8$:$x$$Eu^{2+}$ (a, 0.02; b, 0.16; c, 0.40), and $BaMgAl_{10}O_{17}$:$Eu^{2+}$ (d) ($\lambda_{ex}$ = 400 nm)

With the increase of the $Eu^{2+}$ content, two emission bands were observed at 435 nm and

490 nm. The emission spectrum of $SrAl_2Si_2O_8$:$0.40Eu^{2+}$ has been separated into two components for convenience using the Gaussian function in Fig.4-25. This phenomenon can be presumed that $Eu^{2+}$ ions occupy two types of sites in $SrAl_2Si_2O_8$ lattice separately and form two corresponding emission centers. This result is similar to $Eu^{2+}$ ions in the $CaAl_2Si_2O_8$ lattice.

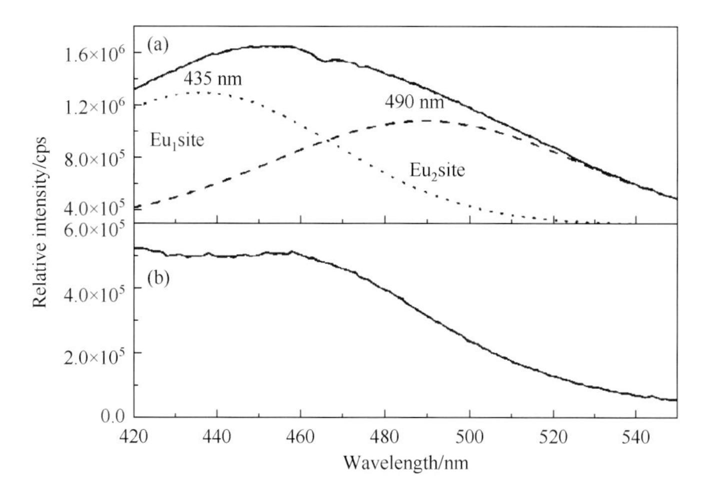


Fig.4-25 The two emission spectra of $SrAl_2Si_2O_8$:$xEu^{2+}$ separated by the Gaussian function ($\lambda_{ex}$ = 400 nm)

The low-temperature emission spectrum (77 K) of the sample $SrAl_2Si_2O_8$:$0.40Eu^{2+}$ under the 325 nm laser excitation is shown in Fig.4-26. Compared with the emission spectrum at room temperature, the emission band shows obvious blue-shift, and the full width at half maximum (FWHM) of the emission peak at 77 K (73 nm) are much narrower than that of the peak at 297 K

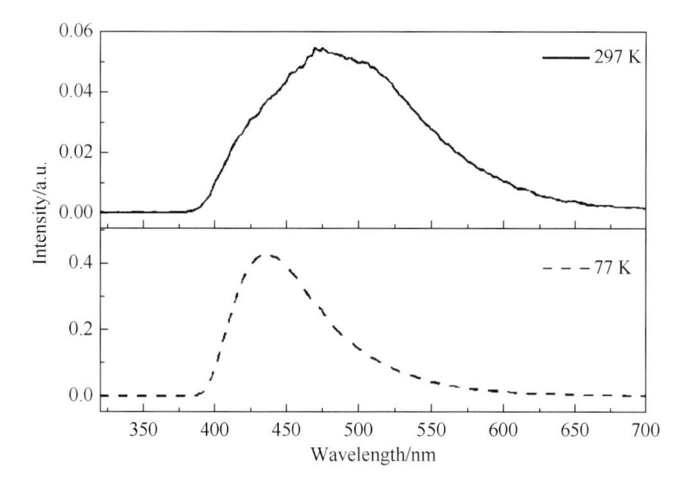


Fig.4-26 The emission spectra of $SrAl_2Si_2O_8$:$0.40Eu^{2+}$ at different temperatures [$\lambda_{ex}$ = 325 nm(laser)]

(130 nm). This may be due to the thermal quenching at the configurational coordinate diagram. The thermally activated luminescent center is strongly interacted with thermally active phonon,

contributing to the FWHM of the emission spectrum. At a higher temperature, the population density of phonon is increased, and the electron-phonon interaction is dominant, then the FWHM of the emission spectrum is broadened. This result is consistent with that of some other phosphors in Ref.

The lifetimes of the phosphors $SrAl_2Si_2O_8:xEu^{2+}$ ($x = 0.02$, 0.16, 0.40) are shown in Fig.4-27. They can be fitted into the single-exponential function [$I = I_0\exp(-t/\tau)$ ], where $I$ and $I_0$ are the luminescence intensities at time $t$ and 0, and $\tau$ is the luminescence lifetime. The lifetime values are calculated to be 595 ns, 425 ns, 262 ns, with the $Eu^{2+}$ content $n = 0.02$, 0.16, 0.40 respectively. The inset of Fig.4-27 shows the dependence of the lifetime of the phosphors on $Eu^{2+}$ content $n$. With the increase of the $Eu^{2+}$ content, the lifetime of the phosphor is decreasing. This decrease of the lifetime can be due to energy transfer among $Eu^{2+}$ ions non-radiatively at the higher concentration of $Eu^{2+}$ ions in the host lattices.

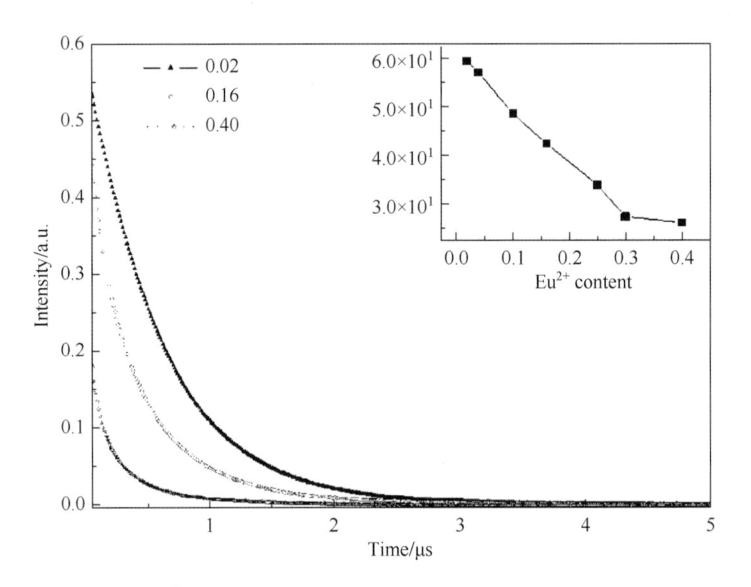


Fig.4-27 Decay curves of the $Eu^{2+}$ emission of the phosphors [$\lambda_{ex} = 337$ nm (laser); $\lambda_{em} = 410$ nm (Xe lamp) ]. The inseted figure is the dependence of the lifetime of the phosphors on $Eu^{2+}$ content $n$

In summary, we have prepared $Eu^{2+}$-activated $SrAl_2Si_2O_8$ phosphors and investigated their PL properties at different temperatures. Two emission bands can be observed in their emission spectra, which means that $Eu^{2+}$ ions maybe occupy two types of sites in the crystal lattice. The excitation and emission spectra of the phosphor $SrAl_2Si_2O_8:xEu^{2+}$ are compared with that of the phosphor $BaMgAl_{10}O_{17}:Eu^{2+}$. The excitation band of the sample $SrAl_2Si_2O_8:0.4Eu^{2+}$ in the range of 350 nm to 390 nm is obviously broader and higher than that of BAM, and its emission intensity is also higher than that of $BaMgAl_{10}O_{17}:Eu^{2+}$. Hence, the phosphor $BaMgAl_{10}O_{17}:Eu^{2+}$ may be an excellent substitute for the near-UV InGaN-based w-LEDs.

# References


Angel R J. 1988. High-pressure structure of anorthite. American Mineralogist, 73 (9-10): 1114.

Burrus H L, Nicholson K P. 1971. Fluorescence of $Eu^{2+}$-activated alkaline earth halosilicates. Journal of Luminescence, 3 (6): 467.

Dalmasso S, Damilano B, Pernot C, et al. 2002. Injection dependence of the electroluminescence spectra of phosphor free GaN-based white light emitting diodes. Physica Status Solidi (a), 192 (1): 139.

Dong Q, Xiong P X, Yang J J, 2021. Bismuth activated blue phosphor with high absorption efficiency for white LEDs. Journal of Alloys and Compounds, 885: 160960.

Dorenbos P. 2003. Energy of the first $4f^7 \rightarrow 4f^6 5d$ transition of $Eu^{2+}$ in inorganic compounds. Journal of Luminescence, 104 (4): 239.

Hao Z D, Zhang X, Luo Y S, et al. 2013. Preparation and photoluminescence properties of single-phase $Ca_2SiO_3Cl_2$: $Eu^{2+}$ bluish-green emitting phosphor. Materials Letters, 93: 272.

Im W B, Kim Y, Jeon D Y. 2006. Thermal stability study of $BaAl_2Si_2O_8$: $Eu^{2+}$ phosphor using its polymorphism for plasma display panel application. Chemistry Materials, 18 (5): 1190.

Jia W, Perez-Andújar A, Rivera I, 2003. Energy transfer between $Bi^{3+}$ and $Pr^{3+}$ in doped $CaTiO_3$. Journal of Electrochemical Society, 150 (7): H161.

Jung K Y, Lee D Y, Kang Y C, et al. 2003. Improved photoluminescence of $BaMgAl_{10}O_{17}$ blue phosphor prepared by spray pyrolysis. Journal of Luminescence, 105 (2): 127.

Kim J S, Jeon P E, Park Y H, et al. 2004. White-light generation through ultraviolet-emitting diode and white-emitting phosphor. Applied Physics Letters, 85 (17): 3696.

Kim J S, Park Y H, Kim S M, et al. 2005. Temperature-dependent emission spectra of $M_2SiO_4$: $Eu^{2+}$ (M = Ca, Sr, Ba) phosphors for green and greenish white LEDs. Solid State Communication, 133 (7): 445.

Kim S W, Zuo Y, Masui T, et al. 2013. Synthesis of red-emitting $Ca_{3-x}Eu_xZrSi_2O_9$ phosphors.ECS. Solid-State Letters, 2 (9): R34.

Kovac J, Peternai L, Lengyel O. 2003. Advanced light-emitting diodes structures for optoelectronic applications. Thin Solid Films, 433 (1): 22.

Leng Z H, Bai H, Qing Q, et al. 2022. A zero-thermal-quenching blue phosphor for sustainable and human-centric WLED lighting. ACS Sustainable Chemistry & Engineering, 10 (33): 10966.

Liu C M, Qi Z M, Ma C G, et al. 2014. High light yield of $Sr_8(Si_4O_{12})Cl_8$: $Eu^{2+}$ under X-ray excitation and its temperature-dependent luminescence characteristics. Chemistry of Materials, 26 (12): 3709.

Lv W, Hao Z D, Zhang X, et al. 2011. Near UV and blue-based LED fabricated with $Ca_8Zn(SiO_4)_4Cl_2$: $Eu^{2+}$ as green-emitting phosphor. Optical Materials, 34 (1): 261.

Mueller-Mach R, Mueller G O, Krames M R, et al. 2002. High-power phosphor-converted light-emitting diodes based on III-Nitrides. IEEE Journal of Selected Topics in Quantum Electronics, 8 (2): 339.

Nakamura S, Senoh M, Mukai T. 1993. High-power InGaN/GaN double-heterostructure violet light-emitting diodes. Applied Physics Letters, 62 (19): 2390.

Neeraj S, Kijima N, Cheetham A K. 2004. Novel red phosphors for solid-state lighting: the system $NaM(WO_4)_{2-x}(MoO_4)x$:$Eu^{3+}$ (M = Gd, Y, Bi). Chemical Physics Letters, 387 (1): 2.

Ohgaki T, Higashida A, Soga K, et al. 2007. Eu-doped $CaAl_2Si_2O_8$ nanocrystalline phosphors crystallized from the $CaO$-$Al_2O_3$-$SiO_2$ glass system. Journal of Electrochemical Society, 154 (5): J163.

Park J K, Kim C H, Park S H, et al. 2004. Application of strontium silicate yellow phosphor for white light-emitting diodes. Applied Physics Letters, 84 (10): 1647.

Park J K, Kim J M, Oh E S, et al. 2005. Luminescence properties of $Eu^{2+}$-activated $CaAl_2Si_2O_8$ by photoluminescence spectra. Electrochemical Solid-State Letters, 8 (1): H6.

Park J K, Lim M A, Kim C H, et al. 2003. White light-emitting diodes of GaN-based $Sr_2SiO_4$: Eu and the luminescent properties. Applied Physics Letters, 82 (5): 683.

Park S H, Yoon H S, Boo H M, et al. 2012. Efficiency and thermal stability enhancements of $Sr_2SiO_4$: $Eu^{2+}$ phosphor via $Bi^{3+}$ codoping for solid-state white lighting. Journal of Applied Physics, 51 (2R): 022602.

Poort S H M, van Krevel J W H, Stomphorst R, et al. 1996. Luminescence of $Eu^{2+}$ in silicate host lattices with alkaline earth ions in a row. Journal of Alloys and Compounds, 241 (1-2): 75.

Qiu J, Miura K, Sugimoto N, et al. 1997. Preparation and fluorescence properties of fluoroaluminate glasses containing $Eu^{2+}$ ions. Journal of Non-Crystal Solids, 213: 266.

Sheu J K, Chang S J, Kuo C H, et al. 2003. White-light emission from near UV InGaN-GaN LED chip precoated with blue/green/red phosphors. IEEE Photonics Technology Letters, 15 (1): 18.

Shionoya S, Yen W M. 1998. Phosphor handbook. New York: CRC Press.

Steigerwald D A, Bhat J C, Collins D, et al. 2002. Illumination with solid-state lighting technology.IEEE Journal on Selected Topics in Quantum Electronics, 8 (2): 310.

Sun J Y, Zhang X Y, Xia Z G, et al. 2012. Luminescent properties of $LiBaPO_4$: RE (RE = $Eu^{2+}$, $Tb^{3+}$, $Sm^{3+}$) phosphors for white light-emitting diodes. Journal of Applied Physics, 111 (1): 013101.

Tang Y S, Hu S F, Lin C C, et al. 2007. Thermally stable luminescence of $KSrPO_4$: $Eu^{2+}$ phosphor for white light UV light-emitting diodes. Applied Physics Letters, 90 (15): 151108.

Tian L H, Yang P, Wu H, et al. 2010. Luminescence properties of $Y_2WO_6$:$Eu^{3+}$ incorporated with $Mo^{6+}$ or $Bi^{3+}$ ions as red phosphors for light-emitting diode applications. Journal of Luminescence, 130 (4): 717.

Van Haecke J E, Smet P F, Poelman D. 2007. Luminescent characterization of $CaAl_2$ $S_4$: Eu powder. Journal of Luminescence, 126 (2): 508.

Van Uitert L G. 1967. Characterization of energy transfer interactions between rare-earth ions. Journal of Electrochemical Society, 114 (10): 1048.

Van Vliet J P M, Blasse G, Brixner L H. 1988. Luminescence properties of alkali europium double tungstates and molybdates $AEuM_2O_8$. Journal of Solid State Chemistry, 76 (1): 160.

Wu H, Zhang X, Guo C, et al. 2005. Three-band white light from InGaN-based blue LED chip precoated with green/red phosphors, IEEE Photonics Technology Letters, 17 (6): 1160.

Wu Z C, Shi J X, Wang J, et al. 2006. A novel blue-emitting phosphor $LiSrPO_4$: $Eu^{2+}$ for white LEDs. Journal of Solid State Chemistry, 179 (8): 2356.

Xia Z, Sun J, Du H. 2004. Study on luminescence properties and crystal-lattice environment of $Eu^{2+}$ in $Sr_{4-x}Mg_xSi_3O_8C_{l4}$:$Eu^{2+}$ phosphor. Journal of Rare Earths, 22 (3): 370.

Yang P, Yao G Q, Lin J H. 2004. Energy transfer and photoluminescence of $BaMgAl_{10}O_{17}$ co-doped with $Eu^{2+}$ and Mn $^{2+}$. Optical Materials, 26 (3): 327.

Yang W J, Luo J, Chen T M, et al. 2005. Luminescence and energy transfer of Eu-and Mn-coactivated $CaAl_2Si_2O_8$ as a potential phosphor for white-light UVLED. Chemistry of Materials, 17 (15): 3883.

Ye S, Wang C H, Jing X P. 2009. Long wavelength extension of the excitation band of $LiEuMo_2O_8$ phosphor with $Bi^{3+}$ doping. Journal of Electrochemical Society, 156 (6): J121.

Yoo J S, Kim S H, W. T. Yoo, et al. 2005. Control of spectral properties of strontium-alkaline earth-silicate-europium phosphors for LED applications. Journal of Electrochemical Society, 152 (5): G382.

Zhang S H, Hu J F, Wang J J, et al. 2010. Effects of $Zn^{2+}$ doping on the structural and luminescent properties of $Sr_4Si_3O_8Cl_4$: $Eu^{2+}$ phosphors. Materials Letters, 64 (12): 1376.

Zhang X G, Zhou F X, Shi J X, et al. 2009. $Sr_{3.5}Mg_{0.5}Si_3O_8Cl_4$: $Eu^{2+}$ bluish-green-emitting phosphor for NUV-based LED. Materials Letters, 63 (11): 852.